AF566863

Redesigning Animal Agriculture

The Challenge of the 21st Century

Redesigning Animal Agriculture

The Challenge of the 21st Century

Edited By

David L. Swain

Chair, Horizons in Livestock Sciences 2005
CSIRO Livestock Industries, Australia

Ed Charmley

CSIRO Livestock Industries, Australia

John Steel

CSIRO Livestock Industries, Australia

And

Shaun Coffey

Industrial Research Limited,
New Zealand

CABI is a trading name of CAB International

CABI Head Office
Nosworthy Way
Wallingford
Oxfordshire OX10 8DE
UK

Tel: +44 (0)1491 832111
Fax: +44 (0)1491 833508
E-mail: cabi@cabi.org
Website: www.cabi.org

CABI North American Office
875 Massachusetts Avenue
7th Floor
Cambridge, MA 02139
USA

Tel: +1 617 395 4056
Fax: +1 617 354 6875
E-mail: cabi-nao@cabi.org

A catalogue record for this book is available from the British Library, London, UK.

A catalogue record for this book is available from the Library of Congress, Washington, DC.

ISBN-13: 978 1 84593 223 7

Typeset by SPi, Pondicherry, India
Printed and bound in the UK by Biddles Ltd, Kings Lynn

Contents

Contributors

Margaret **Alston**, Director, Centre for Rural Social Research, Charles Sturt University, Wagga Wagga, NSW 2650, Australia. Email: malston@csu.edu.au

Richard **Bawden**, Distinguished Visiting University Professor, Michigan State University, Michigan 48824, USA. Email: bawden@msu.edu

Ed **Charmley**, CSIRO Livestock Industries, PO Box 5545, Rockhampton MC, QLD 4702, Australia. Email: ed.charmley@csiro.au

Shaun **Coffey**, Industrial Research Limited, 69 Gracefield Road, Lower Hutt 5040, New Zealand. Email: s.coffey@irl.cri.nz

Alex **Cook**, School of Mathematical and Computer Sciences, Heriot Watt University, Edinburgh EH14 4AS, UK.

Ina **Dobrinski**, Center for Animal Trangenesis and Germ Cell Research, School of Veterinary Medicine, University of Pennsylvania, Pennsylvania, USA. Email: dobrinsk@vet.upenn.edu

Tim **Doran**, CSIRO Livestock Industries, Private Bag 24, Geelong VIC 3220, Australia. Email: timothy.doran@csiro.au

Margaret **Gill**, School of Geosciences, College of Physical Sciences, University of Aberdeen, St Mary's, Elphinstone Road, Aberdeen AB24 3UE, Scotland. Email: Maggie.Gill@scotland.gsi.gov.uk

Iain J. **Gordon**, Sustainable Ecosystems CSIRO, Davies Laboratory, PMB PO Aitkenvale, Qld 4814, Australia. Email: iain.gordon@csiro.au

Jonathan R. **Hill**, CSIRO Livestock Industries, New England Highway, Armidale, Australia. Email: Jon.Hill@csiro.au

Mike R. **Hutchings**, Scottish Agricultural College, West Mains Road, Edinburgh EH9 3JG, UK. Email: Mike.Hutchings@sac.ac.uk

Penny **Johnes**, Professor of Freshwater Science, Aquatic Environments Research Centre, University of Reading, Whiteknights, Reading RG6 6AB, UK. Email: p.j.johnes@reading.ac.uk

Goetz **Laible**, Reproductive Technologies, AgResearch Ltd, PB 3123, Hamilton, New Zealand. Email: Goetz.laible@agresearch.co.nz

Luke **Lambeth**, CSIRO Livestock Industries, Private Bag 24, Geelong, VIC 3220, Australia. Email: luke.lambeth@csiro.au

Glenn **Marion**, Biomathematics & Statistics Scotland, JCMB, The King's Buildings, Mayfield Rd, Edinburgh EH9 3JZ, Scotland, UK. Email: glenn@bioss.sari.ac.uk

Brigid **Nelson**, Department of Primary Industries & Fisheries, PO Box 976, Charters Towers, QLD 4820, Australia.

Kishore **Prayaga**, CSIRO Livestock Industries, PO Box 5545, Rockhampton MC, QLD 4702, Australia. Email: Kishore.Prayaga@csiro.au

Antonio **Reverter**, CSIRO Livestock Industries, Queensland Bioscience Precinct, 306 Carmody Road, St Lucia, QLD 4067, Australia. Email: toni.reverter-gomez@csiro.au

John **Steel**, CSIRO Livestock Industries, 11 Julius Avenue, North Ryde, NSW Australia, PO Box 136, North Ryde, NSW1670, Australia. Email: john.steel@csiro.au

David L. **Swain**, Chair, Horizons in Livestock Sciences 2005. CSIRO Livestock Industries, PO Box 5545, Rockhampton MC, QLD 4702, Australia. Email: dave.swain@csiro.au

Ross **Tellam**, CSIRO Livestock Industries, Queensland Bioscience Precinct, 306 Carmody Rd, St Lucia, QLD 4067, Australia. Email: ross.tellam@csiro.au

Paul **Thompson**, Professor of Philosophy, W. K. Kellogg Chair in Agricultural, Food and Community Ethics, 526 S. Kedzie Hall, Michigan State University, East Lansing, MI 48824-1032. USA. Email: thomp649@msu.edu

David M. **Walker**, Department of Applied Mathematics, Hong Kong Polytechnic University, Hung Hom, Kowloon, Hong Kong.

David **Wells**, Reproductive Technologies, AgResearch Ltd, PB 3123, Hamilton, New Zealand. Email: david.wells@agresearch.co.nz

Acknowledgements

"The editors gratefully acknowledge the support and assistance provided by Howard Philpott who painstakingly checked bibliographies, Annie Grace for coordinating author submissions and Annabelle Davidson for annotating the corrections on the final proofs. In addition the editors especially thank Anne Tuppack who has provided invaluable guidance and support throughout the project."

Introduction to Redesigning Animal Agriculture: Making the Connection Between Holistic and Reductionist Science and the Implications for Delivering Sustainable Solutions

David L. Swain,[1] Ed Charmley,[1] John Steel[2] and Shaun Coffey[3]

[1]*CSIRO Livestock Industries, Australia;* [2]*CSIRO Livestock Industries, Australia;*
[3]*Industrial Research Limited, New Zealand*

This monograph comprises selected papers from the third conference in the series 'Horizons in Livestock Sciences'. Keynote speakers were approached to provide a written record capturing the salient points that came out of the meeting. The conference series was established in 2003 to provide an international forum for scientists, science managers and policy makers to explore some of the emerging issues in the livestock industries. This conference aimed to explore the global challenges and some of the emerging science solutions that will be important for future livestock production systems under the banner of 'Redesigning Animal Agriculture'. Its programme provided a unique opportunity to bring together a diverse array of key international speakers with expertise in a number of core areas that will be important for the future of livestock industries globally. The conference sessions were designed to move from the global drivers through to redesigning the animal and finally redesigning animal systems.

The holistic systems perspective was an important element of the conference programme. Richard Bawden (Chapter 1) emphasizes the importance of the cognitive system that allows us to make sense of, and take appropriate actions as participants in, the systems out there. Cognition itself is the inner processing of external inputs to derive interpretation and meaning. Richard Bawden refers to the move from a reductive, dualist and inherent state to a holistic, contextual and emergent state through epistemic development. Furthermore Richard suggests that the redesigning of animal agriculture will demand the further epistemic cognitive development of the scientists involved, so what does this mean?

Paul Thompson (Chapter 3) gives a glimpse of what epistemic cognitive development might mean through the philosophical and ethical challenges of livestock welfare. Paul explains the philosophical differences between utilitarian, moral rights and virtue ethics. He suggests there is no singular right form of ethics but we need ethical pragmatism to reconcile the opposing reductionist and holistic approaches to dealing with animal welfare. He states that the goal is to develop and institutionalize a practice of moral deliberation that reflects the complexity of human ethics.

Ethical considerations as part of redesigning animal agriculture are linked to a philosophical consideration of what we ought to do in an ethical sense. Systemic considerations are built on an innate sense of right and wrong. However, reconciling a reductionist scientific view of the animal and its welfare under a utilitarian model with the more holistic virtue-based approach is not easy. Herein lies the challenge that both Paul Thompson and Richard Bawden refer to in considering how we can develop a unified model for redesigning animal agriculture. Cognition and the functioning of the brain provide an interesting metaphor for understanding the challenge. The left hemisphere of the brain processes detailed information in a sequential way whereas the right hemisphere simultaneously synthesizes information, creating a 'big picture' view. Unification is derived by balancing the left and right brain modes of operation. The challenge is therefore to make links between the information (reductionist) and conceptual (holistic) systems. So what does this mean in the context of delivering research solutions?

Richard Bawden identifies that we need to take appropriate actions as participants in the systems out there. Science solutions need to be relevant and ethically acceptable and as Margaret Alston states when considering vibrant rural communities that 'when we assess the amenity, heritage and cultural value of rural areas we move beyond the notion of economics as determinant' (Chapter 2). Holism requires understanding and interaction at all levels, which can be achieved through a reductionist methodology. Therefore conceptual reasoning should balance knowledge through interactions that meet the ethical needs of communities. Success will not be defined by the process or the outputs but by the degree of interaction which leads to ethically responsible outcomes.

To develop a programme of research that can interact with existing activities built around a 'left brain' view of the world requires a rethink of livestock sciences. Many of the opportunities can be found in the 'softer' science areas, for example the social sciences. However, a disconnection exists in both philosophy and practice between left brain reductionist science and right brain holistic science. To balance the left and right brain activities requires opportunities that can link the two and ultimately close the loop. This monograph explores the opposing elements of livestock sciences and considers what and how these elements can be connected and balanced to provide a framework to deliver to the challenge of 'Redesigning Animal Agriculture'.

Ross Tellam provides an overview of the impact of genomics on livestock production (Chapter 4) and highlights the challenge of 'the integration of new technological capabilities into existing breeding programmes and production systems'. As an overview Ross puts the molecular biology into a systems context. A more detailed evaluation of some of the emerging molecular techniques and the opportunities for redesigning animals is given by Kishore Prayaga and Antonio Reverter (Chapter 5), Ina Dobrinski and Jonathan Hill (Chapter 6), David Wells and Goetz Laible (Chapter 7) and Tim Doran and Luke Lambeth (Chapter 8). These four chapters explore some of the emerging techniques and technologies for building better animals. In all cases the authors explore how detailed knowledge of the fundamental biology can help deliver production benefits, from transgenic chickens that are resistant to bird flu to transgenic cows that are able to produce health-promoting milk proteins. The value of molecular data in enabling precision animal breeding and the ability to transplant male germ line stem cells to the testes of infertile recipient animals. In each of the examples the authors review the current state of knowledge and demonstrate the value of redesigning animals.

Glenn Marion and colleagues (Chapter 9) consider a complex systems approach using stochastic modelling methods and use a number of examples that emphasize spatio-temporal interactions within a systems framework. In the context of the system Maggie Gill with David Swain (Chapter 12), Iain Gordon and Brigid Nelson (Chapter 10) and Penny Johnes (Chapter 11) develop three chapters that contrast the scale and regionality of agriculture and its impact on the triple bottom line (social, environmental and economic factors). In redesigning animal agricultural systems there is a need to build partnerships and create meaning. Entropic tendencies whether between science disciplines or between farmer and consumer can only be overcome through energy and effort. This monograph aims to contribute to the linking of ideas, and in so doing generate dialogue and debate that will hopefully shape future livestock production systems.

1 Redesigning Animal Agriculture: a Systemic Perspective

Richard Bawden
Michigan State University, USA

Abstract

Faced with a plethora of ever-increasing demands from citizens and producers alike for the design of more responsible and sustainable livestock production practices, animal scientists are becoming increasingly interested in exploring 'systems orientations' to their research and development work. However, two interrelated factors are proving to be especially significant impediments to the successful adoption of what might be termed 'systemic animal science': (i) the widespread confusion about what actually constitutes such research; and (ii) a lack of appreciation of the cognitive demands that the adoption of a 'systemic worldview' on research and development dictates.

Unfortunately – and ironically – the word 'system' itself is used in such a wide variety of contexts within animal science, as indeed it is in virtually every domain of human activity, that confusion reigns about what a systems-oriented research programme actually looks like and what systems-oriented animal scientists actually do. A range of intellectual and practical differences exist, for instance, between work in expert systems, in decision-support systems, in viable systems, in farming systems, in livestock production systems, in agro-ecological systems, in soft systems, and in critical systems, which are all being employed currently within the livestock sciences. In this present chapter, shape is given to the systems movement in the livestock sciences with the articulation of a typology that allows discriminations to be made between first, second and third order systemics, as well as between these three and what is referred to as 'proto-systemics' (which are more systematic than systemic).

This framework is also used to illustrate and explore the contention that the most significant systemic focus within livestock sciences – and yet one that apparently receives very little attention – is the nature and development of the cognitive systems that are fundamental to any systemic comprehension. The fundamental argument here is that acts of systemic development in the material and social world of livestock 'systems' are functions of the systemic intellectual and moral development of those who ought to be involved in those acts – including, and especially, scientists themselves. And while the logic of this submission is clear, the practical challenge is no less daunting.

Introduction

At least as judged by the frequency with which one encounters it in such a wide variety of contexts, there is something clearly beguiling about the notion of systems as an organizing principle or conceptual framework for approaching the world and for guiding concrete actions in, and indeed to it. At the very heart of such a systems approach to reality are three deceptively simple ideas: (i) that systems are whole entities that are separated by their own boundaries from the environments in which they are embedded

 Redesigning Animal Agriculture
(eds D. Swain, E. Charmley, J. Steel and S. Coffey)

and to which they are essentially 'structurally coupled'; (ii) that these systems are composed of interacting component parts that are themselves systems, and thus subsystems of the system in which they are embedded; and (iii) that all such systemic entities have properties that are unique to them and that emerge through the interactions of their component sub-systems. From this it follows that systems are (or can sensibly be assumed to be) organized hierarchically as [sub]systems (ss) within systems (S) within [supra]systems (SS), with unique properties emerging at each level of that system-of-systems hierarchy – and beyond into further hierarchical dimensions.

Emergence is key – it is the single most important difference between systemic and non-systemic worldviews: to the non-systemist, the interest in water is how the atoms of oxygen and hydrogen combine to form the molecule, while to the systemist, the issue of central interest is water's emergent property of wetness!

To approach something or some issue systemically, therefore, is to assume a perspective or worldview of wholeness, comprehensiveness, integrity, embeddedness, dynamic inter-connectedness – and anticipation. It is to think in at least three dimensions concurrently and, most especially, to keenly anticipate the surprises of emergent properties at each change in dimension. It is to be engaged and participative. It is also to be 'holistically' conscious of the intimacy and dialectics of the interdependencies between the observing self and that which is being observed, between the knower and the known, between the abstract and the concrete, between reflection and action, between the instrumental and the normative, between the technical and the practical, between the system and its environmental supra-system.

In the face of the often confusing complexities and messiness of everyday life, there is thus an obvious heuristic satisfaction to be gained by mentally organizing abstract ideas and existential events and concrete things and problematic issues into hierarchical systemic unities, even if the actual nature of those hierarchies and unities and component interactions remain unspecified and unclarified in their details. In use, these ideas can range in the rigour of their application from loose descriptive metaphor in everyday speech through formal analytical procedures to actual ways of 'being' in the world. The domain of agriculture certainly provides a host of examples across this entire spectrum, with its multitude of often unqualified references to research and development work in farming systems, production systems, cropping systems, livestock systems, agro-ecosystems, agri-food systems, marketing systems, integrated pest management systems, management information systems, expert decision systems, and so on and so forth.

Such has been the proliferation of this habit of 'systemizing' agriculture that, with respect to research at least, one commentator has unsurprisingly posited that there appears to be no agreement at all as to what systems-oriented research in agriculture actually is: 'It clearly means different things to different people' (Thornton, 1997). Thornton was commenting specifically on the proceedings of a conference on Systems-Oriented Research in Agriculture and Rural Development held at Montpelier in France in 1994 (Sebillote, 1995). He could have drawn similar conclusions, however, from reflections on the eclectic proceedings of other major international symposia that have focused on systems approaches to agriculture and rural development over the years, such as those held at Reading in England 20 years earlier (Dalton, 1975), at Hawkesbury in Australia (Remenyi, 1985), in California (Allen and van Dusen, 1988), at Reading again (Jones and Street, 1990), and at Bangkok in Thailand (Teng and Penning de Vries, 1992). A review of monographs devoted to the topic of systems approaches to agriculture, and of articles in past and present specialist journals like *Agricultural Systems*, *Farming Systems Research and Extension* and *Agroecosystems and Environment*, would also have reinforced the apparent confusion and complexity in what might be termed, following Checkland (1981), the *systems movement* within agriculture.

The irony here is that the conceptual clarity that systems perspectives can often bring to the confusions and complexities of life in the 'real world' of agriculture and rural affairs has somehow been ignored or avoided as a principle for organizing the different applications of the systems idea themselves as they appear in practice.

The matter of the redesign of animal agriculture provides an opportunity for this oversight to be addressed, especially within the context of the central claim of this chapter that any comprehensive and fundamental act of systemic transformation in agriculture, such as the redesign of animal agriculture, demands the transformation of the cognitive systems of the intended transformers. Because what we do in this world is a function of the way that we view it, as the neurobiologists Maturana and Varela (1987) insist, then the transformation of our worldviews to embrace systemicity is a necessary prerequisite for the systemic transformation of our work.

Orders of Systemics

A decision to redesign something implies that there is, at minimum, (a) some clarity in some quarters about the identity of the 'something' that is to be the focus of the endeavour, and (b) some rationality for why such a redesign is necessary and/or desirable. From a basic systems perspective then, the first decision relates to the identification of a 'system of interest' while the second relates to the identification of what it is that constitutes an improvement to that system, taking account of any of the implications of any potential impacts this might have on the rest of the system hierarchy. For reasons that will become evident a little later, this 'basic systems perspective' can be referred to as First Order Systemics. Elaborating on the idea of 'systems work' as introduced by Checkland (1981), the First Order Systemics being identified here for agriculture relates to research and development work in 'hard' systems – tangible systems that are presumed to exist and that can be simulated, redesigned and 'engineered' to meet some specified aim or another within agriculture. In the context of animal agriculture, this might be to make some particular livestock system more productive or more adaptable or more sustainable or resilient in the face of the turbulence that can often characterize the environmental supra-system in which it is embedded. Agroecosystem analysis, for instance, has been developed to capture precisely these matters with a commitment to such systems as formal 'well defined systems of cybernetic nature' (Conway, 1987). In contrast to much of the work in farming systems research, human beings are envisaged as regulators of agroecosystems rather than either managers or optimizers, and this allows for the emergence of the key ideas of (a) hierarchies, on the one hand, and the importance of system/environment interrelationships as a function of this, and (b) stability, productivity, equitability, and sustainability as emergent properties of such agroecosystems (Conway, 1987).

As indicated by the rhetoric of CSIRO Livestock Industries, the advantages of taking such a First Order Systemic perspective on the redesign of animal agriculture has already been intellectually and normatively appreciated, at least within that organization. Witness the following illustrative statement taken from the Mission of the Division:

> We will be valued for supporting Australia's meat, dairy, aquaculture and fibre industries to be stronger global competitors in the 21st century by [*inter alia*] enhancing welfare, health and safety of livestock products and *production systems*.

Further reinforcement of an appreciation of such a systemic perspective is found through explicit references to *systems* in the stated goals of three of the five Research Areas of the Division: *farm and business systems* within a context of Enabling Technology Innovation, *disease risk management systems* within a context of Ensuring Product Integrity and Market Access, and *production systems* within a context of Achieving Environmental Sustainability and Social Acceptance.

However, the issue of systemicity here is clearly not confined to the explicit mention of systems *qua* systems, but is refreshingly extended through an apparently pervasive appreciation of the comprehensiveness and indeed interrelatedness and inter-dependencies of the range of different criteria that are here seen within CSIRO Livestock Industries as constitutive of improvements in animal agriculture. These multi-criteria include stated concerns for the welfare, health, safety and quality of products as well of entire production systems. There are also stated concerns for human health and community well-being, for social acceptance, and for environmental sustainability, and these are far more comprehensive, and complex, criteria than most of the First Order Systemics allows. It is also a very long way away from the 'get big or get out' utterly unsystemic policy environment that characterized Australian agriculture for a great many decades. Yet it is the unintended consequences of precisely this latter worldview that now dictates the urgent and critical need for animal scientists to think in systemic terms not just of the comprehensiveness of their enterprise, but also of the consequences of not thinking in that manner.

Such is, and has been, the extent of the unintended consequences of many agricultural practices on social and ecological 'systems' alike that it has led to the growing realization, even among the citizenry, that it is not just the range of negative and destructive outcomes that have been perceived to be associated with many aspects of techno-scientific agriculture with which we should be concerned but the processes of techno-scientific development itself. As Beck (1992) has argued, the concern is no longer exclusively with transforming nature to make it useful, or with relieving mankind from traditional constraints, 'but also and essentially with problems of techno-development itself'. As it is the public at large who will be impacted directly or indirectly and beneficially or otherwise by future developments in animal agriculture, it is imperative that citizens now have a voice in the determination of the nature of those developments. And this is especially so where these are generated through novel scientific initiatives where public scepticism is already understandably high in the face of promise over substance. The increasing appreciation of, and thus concern for, the fact that local agricultural practices can have truly global impacts is leading to public calls for greatly enhanced guarantees of safety with respect is to the introduction of new agricultural technologies such as those involving rDNA. The range of concerns about their use is broad and grounded as much in ethical and aesthetic considerations as in technical and economic terms. There are anxieties about possible new risks to human health, to biodiversity, to equity, to the integrity of different eco-systems on the planet, and indeed to the entire nature of nature as it is now encountered. Genetic engineering, as Hindmarsh and Lawrence (2001) have observed, confronts us all with 'a new medium by which to imagine a future nature' that is very different to the nature that we have long known. Here then we have a novel systemic twist, with 'culture' suddenly assuming the ability to transform 'nature' using technologies like recombinant DNA which themselves represent culturally transformed nature.

All of these motivations together represent the need for a very different order of systemics for animal agriculture, where the focus is not on any particular tangible system, but essentially on the wholeness and integrity of the entire enterprise as a comprehensive human endeavour that is embedded within the environments in which it is conducted. Such a perspective connotes the ways by which scientists need to integrate their work in animal agriculture into the very way they live their lives as 'systemic beings' (Bawden, 1995) with an acute appreciation of the bio-physical and socio-cultural worlds with which they are structurally coupled. This profoundly holistic sense of being represents what might be termed a Third Order of Systemics, where the focus of the systems work is shifted to cognitive systems and on how the development of these greatly impacts on other systemic and non-systemic work in agriculture.

The acceptance of the comprehensiveness and inherent contestability of criteria such as environmental integrity along with social acceptance, and the obvious holistic appreciation of the complexity of the challenges being faced within contemporary animal agriculture, as both are revealed in the language of CSIRO, are very different from the reductionist/objectivist/positivist worldview traditions of animal science in Australia (although perhaps they are much less far from the worldview perspectives of livestock producers themselves, who have to live with the messy and complex realities of 'daily life down on the farm'). The gap between the worldviews, of traditional science on the one hand and this holistic position of Third Order Systemics on the other, is so great that, if the reality matches the rhetoric, this third-order perspective represents nothing less than the emergence of a new and critical paradigm for livestock agriculture that reflects the establishment and practical expression of a new order of systemic worldviews among livestock scientists. The adoption of such a worldview would represent the acceptance of profoundly unconventional beliefs and literally extra-ordinary epistemic assumptions about the nature of nature, the nature of human nature, and the nature of knowledge about both. Research and development work from this third-order perspective will necessarily embrace concerns for, and the development of, systemically integrated methods for dealing with the responsibilities of ecological integrity (in both senses of that world) as comfortably as matters to do with technical feasibility, and with issues of ethical defensibility as well as with matters of economic viability. Criteria for improvement, moreover, would extend to include aesthetic sensibility and cultural amenability in addition to political acceptability and social desirability, while spiritual sensitivity would be as central to judgement as instrumental rationality. And all of this together represents a formidable cognitive challenge.

The inherent complexity of the situation being faced by those who would systemically redesign animal agriculture to improve it is thus further 'complexified' by the nature of the process by which such improvements are to be evaluated, established, and eventually achieved. This in turn reflects the emergence of holistic worldviews that differ very considerably not just from the reductionistic, mechanistic, monist, techno-centric perspectives that have characterized the development of Australian agriculture for more than two centuries, but also from First Order Systemic thinking and practices.

The relatively simple, but essentially revolutionary, small cluster of systems ideas, along with the systemic perspective that they impute, become quickly complicated when epistemic matters are included with them. Systemists – those concerned with matters systemic – are now forced to confront profound epistemic questions like: Is nature 'really' systemically organized in the form of cybernetic systems-of-systems, and if so, are emergent (unpredictable) properties characteristic of all natural phenomena? Where do humans sit with respect to such 'natural systems', and how does their nature influence the systemic nature of nature? Do 'social systems' exist in reality, and if so, how are these to be reconciled with 'natural systems'? What influences do different assumptions about knowledge and about knowing have on beliefs about the systemic nature of nature and about human nature? Who should be involved in decisions about what it is that now constitutes improvements, given the greatly expanded and diversified range of criteria for 'betterment' that must now be embraced? And how are their different value assumptions to be assessed and translated into consensual actions for 'betterment'?

At the centre of appeals for the adoption of systemic perspectives on, and practical systems approaches to, the redesign of animal agriculture, therefore, lies the recognition that this needs to include epistemic concerns for the process of cognition itself.

Thus the redesign of animal agriculture, if it is to satisfy the multi-criteria that have been identified as vital to the endeavour, needs to commence with a redesign of the cognitive systems of the redesigners. In other words while the systemic attention to the redesigning process to date appears to have been mainly confined to First Order Systemics, logic dictates that much more emphasis needs to be given to Third Order Systemics to allow the attributes of former order to be more fully appreciated and more critically expressed.

There is justification for concern about the extent to which these epistemic dimensions of systemicity have been appreciated so far in the adoption of a putative systems perspective on the redesign of animal agriculture in Australia. In spite of a growing interest in systems approaches within agricultural research and development communities over the past three decades or so in this country, as elsewhere, those adopting them in their work appear to remain essentially unappreciative of the Third Order dimensions – or at least uncertain as to how to express them in it.

Before further pursuing these assertions and their implications, however, it is necessary to first fill in the gap and introduce the 'missing link' between First and Third Order Systemics as they have been articulated here.

Essentially, Second Order Systemics is concerned with the methodologies/methods of systemic research and development which involves work with 'soft' or enquiring systems. To paraphrase Checkland (1981), 'soft' systems approaches to agriculture, as for other human endeavours, reflects a shift in systemicity from the systems of the world to systemic ways of enquiry into issues that are identified as being in some way problematic in that world. In essence, there is a shift in focus here, away from researched systems and on to the processes by which they are researched – from researched to researching systems as it were.

The work in Second Order Systemics is thus concerned with the nature, development and employment of different systemic methodologies. In this manner, it truly provides a link between the first-order focus on the development of particular systems, and the third-order focus on the epistemic assumptions which will be brought to bear in any particular 'systemic intervention'. Strictly speaking, as Midgley (2000) has emphasized, systemic methodologies present the epistemic (theoretical and philosophical) foundations that justify the choice of any particular systems research method or development practice. In other words, the conscious choice of any particular systems method should, in practice, reflect a set of intellectual and moral foundations that are appropriate to the purpose and nature of the work. In actuality, as Midgley also points out, it is common for there to be a blurring of the boundary between method and methodology, as the former is really the expression of the latter, in action!

While a very considerable attention has been given to the development of systems methodologies over the years – and indeed to the internecine wars that have characterized these developments (Jackson, 2000) – little of this has extended into agriculture, at least in a manner that reflects third-order concerns.

To further expand on this issue of systems methodologies in use in agriculture, as well as to illustrate a number of the other details raised above, it is now appropriate to give some shape to the systems movement in agriculture (Fig. 1.1).

The Shape of the Systems Movement in Agriculture

As indicated on the far right-hand side of the diagram in Fig. 1.1, a number of different systems methods/methodologies can be identified for the three different orders of systemics along with their respective domains of work – in 'hard' engineered systems (seven methodologies), in 'soft' inquiring systems (five methodologies), and in cognitive systems (one methodology with clear links to the 'soft' domain). There are clearly some significant interactions between the domains while the logic of their origins is indicated through following the connections through to the left of the diagram. This is not by any means an exclusive list of all of the methodologies that are employed in systems work in agriculture. Moreover, each of these methodological positions typically represents a number of different methods which have been omitted here for the sake of clarity. There are, for instance, as many as six different farming systems research (FSR) methods spread across the anglophone and franco-

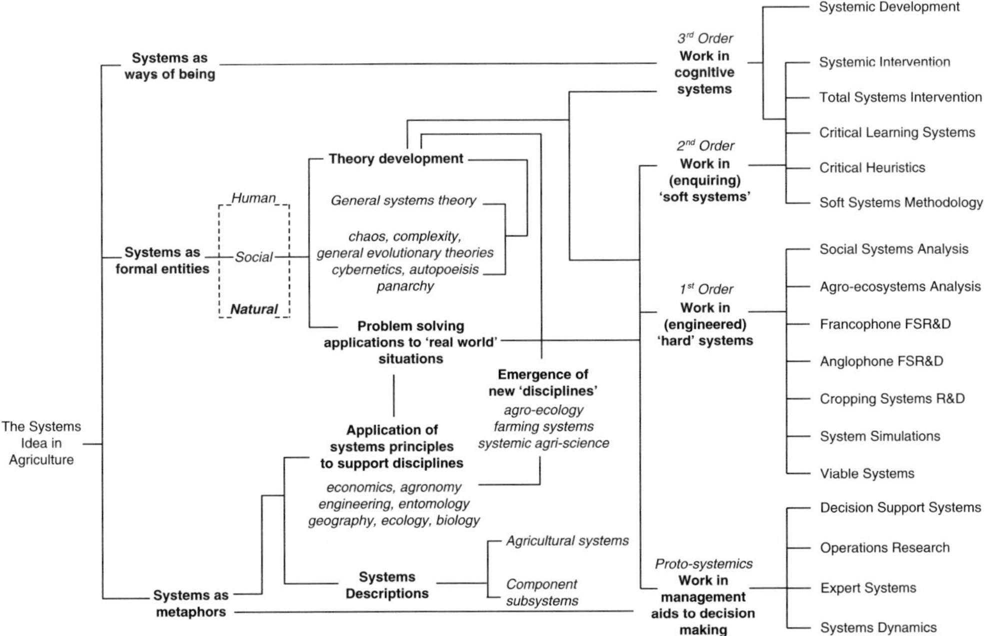

Fig. 1.1. The systems movement in agriculture.

phone methodologies indicated here (Sands, 1986). In a similar vein, systems simulations can be conducted for at least four distinctly different purposes (Rykiel, 1984), each of which demands slightly different methods of conduct. There is also a variety of methods within agro-ecological approaches to agriculture.

A fourth domain of work, associated with management aids to decision making, has also been included in this diagram, with indications of two sources of its origin. The first identifies it as another branch of problem-solving applications of the formal entity systems idea in agriculture, in common with the work in 'hard' and 'soft' systems, while another links it back directly to the notion of system as a metaphor in contrast to a formal entity. This ambiguity reflects the argument that endeavours such as expert systems, operations research, systems dynamics and decision support systems are more accurately regarded as *systematic* procedures rather than *systemic* ones (Jackson, 2000). A key aspect of that argument relates to the typical lack of explicit recognition of emergent properties in any of this work, while the notions of boundary and of environmental embeddedness are also often less than clearly articulated. Because of these features, the work in this domain is regarded here as being proto-systemic, which in no manner is intended to diminish the importance of the work, but is introduced simply to classify it within a 'systems movement'. The same can be said for the other areas within the 'map' that relate to the expression of system as a metaphor, with the importance of systems descriptions and applications to other disciplines again not to be denied, especially in the context of a consciousness-raising orientation to interconnectedness and wholeness, while nurturing at least a naïve sense of systemicity.

It must be emphasized that the typology of systems approaches illustrated here itself has systemic characteristics! What are shown as simple linear connections and causal pathways are in reality much more interconnected and non-linear than they appear. Moreover, there are also many interconnections which are not shown here but which are significant as part of a complicated network of influences right across the whole map. Finally, the clear distinctions between different endeavours that are illustrated in the model are, in practice, much more indistinct, including as they do the relationships between the generation of theories relevant to systemics and their application in practice.

The intention at this point is not to further explore the details of the distinctions between all of the different methods and methodologies that have been recognized. Rather, the schema has been introduced to give an indication of the relationships that exist between First, Second and Third Order Systemics (and proto-systemics) while also illustrating how systemic organization can be brought to the many aspects of systems approaches to agricultural and rural development.

Cognitive Systems

The logic that has been used here to discriminate between three different 'orders' of systemicity has its origins in a three-level model of cognitive processing that has been proposed by Kitchener (1983) and the interpretation of that model into a three-dimensional learning/researching system by Bawden (1998). Kitchener suggests that different levels of cognitive processing allow individuals to monitor, evaluate and potentially transform one level of cognitive tasks at another. In this manner she distinguishes between cognition, meta-cognition and epistemic cognition as a hierarchical sequence through which individuals can monitor the way by which they conduct their own basic cognitive tasks, like dealing with the immediate matters to hand, through a meta-cognitive process, which is then itself monitored by a process of epistemic cognition that concerns itself with 'the nature of problematic matters to hand' and 'the truth values of alternative solutions' (Kitchener, 1983). Epistemic cognition is thus essentially a reflexive focus on worldview perspectives.

The implication of the very deliberate congruence between this model of cognitive processing and the three orders of systemics being proposed within the context of the redesign of animal agriculture is that there is a vital need to provide opportunities for the different systemic orders to be reflexively monitored, evaluated and transformed as a prerequisite of the systemic transformation of animal agriculture. And this in turn is congruent with the notion of a three-dimensional learning or researching system. To put it into researching or learning terms, we are not only capable of researching into the world about us, but also into the process of researching itself (meta research), while, at a third (epistemic research) level of cognition, we can research into the limits to both 'lower order' forms of researching that are being imposed by, or at least profoundly influenced by, the particular worldviews that are prevalent within the systems researching community. In this manner the three-level model of cognitive processing is thus transformed into a three-level hierarchy of a researching system-of-systems in which each level is contextualized and profoundly influenced by the next higher system. Researching the matter is clearly contextualized by the methods/methodologies of research adopted, which in turn are contextualized by the epistemic worldviews of the researchers involved which provide the intellectual, moral, aesthetic and spiritual foundations for the two 'lower order' levels in this systems hierarchy.

This emphasis on the significance of worldviews is certainly consistent with the claims of the cognitive psychologist Koltko-Rivera (2004) that, in addition to defining what goals can be sought in life, a worldview also defines what goals should be pursued. As he further elaborates, worldviews involve sets of beliefs about what does and what does not exist (either in actuality or in principle), what assumptions are held, what value judgements are made about the goodness or badness of objects or experiences, and what objectives, behaviours, and relationships are deemed desirable or undesirable. And as he finally posits, worldviews also define what can be known or done in the world, and how it can be known or done.

These are all matters that are entirely relevant to the criteria that were articulated earlier with respect to the redesign of an animal agriculture that is comprehensive in its embrace. To borrow from philosophy and cognitive psychology, redesigning animal agriculture from a systemic perspective will necessarily involve the redesign of the composite ontological beliefs (about reality and being in the world), axiological assumptions (about what is good, virtuous, beautiful, etc.), and epistemological beliefs and assumptions (about knowledge and how it can be known) that currently prevail within the livestock science communities as well as within the livestock industries as a whole.

The connection between worldview and action is both crucial and, by its very systemic interconnectedness, 'charged with emergence'. Thus, in this context it is not a matter of whether ecosystems actually exist as self-organizing, cybernetically regulated, dynamic equilibrium-seeking entities, but the emergent impacts that these ontological beliefs have on the activities of those who hold to them. Similarly, it is not a matter of whether human beings are intrinsically good or evil, extravagant or mean spirited, virtuous or otherwise, but the influence that the particular axiological assumptions that are held under any particular circumstance have on behaviours. And finally, it is not a matter of whether objective, empirically generated, and experimentally validated scientific knowledge is the only form of truth, but the impact that that epistemological belief has on the way people accept and use knowledge in their everyday affairs.

The relevance of the work of Kitchener (1983) in this context is that we are indeed capable of clarifying the nature and details of our worldviews and of appreciating their impact on our behaviours, through focused attention on epistemic cognition. This is, however, not a trivial task for, as Salner (1986) submits, it involves the uncommon activities of thinking about, and evaluating, the very foundations of thought itself. Salner's particular interest in this issue is of signal importance to the arguments being developed in this chapter, for the essential thesis that she mounts is that the development of systemic capabilities is basically a function of the development of epistemic competence. In the present context, this implies that unless we sustainably engage with, and commit ourselves critically to, Third Order Systemics it is highly unlikely that our efforts at First Order and Second Order Systemics will be much different from the non-systemic, essentially mechanistic approaches that have prevailed to this point. Such a commitment, Salner would argue, is often a very demanding developmental process that involves persistent experiential, intellectual and moral confrontation for any development to be achieved in our worldview perspectives: as the aphorism has it, most of us do not even know that we have a worldview, let alone know the influence that it has on our knowing!

Yet this emphasis on epistemic *development* and worldview transformation is of fundamental significance, and both Kitchener and Salner draw heavily on the intellectual and moral development theories of Perry (1968) in support of their respective theses. From his longitudinal research on undergraduate students, Perry postulated that there appeared to be a sequential developmental pattern, which, over the course of their studies, was reflected in changes in the manner by which these young people 'addressed themselves to challenges'. At each step in the sequence of what he saw as 'structural transformations', Perry argued that each student saw him or herself, his or her instructors, and even truth itself, in very different terms. While identifying nine such steps or phases in the sequence, the essence of his findings was that the students 'progressed' – essentially through persistent and consistent challenge – from an initial developmental state of epistemological *dualism,* passing through a state of *multiplicity*, to eventually reach a state of *contextual relativism* which then found expression in *commitment*.

In other words, in response to the constant pressures of the intellectual and moral challenges that they were forced to confront over the years of their study, these young people essentially changed key features of their worldviews in what was carefully observed to be an almost invariant sequence. The initial 'epistemic' stage (of dualism) was characterized by clear distinctions between self and the external world and between right and wrong and true and false, by an acceptance of the objectivity of techno-scientific knowledge as absolute truth, by an unqualified respect for authority, and by palpable discomfort with any ambiguities in any of these arenas. The characteristics of the final stage (contextual relativism) were in almost direct opposition to those of the initial one: there was wide acceptance of the importance of contexts in defining truth as well as value, ambiguity was not only tolerated but accepted as a fertile ground for learning, and there was a much greater awareness of the subjectivity of what might be termed boundary judgements. There was also a much greater recognition of the need for interaction and participative discourse and engagement with others, as well as with issues, as essential to the process of the generation of knowledge and truth.

Importantly, Perry saw the interim *multiplicity* stages of development between *dualism* and *contextual relativism* as periods of significant intellectual and social confusion, with students often showing a strong tendency to revert to dualistic worldviews and to reject authority, rather than either persist with the confusions of multiplicity or move on to explore the more comfortable but much more complex state of contextual relativism. As Salner (1986) reported from her own work, the true/false, right/wrong dichotomies became 'much less serviceable' as students came to face increasing ambiguities in the focus of their studies and increasing conflict in both their intellectual and social circumstances. At the stage of multiplicity, truth was most likely to be seen to reside within the self, rather than either the external world, or in the dialectical interactions between the self and that external world.

West (2004) has recently reviewed the work of a number of other researchers who have subsequently and significantly extended Perry's work with other constituencies and over different timescales that indicate that epistemic development is often a lifelong endeavour. She points out that the results of all of these studies into cognitive development seem to agree on four essential propositions: (i) that the different stages are qualitatively different structures that nevertheless allow knowers, as they develop, to continue the same functions albeit in different ways; (ii) that the stages follow an invariant sequence; (iii) that each stage forms a structured whole based on an internally consistent logic; and (iv) that the stages are hierarchically integrated such that the structures of each successive stage differentiate and reorganize the knowledge constructed at the previous stage. With regard to this last proposition, Colby and Kohlberg (1987) have made the important point that the nature of the hierarchy is additive rather than purely transformational, and, as West (2004) points out, this means that the insights from earlier stages are included within new structures although they may be used differently. This position is consistent with the systemic view that any one 'level' is not better than another, but higher orders contextualize lower orders. Thus Perry's 'contextual relativism' is not 'better' than 'dualism' but it provides a more comprehensive context in which dualistic principles and practices can be sensibly employed.

There also seems to be general agreement that epistemic development is associated with both an increasing complexity in the structure of what we are calling worldviews here, and an associated capacity to deal with complex situations. This, of course, has immediate relevance to the nexus between epistemic development and the development of systemic competencies: as Salner and others essentially argue – albeit not in these precise words – one can't deal with complex issues in any profoundly systemic manner until and unless one develops a complex systemic worldview.

Experiences in systemic agricultural education and in systemic action research in rural development at Hawkesbury over the past two decades or so strongly reinforce this claim (Bawden, 2005); indeed the notion of *systemic development* that emerged at that institution in the late 1980s was intended to capture this vital interconnectedness effectively between systemic 'acts' of development in the material and social worlds and the epistemic development of all of the 'actors' who ought to be engaged in those endeavours. From such a perspective it makes sense to now recast the task of redesigning animal agriculture as an initiative in systemic development that embraces the need to extend the criteria for improvement in agriculture way beyond the mere efficiency and effectiveness of livestock systems. This shift in focus has two essential features: (i) that the emphasis is changed from one which privileges the relatively remote processes of research and of design, to one which now focuses on the participative process of development, with all of the messiness that that connotes; and (ii) that the particular perspective of this development is systemic in both form and function.

Systemic Development in Animal Agriculture

As indicated earlier in the Division's literature, the welfare, health, safety and quality of the products of animal agriculture, as well of entire livestock production systems, must now be regarded as essential features of the new horizons. So too, it is argued, must be concerns for human health and community well-being, for social acceptance, and for environmental sustainability. This last matter alone provides a substantial epistemic challenge, not the least because it is a quintessentially contestable concept that, as Thompson (2004) opines, combines questions about what it is that could be made to persist, with what it is that should be allowed to do so. It is highly unlikely that satisfactory answers to these questions could ever be generated from a dualistic worldview and indeed their very contestability demands the type of dialectical discourse that characterizes the epistemic state of contextual relativism, where meaning and plans for meaningful action only emerge through communicative interactions between people and the environmental circumstances that they face. And if nothing else, this implies types of participation in issues and engagement with the citizenry, and with those in other relevant institutions, that are far removed from the conventions of animal scientists and which demand a greatly heightened epistemic awareness as a prelude to necessary worldview transformations by all concerned, including, and perhaps especially, scientists themselves.

A reinterpretation of the redesign of animal agriculture as systemic development presents those committed to the quest for new horizons with at least three epistemic challenges. The first of these relates to the question of how nature is known through science and how adequate and salient that prevailing epistemology of science is to the development process. The second issue concerns the impact of such matters on citizens as they attempt to engage together with scientists with issues of development that are of profound significance to all. And this then leads to the third issue of what could and should be done from a systemic perspective to remove the constraints to the transformation of the current situation representing the two other issues!

There is, of course, no single epistemology of science, just as there is no single model of development. It is useful, however, to contrast two particular epistemologies that prevail within scientific work in agriculture, which reflect a dualistic and a contextual position respectively. Manicas and Secord (1983), for instance, have identified and contrasted what they refer to as traditional science on the one hand and realist science on the other. Where the former has a foundationalist epistemology in which the test of the truth of a proposition is the correspondence between theory and data and theories are deduced from the factual data, with the latter, it is the practices of science that generate their own criteria for which theory is accepted or rejected. Where conventional science accepts the logic of reductionism, the realist position is that the natural world is non-reductive and emergent. Where traditional science accepts causality as a central feature that leads to symmetry between explanations and predictions, the realist position takes the opposing view and rejects the idea that explanation and prediction are ever symmetrical. Finally, where truth is seen to 'reside out there' in the traditional paradigm, from a realist perspective the source of truth is neither 'out there' nor within the self, but in the dialectical relationship between the knower and the known. This latter position is therefore more likely to admit to normative influences where the former, traditional epistemology of science, eschews that.

While all of these distinctions are significant in one manner or another, perhaps the most constraining element of traditional science, with respect to its role in development at least, is the rejection of the normative dimension, for, as Goulet (1995) submits, development is above all else a question of human values and attitudes. And as he further posits, it is about the self-definition of goals by societies and the collective determination of criteria 'for determining what are tolerable costs to be borne, and by whom, in the course of change'. There is little doubt that this situation presents a significant epistemic challenge to those who would accept the need for the systemic development of animal agriculture but who have traditionally been captives of positivistic science.

The next challenge is, in many ways, even more difficult to reconcile, for it relates not directly to science or to scientists but to the manner by which society has adopted scientific rationality to the apparent exclusion of virtually all other ways of knowing and all other epistemologies. In this manner, as Norgaard (1994) suggests, very significant constraints are being imposed on the process of development through the 'narrowness of accepted patterns of thinking in the modern world', with the instrumentality of techno-science thoroughly embroiled at the heart of the matter. So it is not so much how scientists go about their work of understanding the natural and social worlds, but how the citizenry have come to abdicate their own responsibilities for judgements about the way they live their lives in those worlds in deferring to the authority of scientists. This phenomenon extends beyond instrumental knowledge into the process of moral judgement. As Busch (2000) argues, we have spent several centuries now, secure in the abdication of our individual moral responsibilities to the care of what he refers to as 'one Leviathan or another' such as 'science', or 'the state', or 'the market'. We have been seemingly content, he submits, to place our trust respectively in scientists, authoritarian monarchs, or the market to tell us what is 'good' or 'bad'. Similar arguments can be made for organized religion, with individuals abdicating their individual moral responsibilities to one 'church' or another whose dogmas have led to the loss of the will and/or the capacity for critical reflexivity. That which is 'right and proper' will be found in 'the good book' or will be revealed in some other spiritual manner, and a moral life then becomes that which is led under guidance of such revelations. The call for transformation in this context is based on the premise that there would seem to be too many instances where faith appears 'blind' to contemporary circumstances or where it is being lost without the resumption of moral responsibilities by individuals.

The instrumentalist dominance has thus been accompanied by the elevation of scientists to positions of social dominance as 'public decision makers' by virtue of their status as 'experts'. Regrettably there is little doubt that some scientists have exploited what Gadamer (1975) has called 'this peculiar falsehood of modern consciousness'. And regrettably, as Wynne (1996) has claimed, there are ample examples of situations where 'experts', including animal scientists, have 'tacitly and furtively' imposed prescriptive models of development upon lay people which have subsequently often 'been wanting in human terms'. This has all led, as some see it, to the establishment of what Yankelovich (1991) calls a 'culture of technical control' and to an epistemic climate of 'cognitive authoritarianism', where the rationality of thinking for oneself diminishes as the knowledge-gathering activities within societies expand beyond their capacities to deal with all they gather (Fuller, 1988).

And so to the matter of what can and should be done in the face of these epistemic challenges to the systemic development of an animal agriculture, which both wants and needs to be comprehensive in character in its embrace of a variety of views and opinions, and values and epistemologies consistent with an appreciation of the vital significance of public participation in the entire endeavour.

Participation is indeed the key element where such collective engagement is sensitive to the epistemic challenges that need to be addressed as well as the systemic nature of the development task itself. It thus makes sense to start with a systemic position on participation, and in this regard, Skowlimowski (1985) has put it as well as anyone in arguing that wholeness means 'that all parts belong together' such that they 'partake in each other'. Therefore from the central idea that everything is connected and that each is a part of the whole comes the idea that each participates in the whole. 'Participation is thus an implicit aspect of wholeness.' Observing, analysing, designing, simulating, engineering, developing, and learning about the systems of nature is one thing: 'being systemic', and being a participant in nature as an inherent part of it, is quite another, demanding, as the polymath Goethe insisted, nothing less than a change of consciousness. As Goethe saw it, scientists should not lose themselves in nature, but seek to find nature within themselves – to engage, in a fully conscious manner, in work in which humanity and nature come together in such a way 'that each becomes more fully itself through the other' (Bortoft, 1996).

This is much more than a mere literary image or a metaphor, as this view of 'participation as wholeness' connotes an ontology – a way of being in the world. Participation from this systemic perspective represents a 'cosmology of connectedness': of 'the self' participating with others and with the world at large, and of the participative interrelationships between the individual and the collective as well as between nature and society. Most significantly, perhaps, it intimately connects the 'knower' with the 'known', while extending the notions of knowing and knowledge to incorporate values and emotions as well as facts and theories. It also provides a fresh, participative perspective on the 'moments' of the hermeneutic as a process of dynamic flux between 'wholes' and 'parts'. It involves us in creating knowledge of the world that we experience about us, in a manner that evokes all of the elements of what the physicist Bohm (1994) referred to as our 'thought system', which includes the whole of society – with thought passing back and forth between people in a holo-movement 'by which thought has evolved from ancient times'. There is a continuous 'unfolding' and 'enfolding' of meanings, thoughts, 'felts' and even intentions and 'urges to do things' which cannot be anything other than a dynamic, transformational process of individuals and social groupings alike – and indeed of the mind itself.

These are immensely powerful images, which, while they appear to be utterly beyond the pale, are statements made by an eminent scientist, and while they might give a metaphysical or spiritual flavour to systemicity, they still allow, indeed reinforce, the arguments for epistemic development.

Salner (1986) restores the prosaic with her two vital conclusions about the types of epistemic transformations that she connects with the development of systemic capabilities: (i) the structural reorganization of epistemic assumptions, in the direction of increasing complexity that Perry's model implies, takes place on an individual timetable as a function of confrontation with intellectual and moral challenges which must be confronted; and (ii) while these might be confronted accidentally as a person's life unfolds, they can and should be facilitated by relevant institutions, for there is no evidence to support the idea that epistemic development occurs naturally, without confrontation. Rather, epistemic development occurs in relation to complex, integrated psychological and social development of that person. Most of the workers in the field of cognitive development agree that progression from one 'epistemic state' to another is rarely achieved without a combination of 'social participation' and sustained 'experiential stimuli'. As Salner (1986) has argued, learners must have the opportunity to *experience* the epistemic tensions and dilemmas that characterize each stage 'as his or her own personal dilemmas'. Furthermore, as she sees it, they must be 'emotionally able to contend with the temporary stress induced by such dilemmas' and deal with the mental blocks that are invariably, if unconsciously, erected 'to slow the pace of cognitive change'. In process terms, epistemic development is a function of 'mild' but persistent pressure in the learning environment such that the kinds of confrontations that produce development and transformation cannot be avoided. And this is precisely the design principle that was adopted at the Faculty of Agriculture and Rural Development at the University of Western Sydney Hawkesbury around that same time and interpolated into the working construct of critical learning systems (Bawden, 1998).

Critical Learning Systems

A critical learning system is a group of people that, in the first instance, have decided to collaborate in order to seek improvement in some situation that together the group members regard as problematic – the current status of the livestock industry in Australia for instance! Rather than being encouraged to think of themselves collectively as a group of decision makers or researchers, or a task force, or a committee, they are introduced to the idea of 'being a system'. They are thus encouraged to imagine themselves as a whole entity in which each individual participates and indeed is embedded, while contributing to both the 'form' of that system as well as its functions. The essential function of this system is to learn its way to betterment with respect to: (i) the issue that it has identified as problematic in its environment; (ii) its own integrity and functions as a system and its development in these regards; and (iii) the quality of its relationships with its environment. Learning is accepted as the experiential process by which the experiences of the system are transformed by it into the collective or social knowledge of the system, which is then employed in the design and conduct of knowledgeable action by the system as a whole. The system is also conscious of its own three-dimensionality, in two different regards. First, it sees itself as a sub-system within a system of interest (an institution or a community or a more abstract system of endeavours) that is, in turn, embedded within a higher-order environmental supra-system (and this in effect means that it has two levels of environment with which to deal). Second, because it is a learning system, it identifies three levels of its own cognitive functions as learning, meta-learning, and epistemic learning. In other words it can learn about the matter to hand, learn about the learning processes that it brings to bear to learn about the matter to hand, and it can learn about its own epistemic nature, which represents the collective (systemically emergent) worldview. Essentially it can also learn how to develop itself across all three of those cognitive domains, with the special, and somewhat paradoxical, need to develop to an epistemic state that is sufficiently complex in its character and 'mature' enough to enable it to comprehend systemicity – including its own.

Any critical learning system can be regarded as critical from three points of view. First, it is inherently critical of the conditions of the environments in which it identifies itself as being embedded, and with which it is consciously aware it is 'structurally coupled'. The situation it seeks to improve is often a manifestation of the history (or lack thereof) of such structural couplings. Second, it is reflexive of its own structure and functions and is consistently monitoring itself and adapting its behaviours in response to 'triggers' from its environments as well as from itself across its three cognitive dimensions. And finally, every critical learning system is critically conscious of the character and implications of each boundary judgement that it makes with reference particularly to what and who it includes and excludes from its activities as a learning system.

While all of this might sound incredibly abstract, the logic has proved to be a wonderful example of the systems idea in application. It presents a set of principles which can be used as the conceptual foundations for the design of the organization, structure and activities of self-reflexive groups of people who can journey together towards the level of epistemic development that, as has been argued above, leads to profound systemic appreciation. Experience across a wide diversity of settings, both within the academy and beyond into communities, institutions, and both private and public sector organizations, has shown that those who consciously participate in a critical learning system with a level of commitment not just to the task at hand but to their own 'epistemic development' can indeed learn how to become systemic beings!

Critical learning systems are thus a vital aspect of Third Order Systemics and therefore represent a potentially sound conceptual framework for those interested in pursuing the systemic development of animal agriculture. The intellectually powerful combination of the logic of critical learning systems, of systemic development, and of cognitive systems provides a stark contrast with the great majority of other systems approaches that are employed in the quest of improving agriculture. In essence it represents the seeds of a new paradigm of systemic science which is comprehensive, integrated and, most importantly, critically reflexive.

Towards a Systemic Science

The globalization of the risks which have arisen through the unintended consequences of conventional, reductionist agricultural techno-science is certainly sufficient reason alone for agricultural scientists to question the ethical defensibility of their work. They need especially to become self-critical, not just of the outcomes of their processes of enquiry, but of the very nature of those processes. Arguments can be mounted to support the claim that it is both the non-reflexive nature and the limited epistemic character of the processes of 'normal' agricultural science that leads inevitably to consequences that are as indefensible as they undesirable. It is salutary to reflect on the number of such consequences that, while certainly undesirable, were anything but unpredictable. For this reason alone, and there are indeed many others as has been emphasized throughout this narrative, a focus on the redesign of an endeavour as important as animal agriculture must include a focus on the redesign of the science that will be used to inform it. Busch (1994) reinforces this need with his passionate argument that the attempt to attain perfect knowledge by reductionist means 'can only lead us to madness' not just of individuals but of entire societies. Such knowledge, detached from caring and valuing, he claims, 'is as likely to create new problems as it is to help resolve old ones'.

A shift to systems approaches to agricultural science is clearly indicated here: as Churchman (1971) observed, 'the most important feature of any systems approach is that, in addition to ascertaining whether or not the choices that decision makers make lead to ends that they consider desirable, they also embrace the need to assure ends that are ethically defensible'. Regrettably, most of the systems approaches that are currently used within agricultural research and development have not embraced the need for ethical and aesthetic evaluations or established procedures through which they might be secured. It is for this reason that this particular narrative has taken the position that those systems approaches themselves need to be subjected to systemic transformation. Thus the argument here, with respect to what is being referred to as Third Order Systemics, is that a 'higher order' of systemic enquiry and development is necessary in order to improve the quality and relevance of the current first- and second-order systemic approaches to agriculture.

The emergence of so-called 'sustainability science' (NRC, 1999) is an example of a movement away from the reductionism of conventional science to embrace a much more systemic perspective that does appreciate the need for new perspectives on the dynamic interactions between nature and society. This movement seeks to address the essential complexity of these interactions, recognizing that understanding the individual components of nature–society systems does not yield sufficient understanding about the behaviour of the systems themselves (Clark and Dickson, 2003). The panarchy theories that attempt to inform transformations in human and natural systems (Gunderson and Holling, 2002) have a similar intent. There is the urgent need to transcend old dichotomies here, such as that between the natural sciences and the social sciences for instance, which bedevil attempts to gain systemic understanding. There is need for the beta and the gamma sciences, as Röling (2000) calls the natural and social sciences, respectively, to be synthesized, as he argues, into a new beta/gamma systemic complex. The call is essentially for a new 'post-normal' science which will be based on the assumptions of unpredictability, of incomplete control and of a plurality of legitimate perspectives (Funtowicz and Ravetz, 1993) that translates into a form of systemic development that appreciates the need to accommodate multi-criteria, multi-functionality, and sheer complexity. Such post-normal science will also reject the fact/value dichotomy and seek, instead, their integration into a systemic science which is self-reflexively aware of its own systemic character (Alrøe and Kristensen, 2002).

The new systemic science must recognize that values play an important role in science – as indeed they ought to do – not just with respect to constitutive values such as the norms of 'good science', but also in the form of contextual values that 'enter into the very process of science' (Alrøe and Kristensen, 2002) and reflect significant epistemic maturation. All of these matters strongly reinforce the call for much more attention to be paid to the development of cognitive systems and for epistemic transformations as the foundations for very different approaches to developments in 'systems' in the material and social worlds – and that perforce includes the development of animal agriculture.

References

Allen, P. and van Dusen, D. (1988) *Global Perspectives on Agroecology and Sustainable Agricultural Systems. Proceedings of the Sixth International Conference of the International Federation of Organic Agriculture Movements, Agroecology Program*. University of California, Berkeley, California.

Alrøe, H.F. and Kristensen, E.S. (2002) Towards a systemic research methodology in agriculture: rethinking the role of values in science. *Agriculture and Human Values* 19, 3–23.

Bawden, R.J. (1995) I as in academy: learning to be systemic. *Systems Research* 12, 229–238.

Bawden, R.J. (1998) The community challenge: the learning response. *New Horizons* 99, 40–59.

Bawden, R.J. (2005) Systemic development at Hawkesbury: some personal reflections from experience. *Systems Research and Human Behaviour* 22, 151–164.

Beck, U. (1992) *Risk Society: Towards a New Society*. Sage, London.

Bohm, D. (1994) *Thought as a System: A Seminar Transcription*. Routledge, London.

Bortoft, H. (1996) *The Wholeness of Nature: Goethe's Way Toward a Science of Conscious Participation in Nature*. Lindisfarne Books, New York.

Busch, L. (1994) Bringing nature back in. Paper presented to the annual meeting of the Michigan Sociological Association. MSA, Kalamazoo, Michigan.

Busch, L. (2000) *The Eclipse of Morality*. Aldine de Gruyter, New York.

Checkland, P.B. (1981) *Systems Thinking Systems Practice*. John Wiley, London.

Churchman, C.W. (1971) *The Design of Inquiring Systems*. Basic Books, New York.

Clark, W.C. and Dickson, N.M. (2003) Sustainability science: the emerging research program. *Proceedings of the National Academy of Sciences of the United States of America* 100, 8059–8061.

Colby, A. and Kohlberg, L. (1987) *The Measurement of Moral Judgment: Volume 1. Theoretical Foundations and Research Validation*. Cambridge University Press, Cambridge.

Conway, G. (1987) The properties of agroecosystems. *Agricultural Systems* 20, 31–48.

Dalton, G.E. (1975) *The Study of Agricultural Systems*. Applied Science Publishers, London.

Fuller, S. (1988) *Social Epistemology*. Indiana University Press, Bloomington, Indiana.

Funtowicz, S.O. and Ravetz, J.R. (1993) Science for the post-normal age. *Futures* 25, 739–756.

Gadamer, H-G. (1975) Hermeneutics and social science. *Cultural Hermeneutics* 2, 312–321.

Goulet, D. (1995) *Development Ethics: A Guide to Theory and Practice*. Apex Books, New York.

Gunderson, L.H. and Holling, C.S. (2002) *Panarchy: Understanding Transformations in Human and Natural Systems*. Island Press, Washington DC.

Hindmarsh, R. and Lawrence, G. (2001) *Altered Genes II*. Allen and Unwin, London.

Jackson, M.C. (2000) *Systems Approaches to Management*. Plenum, New York.

Jones, J.G.W. and Street, P.R. (1990) *Systems Theory Applied to Agriculture and the Food Chain*. Elsevier Applied Science, London.

Kitchener, K.S. (1983) Cognition, meta-cognition, and epistemic cognition: a three level model of cognitive processing. *Human Development* 26, 222–232.

Koltko-Rivera, M.E. (2004) The psychology of worldviews. *Review of General Psychology* 8, 3–58.

Manicas, P.T. and Secord, P.F. (1983) Implications for psychology of the new philosophy of science. *American Psychologist* 43, 399–413.

Maturana, H.R. and Varela, F.J. (1987) *The Tree of Knowledge: The Biological Roots of Human Understanding*. Shambhala, Boston.

Midgley, G. (2000) *Systemic Intervention: Philosophy, Methodology and Practice*. Kluwer Academic, New York.

Norgaard, R.B. (1994). *Development Betrayed: The End of Progress and a Co-Evolutionary Revisioning of the Future*. Routledge, London.

NRC (National Research Council) (1999) *Our Common Journey*. National Academy Press, Washington, DC.

Perry W. (1968) *Forms of Intellectual and Ethical Development in the College Years*. Holt, Rinehart and Winston, New York.

Remenyi, J. (1985) *Agricultural Systems Research for Developing Countries*. ACIAR, Canberra.

Röling, N. (2000) Gateway to the global garden: beta/gamma science for dealing with ecological rationality. *Eighth Annual Hopper Lecture*. University of Guelph Canada.

Rykiel, R. (1984) Modelling agroecosystems. In: Lowrance, R., Stinner, B.R. and House, G.J. (eds) *Agricultural Ecosystems: Unifying Concepts*. John Wiley, New York, pp. 157–178.

Salner, M. (1986) Adult cognitive and epistemological development. *Systems Research* 2, 225–232.

Sands, D.M. (1986) Farming systems research; clarification of terms and concepts. *Experimental Agriculture* 22, 87–93.

Sebillote, M. (1995) *Proceedings of the International Symposium on Systems-oriented Research in Agriculture and Rural Development*. ASFRE, Montpellier.

Skowlimowski, H. (1985) The co-operative mind as a partner of the creative evolution. *Proceedings of the First International Conference on the Mind–Matter Interaction*. Universidad Estadual de Campinas, Brazil.

Teng, P.S. and Penning de Vries, F. (1992) Systems approaches for agricultural development. *Agricultural Systems* 40, 1–19.

Thompson, P.B. (2004) Sustainable agriculture: philosophical framework. In: Goodman, R.M. (ed.) *Encyclopedia of Plant and Crop Science*. Marcel Dekker, New York, pp. 1198–2000.

Thornton, P. (1997) Book reviews. *Agricultural Systems* 54, 261.

West, E.J. (2004) Perry's legacy: models of epistemological development. *Journal of Adult Development* 11, 61–70.

Wynne, B. (1996) May the sheep safely graze? A reflexive view of the expert–lay knowledge divide. In: Lash, S., Szerszynski, B. and Wynne, B. (eds) *Risk, Environment and Modernity: Towards a New Ecology*. Sage Publications, London.

Yankelovich, D. (1991) *Coming to Public Judgment: Making Democracy Work in a Complex World.* Syracuse University Press, Syracuse, New York.

2 Maintaining Vibrant Rural Communities

Professor Margaret Alston
Centre for Rural Social Research, Charles Sturt University, Australia

Abstract
Agricultural productivity has trebled in Australia over the last 50 years, suggesting that all is well in the heartland of agriculture – the inland rural communities around which much of our production is based. None the less in this chapter I sound a note of warning, arguing that an over-reliance on economic parameters as the only indicator of success will inevitably compromise the industry. Social sustainability is critically important to an industry that relies so heavily on farm families. Yet access to education, employment, health and welfare service and transport and telecommunications infrastructure is reducing the attractiveness of rural areas. Sustaining families and attracting young people into the industry are vital to its survival. In this chapter I introduce the notion of multifunctionality, a concept in favour outside Australia, as a way of valuing rural areas beyond their productive capacity. Thus when we assess the amenity, heritage and cultural value of rural areas we move beyond the notion of economics as determinant. The concept opens up the need for wider investment in human, institutional, environmental and social capital as ways of revitalizing rural communities and of ultimately ensuring the future viability of agriculture.

Introduction

The long-term viability of Australian agriculture and sustainable livestock industries are highly dependent on vibrant rural communities. Yet, because agricultural productivity continues to grow (Productivity Commission, 2005), the link between production, people and the communities in which they live is often overlooked. For example the 2006 Corish[1] report (Agriculture and Food Policy Reference Group, 2006) – *Creating our Future: Agriculture and Food Policy for the Next Generation*, a report developed under the auspices of the Australian Government, makes limited reference to the importance of thick institutional structures within rural communities, preferring to focus on freeing the marketplace to facilitate even greater production. While this approach implies that economic growth will lead to a revitalization of rural communities, none the less, it is timely to evaluate the importance of communities and to assess the *social sustainability* of agriculture in the long term because a failure to do so will undoubtedly compromise agricultural productivity into the future.

Over 95% of Australian agricultural producers are farm families (Barr, 2004), families who are highly dependent on local educational facilities, employment, transport and telecommunications infrastructure and health and welfare services. If these institutions and infrastructure are not readily available or are inadequate in some way, attracting people into farming and/or retaining existing families and their skills base will be undermined. In this chapter I endeavour to add substance to this link between a healthy agricultural sector and vibrant communities. Further, I seek to underline the importance of the social relations of agriculture in order to override the blinkered view that the sector is somehow self-sustaining into the future if economic indicators and market freedom prevail.

Maintaining viable communities to support agriculture is challenging given the volatility in rural Australia. There are significant ongoing processes of restructuring and productive units are undergoing major change, a process that has been underway in Australian agriculture since the 1950s. The introduction of capital-intensive to replace labour-intensive agriculture, processes of globalization, a decline in commodity prices, a spiralling cost–price squeeze and, more recently, various industry deregulations and the adoption of a free-market mantra have placed extraordinary pressures on farm families and the communities that serve them (Lawrence, 1987; Gray and Lawrence, 2001). Added to this has been the extensive drought of the early years of the new century, a drought that has had a devastating effect on production, the landscape and rural people (Alston *et al.*, 2004; Alston, 2006).

These pressures, together with changes in community perceptions about lifestyle factors, have seen three trends emerging – broadacre farms have been expanding in size, becoming more capitalized and responsive to liberalization (Potter and Tilzey, 2005); at the other end of the scale, peri-urban land with high amenity value is being broken into smaller parcels and sold for lifestyle rather than production reasons; and, in the middle are the large majority of farm families, who are increasingly dependent on off-farm income to continue farming (Alston, 2000, 2006; Alston *et al.*, 2004; Potter and Tilzey, 2005).

The ongoing restructuring in agriculture has inevitably impacted on rural communities, particularly those small inland towns whose very existence is highly dependent on agricultural production. While regional centres continue to prosper and mining communities have found new sources of capital, many of the small inland communities are under stress with static or declining populations, an ageing population and reduced service infrastructure.

Yet Australian governments are committed to neoliberal policies that espouse market primacy and reduced government intervention, policies that have paradoxically assisted the higher-end producers, at the same time as they have resulted in a reduction in services into many rural communities, eroded service infrastructure, induced a greater use of volunteers and accelerated the privatization of some critical services (Cheers and Taylor, 2005). It is evident that rural Australians, who have higher unemployment, lower incomes, less education access, and poorer health (Cheers, 1998; Cheers and Taylor, 2005; Hugo, 2005; Alston and Kent, 2006), are compromised by processes of restructuring and neoliberal policy responses. Although agricultural production in Australia has risen nearly threefold since the 1960s (Productivity Commission, 2005), I would argue that the over-reliance on economic growth as the determinant of rural community viability is moving us to a crisis point where production may also be compromised unless circumstances change. If rural communities offer little to attract young people or to sustain the families currently in agriculture, then the long-term viability of agriculture based on farm family production units must necessarily be at risk in the foreseeable future. There is also ample evidence of an ageing demographic of farmers in Australia and a major (approximately 60%) decline in the number of farmers aged less than 35 years of age over the last 30 years (Barr, 2004), adding weight to my argument that agricultural production is at risk.

These issues are not unique to Australia – policy makers across the world are struggling with ways of supporting rural people and communities. In the European Union and United States, where similar processes of restructuring are under way, farming families have been protected to a degree by policies that recognize and reward the multifunctionality of rural areas – the various heritage, cultural and environmental aspects of landscape.

In this chapter I present an analysis of farming families and rural communities, noting their interdependence, and I argue that free-market policies may result in unsustainable social and environmental practices. I present data on Australian farm families and rural communities before examining multifunctionality and its usefulness in the Australian context. The Corish Report goes some way towards a valuing of multifunctionality by introducing the notion of environmental incentives and the Minister for Agriculture's response is supportive of a limited move in this direction.[2] However, I argue that an expanded view of multifunctionality is necessary to support rural people and rural landscapes, and ultimately agricultural production. My suggestions for ways to facilitate the growth of rural areas is to move beyond a focus on agriculture and productive units to encompass the rural communities themselves, a 'decoupling' that has found acceptance in the EU (Bryden, 2000).

Agriculture in Australia

The links between agriculture and rural communities become more transparent when we assess the diversity of the agricultural workforce. There are approximately 630,000 people living on Australian farms (Garnaut and Lim-Applegate, 1998) and approximately 131,000 farm units (Productivity Commission, 2005; ABS, 2005). While this number appears healthy, it none the less represents a significant decline in farm numbers of approximately 40,000 over the last quarter of a century (Gray and Lawrence, 2001) and there is some speculation that an ongoing process of restructuring will see the numbers of farms further decline by between 10 and 50% over the next 30 years (Barr, 2004). Yet despite the area of land under production declining by 9% over the last 20 years, agricultural output has doubled in the past 40 years (Productivity Commission, 2005). Yet, as Losch (2004) notes, increased productivity is occurring at the expense of overproduction, erosion of the natural environment and reduction in farm numbers and rural conditions. An analysis of the demographics of farming hints at future instability in the industry (Barr, 2004).

While over 95% of farms are run by families (and as many as 99.6% in broadacre and dairy farms) (Barr, 2004), the average age of owner-managers is increasing. In 1998 it was 52 (Garnaut and Lim-Applegate, 1998) and is now 55 for broadacre farms (ABARE, 2006). Older couples are delaying their exit/retirement and younger people are more reluctant to take on farming (Barr, 2004). There has, in fact, been a general decline in the numbers of young people entering farming since the 1970s and, although patrilineal inheritance dominates family farm agriculture, there is now more limited entry by inheritance as young people choose further education and alternative careers because of the insecurity of agriculture (Alston *et al.*, 2004; Barr, 2004). None the less there is a growing trend for family members to return mid-career (Barr, 2004).

The sector is significantly stratified, with the top 10% producing 50% of product and the bottom 50% producing 10% of product (Barr, 2004). The trend at the top end is favouring the development of large properties benefiting from liberalization. Dramatic changes are evident in some industries and have led to a significant shake up in farm numbers and structures. For example the decline of the wool industry, coupled with drought circumstances, has resulted in the amalgamation of properties into larger holdings and/or a switch to another industry in many remote areas of the country (Alston and Kent, 2006). Properties are being bought up by neighbours or absentee landlords, and run by a reduced staff, often facilitated by a move from a more intensive industry like sheep into a less intensive industry like cattle (Alston and Kent, 2006). Similarly the deregulation of the dairy industry has resulted in amalgamations of dairy farms and a loss of families from rural communities (Alston and Kent, 2006). This process of consolidation of farms into larger holdings is particularly evident in less populated areas, and has accelerated under drought circumstances (Alston and Kent, 2006), resulting in a further loss of farm families, a loss of employment for shearers and workers in more remote areas, a significant rise in the numbers of empty farm houses and a closure of small schools and bus routes (Alston and Kent, 2006).

Meanwhile a significant proportion of smaller family farms are dependent on other strategies to survive in agriculture and many are finding survival challenging. The numbers of farm families living in poverty has risen from 20% in the 1990s (Garnaut *et al.*, 1997), given the long-standing drought during the early years of the new century (Alston *et al.*, 2004; Alston and Kent, 2006). ABARE (2006) figures suggest 31% of broadacre farms and 15% of dairy farms had a negative income in 2004/05. It is not surprising then that over 50% of farm families are now reliant on pluriactivity, in particular off-farm income generation, to maintain their precarious attachment to farming (Alston, 2000). Farm women, who are twice as likely to have higher education as their partners, undertake over 80% of this sourcing of alternative income off-farm (Board, 1997; Alston, 2000).

The critical significance of rural communities for farm families is revealed in this statistic as it indicates that Australian farm women, like their overseas counterparts (Shortall, 2006), are reliant on employment in rural areas to support their families in farming. Yet the sourcing of this income is made more difficult by policies resulting in government withdrawal from rural community service delivery and service infrastructure decline, a process that has removed many jobs traditionally held by women. The closure of small hospitals, small schools, banks, health and welfare services in recent years has made the task of finding employment that much harder for rural women. As a result women report having to travel significant distances for work and/or many report having to involuntarily separate – moving away from their farm and home – in order to source the income needed by the family to remain farming (Alston, 2000; Alston *et al.*, 2004).

Meanwhile back on the farm, over 85% of labour is provided from within the family (Barr, 2004), and this includes increased farm work for women and children, who have been drawn more into the physical farm labour during drought (Alston and Kent, 2006), as families struggle to reduce costs. The need to reduce the use of hired labour has been highlighted during the drought as a significant reason for the loss of workers and their families from rural and remote communities (Alston and Kent, 2006). While reducing the labour pool to the immediate family preserves some autonomy for farm families, the loss of workers destabilizes the viability of rural communities by reducing the population and hence the community's ability to provide the breadth of services needed by farm families.

The third discernible group are those engaged in peri-urban post-productivist development and located on the urban fringes, along the coast and around regional centres (Barr, 2004). In these areas, high amenity farm land is being taken out of serious production, subdivided and sold to buyers seeking lifestyle options (Barr, 2004). This trend is facilitated in part by a 'tree-change' phenomenon – people escaping the stresses of city living and moving to the country for lifestyle reasons (Salt, 2005).

In summary – farm units are undergoing restructuring as particularly those in more remote areas are getting bigger, those on the peri-urban fringe move into small-scale post-productivism and the vast majority survive by adopting pluriactive strategies. Farm families still dominate the agricultural landscape and, while the top 10% of farm units are doing very well, there are a rising number of farm families living in poverty. Yet all farm units, large or small, rely on a healthy service infrastructure for their families and workers and good telecommunications and transport for business profitability. Despite this, an examination of Australia's rural communities suggests that many rural communities are under stress and unable to service the needs of rural producers.

Rural Communities

Two-thirds of Australians live in capital cities and 84% of the population live within 50 km of the coast (Hugo, 2005). Only 16% of Australians inhabit the vast inland areas of this large continent and these are the people most likely to be critically engaged in or dependent in some way on agriculture. A sharp distinction needs to be made between rural coastal communities, which are experiencing population increase and the associated pressures of rapid growth, and rural inland communities dependent on agriculture, 40% of which are in stasis or decline (Tonts, 2000). While regional 'sponge cities' (Salt, 1992) are growing beyond a dependence on agriculture and mining communities are experiencing some prosperity (Hugo, 2005), rural inland communities, particularly those in areas where mixed cropping and livestock farming are practiced away from major cities are under significant stress (Hugo, 2005). In these communities the pressures on farm families, including drought, a changing policy environment and the impacts of globalization, rebound, creating unemployment, a loss of families, an erosion of services, a loss of professionals and a widespread general community malaise (Alston and Kent, 2006).

The loss of population in many rural areas occurs for complex reasons including the reduced employment options on farms, the loss of professionals and their families as services close down, a drift of population to the coastal regions for lifestyle reasons and a decline in the size of rural families. However, one of the more critical factors associated with rural depopulation is the loss of young people from rural communities as they leave in search of work and higher education (Alston and Kent, 2001, 2006; Alston *et al.*, 2004). This rural out-migration of young people is highly gendered as more young women leave, their local choices more constrained than those of young rural men. Yet this gendered drift leaves a higher proportion of young men in rural communities unable to find partners. This social issue is compounded by the decline in the numbers of young professionals – teachers, nurses, allied health professionals (usually women) – entering rural communities, as rural health and education services are reduced. Keeping and attracting rural young people remains a vexed issue when employment options are limited and social networks sparse.

None the less there is a small but counter-trend of people moving to rural areas – the 'tree-changers' noted by Salt (2005), or the 'downshifters' noted by Hamilton and Mail (2003, p. 8), who are defined as 'those people who make a voluntary, long term lifestyle change that involves accepting significantly less income and consuming less'. Many are retirees who may move to small rural communities for low-cost living and lifestyle. For example several houses in Blackall, Queensland, have been purchased by Victoria retirees who spend their winters in the hotter climate (Alston and Kent, 2006). Burnley and Murphy (2004) make an important distinction between downshifters who are 'free agents' – those retirees and workers who make the choice for lifestyle reasons – and the 'forced relocators', those often welfare-dependent people who must move to escape the high cost of city living. Whatever the reason for the escape from the cities, these people offer some hope of a turnaround in population drift. For agriculture, a potential bright spot is the 'tree-changers' who return to take over the family farm mid-career or those relocators who move into farming for the first time. Capturing this demographic depends on the attractiveness of the area, the services such as education for children and the lifestyle offered to families. However, demographic data suggests that health, welfare, education and employment are areas of disadvantage in rural areas.

Representing ongoing stress in the hinterland, rural Australians continue to experience poorer health, mortality rates from 10 to 40% higher than urban Australians (NRHPF and NRHA, 1999, p. 38; ACSWC, 2000), higher morbidity rates (Cheers and Taylor, 2005) and yet have precarious access to health services (Alston *et al.*, 2006). The overwork of many farm family members is indicated in higher rates of farm accidents as families struggle to accommodate farm labour needs within the family (AIHW, 1998; Alston and Kent, 2006). Women in rural and remote areas are up to 25 times more likely to be hospitalized as a result of domestic violence (AIHW, 1998; ACSWC, 2000) and are in danger of losing local maternity services entirely with the closure of over 120 maternity services in rural areas in the last 10 years (Rural Doctors' Association, 2005). Access to bulk-billing doctors and specialists is precarious indeed, as is cancer care, counselling and family planning services (Alston *et al.*, 2006). These indicators suggest that attracting and retaining families in rural areas is compromised by poor health and welfare access and that farm family viability is similarly compromised. Attracting women to marry into farm families may become even more difficult if they are unable to give birth, see a doctor or access family planning and counselling advice in their local community, suggesting again a looming threat to the social sustainability of family-farm based agriculture.

The disturbing high incidence of male suicides in rural areas is another indicator that all is not well for rural men. Official figures suggest the rate for rural men is at least twice that of urban areas and that this is related to lack of employment, financial stress, alcohol abuse, greater access to firearms, an uncertain future and lack of access to education (ACSWC, 2000; Bourke, 2001). The drought has escalated the mental health issues in rural areas and yet there is a dearth of mental health facilities in rural areas[3] (Alston and Kent, 2006).

Educational access for rural Australians appears to be declining, with young people in rural and remote areas less likely to complete high school than urban young people and less likely to be able to afford to attend tertiary education institutions (Alston and Kent, 2006). This issue was recognized by the Human Rights and Equal Opportunities Commission report in 2000, *National Inquiry into Rural and Remote Education* (HREOC, 2000). Our more recent report (Alston and Kent, 2006) indicates that little has changed, and that if anything, access has been reduced by years of drought and financial hardship. Small schools continue to close, school bus routes are shut down, young people are being withdrawn from boarding schools and access to tertiary education is increasingly determined by ability to pay rather than merit, leaving many rural young people at a significant disadvantage. This declining access is a critical indicator of the precariousness of agriculture's future as its ongoing viability and innovation would appear to depend on an educated workforce. If the industry continues to rely on a locally based workforce, education of rural people is a critical factor in its future. The alternative is to bring people from outside the communities – contract managers and labourers for example. However, this issue is complicated by the amenity value of an area and its intrinsic attractiveness to outsiders. If we rely on importing labour then some regions will be marginalized by their lack of amenity. The notion of a fly-in fly-out workforce is one that also might be considered but again there are social issues associated with a transient workforce that may mitigate against stability in the industry. Providing education for locals would appear the most appropriate way of ensuring the future of the industry.

Meanwhile the loss of services, and hence the loss of locally based employment, has created greater stress for families, who must travel to access even basic services. With the high cost of petrol, families report their access is reduced and their quality of life impacted by the erosion of service infrastructure (Alston and Kent, 2006), an issue that will affect the ability of farm businesses to operate efficiently and effectively.

How do the people who live in rural Australia respond to the significant restructuring going on around them? Our research on drought and its social impacts indicates ongoing tensions created by the downgrading of rural Australia, a situation recognized by the then Deputy Prime Minister and Leader of the National Party in a speech to the National Press Club when he notes the possibility of Australia becoming two nations:

> The sense of alienation, of being left behind, of no longer being recognized and respected for the contribution to the nation being made, is deep and palpable in much of rural and regional Australia today. While there are areas and industries that are doing very well, there are many that are not. This issue must be addressed by all of us who collectively make up Australia, if we are to be a whole nation, because we can and must do everything we can to draw alongside those facing great challenge.
>
> (Anderson, 1999)

a view shared by the Australian Catholic Social Welfare Commission in their 2000 report:

> There is now such a deep, almost pathological, cynicism amongst rural Australians about the nature of government interest in them.
>
> (ACSWC, 2000)

A lack of trust in institutions, and in policies and political processes, continues to frame the views of many in rural Australia as they struggle to cope with significant changes (Alston *et al.*, 2004; Alston and Kent, 2006). Increasing levels of social isolation and declining levels of social capital[4] are evident in rural communities under stress (Alston *et al.*, 2004; Lawrence, 2005; Alston and Kent, 2006). Yet there is also strong evidence of resilience amongst rural people (Stehlik *et al.*, 1999). Can this resilience maintain agricultural productivity if rural Australians are not supported and valued; if their services continue to erode and if their trust in the Australian community and its institutions is corrupted?

Why Do We Need Rural Communities?

Given the issues facing inland rural areas, it is useful to ask why do we need rural communities? Drawing on the issues raised already in this chapter, it is clear that rural communities offer services to agriculture – the infrastructure, commercial hubs and educational and health facilities so vital to the economic viability of agriculture. The removal of services from rural communities also impacts on the ability of farm families to access input supplies, technical advice, loans and marketing information (Losch, 2004, p. 348). What is also evident from the previous discussion is that they also offer loci of much needed employment for farm family members to source the income that underpins agriculture. Less clear is that rural communities can provide the sites of social integration and social capital generation so vital to farm family functioning and the health of rural people. They offer quality of life facilities – sporting fields, perhaps a movie theatre, a community centre and adult learning facilities. Even less clearly articulated is that they have heritage value and hold a special place in the national psyche. Becoming more critical in the new century is the fact that they offer front-line defence against disease and biosecurity risks. There is no doubt that rural communities are significant to the future of agriculture and the country, but providing for rural communities and the people who live in them requires a change in policy drivers. One area where we might find inspiration is in the language of multifunctionality.

Multifunctionality

Neoliberalism has gained precedence in Australian government policy circles, a discourse that prioritizes market primacy and reduced government intervention (Gray and Lawrence, 2001). This discourse also dominates World Trade Organization (WTO) negotiations and thus international agricultural policy (Potter and Tilzey, 2005) and hence has a significant impact on papers such as the Corish report. None the less, as I have argued, neoliberalism has had a significant impact on Australia's rural areas, resulting in a withdrawal of services and infrastructure and significant stresses for rural people. In Europe and the USA, resistance to this discourse has developed as a result of the consequences productivist agriculture has had in rural areas – environmental degradation, overproduction and loss of farm families and workers (Losch, 2004). There is also a growing community consciousness of environmental issues, an increasing anxiety about food security in many countries and an awareness that subsidies in developed countries mean there is no level playing field in international markets (Losch, 2004). As in Australia, trade liberalization facilitated by neoliberal policies has seen the emergence of an elite group of family farms that have expanded and adopted a business model, and, again as in Australia, there are also growing numbers of family farms that are marginalized, reliant on off-farm income and/or subject to poverty, particularly when hit by market instability or natural disasters (Losch, 2004).

Multifunctionality is essentially a term coined to describe the wider non-commodity functions of agriculture beyond production – for example the protection of biodiversity, the landscape and environment, amenities and aesthetics and the preservation of cultural heritage (McCarthy, 2005; Potter and Tilzey, 2005). The recent articulations of the concept have been shaped in response to a growing awareness of the need to intervene in rural areas to underpin the incomes of farm families, to preserve agricultural production as a legitimate activity, to ensure food security, but also in response to wider community concerns about the environmental consequences of intensive agriculture (McCarthy, 2005; Potter and Tilzey, 2005).

> Thus advocates of strong multifunctionality position their case firmly within . . . the moral economy of the European Community by regarding the activity of farming as one of the defining conditions of rural space, the purpose of state assistance being to create the conditions under which family farming, rural landscapes and society can flourish. Agricultural liberalization, and the withdrawal of state support that this would entail, is seen as deeply antipathetic to these objectives, threatening the existence of large numbers of small and marginal family-run holdings and fashioning a countryside geared to mass production.
> (Potter and Tilzey, 2005, p. 590)

Supporters of multifunctionality also argue that those producers most likely to protect wildlife corridors and cultural sites are those families not necessarily geared to large-scale production, who recognize the alternative value in rural landscapes (McCarthy, 2005).

A discourse of resistance to the dominance of market liberalization and its consequences for farm families and the rural countryside is muted in Australia. Market liberalization dominates government and industry policy and is evident in publications like the Corish report and in welfare and workplace reforms which have undermined the wider social supports Australians have enjoyed. Perhaps this is understandable given Australia's reliance on export agriculture, its minnow status in world affairs and its membership of the Cairns group of agricultural exporting countries (Losch, 2004). None the less there is emerging evidence of alternative discursive constructions of land use within the community – a growing community awareness of ecology and the environment and a challenging of the right to farm. These discourses represent a move to recognize the multiple uses of land extant from agricultural production.

Multifunctionality offers an alternative discourse to market primacy for Australia that, none the less, incorporates recognition of agriculture. A useful start is that acknowledged in the Corish report – the need to invest in environmental management by farm families, an idea being taken seriously by government. However, we need to move much further, adopting an interpretation of multifunctionality that allows us to value the social aspects of rural landscapes and addresses the growing disparity between rural and urban areas. An investment in rural people and services is required to create rural environments where people are drawn not just for the aesthetic value or because they are escaping from the stresses of urban environments, but because they are moving to areas where people are valued and supported. This requires a significant investment in rural development and entails attention to the human, institutional, environmental and social capital aspects of rural areas and people. The following suggestions are offered as ways we might improve the viability of rural communities and, in the process, agriculture's future.

Human capital investment would include at least the following:

- Ensuring accessible, available and affordable education – making sure that no rural children miss out on education because of where they live or their income.
- Investing in job creation – providing incentives to attract businesses to rural areas and moving government positions and departments to the country.
- Ensuring rural people have access to service infrastructure, including telecommunications and transport, that enables them to participate fully in the life of Australia.

Institutional capital investment requires:

- Attention to services to ensure that rural people have the best possible, and safest, access to health and welfare care, education and training, telecommunications and transport.
- Attention to rurally embedded models of service delivery.
- Governance structures that empower rural people and shift decision making as close to the ground as possible.
- Infrastructure that is accessible, professionally managed and 'thick'.
- Decentralization of services to rural and regional areas.
- Public service middle-management positions being shifted to the country to allow rural experience within organizations and career-oriented professional jobs to be rurally located.
- Investment in alternative options of employment such as teleworking.

Investment in environmental capital requires:

- Financial reward for environmental stewardship.
- Investment in environmental and water management systems.
- Investment in alternative, environmentally friendly industries in rural areas.
- Recognition of the value of the services that farm families provide to the environment.

Finally, and perhaps most importantly, social capital is the key to vibrant rural communities (Dibden and Cocklin, 2005). Creating the conditions where social capital in rural communities can expand is the task of government, industry and community. Social capital comes from certainty in the future, from security about one's financial circumstances, from knowing that even in difficult times there are community supports and welfare provisions for families. While investment in human capital contributes to economic development (Stephens and Laughton, 2003), it is also arguable that enhanced social capital will assist the growth of vibrant communities, build networks and attract jobs and infrastructure through a facilitation of economic development (BTRE, 2003).

Conclusion

Despite severe cost–price pressures, commodity price fluctuations and drought, agriculture production continues to grow. Major restructuring has created a leaner, more efficient industry, with fewer farm families and workers producing more product on larger farms. Government policies have acted to reduce government involvement, allowing market primacy, and the top 10% of producers have responded very effectively, suggesting on paper at least that agriculture will continue to prosper. However, all is not rosy – in this chapter I alert to an impending crisis brought about by the very conditions that have favoured increased productivity. Such a crisis is foreseeable based on currently available data and will take decades to reverse unless sustained action is undertaken. The destabilization in many rural communities, the lack of access to services and infrastructure and inadequate access to education, health services and employment create conditions where the future of agriculture will be undermined.

Agricultural demographics reveal that farm families, who are our major agricultural producers, are a stratified group, and that at least half now rely on off-farm income to survive in agriculture. At the same time young people are leaving farms and communities in droves, turning their backs on agriculture, lured by the opportunities in the cities, and in the process eroding the skills base of agriculture. Rural communities are under pressure and the services vital to agricultural producers are being withdrawn.

In this chapter I have argued for a shift in policy to incorporate the notion of the multifunctionality of rural areas – the valuing of rural areas and landscapes for non-agricultural purposes and hence support for the idea of stewardship payments and rewards for environmental and ecological management. I argue that the definition of multifunctionality should extend also to the valuing of rural people and communities through investment in human, institutional and social capital. What is required is a softer approach to policy – a move away from economic fundamentalism, a stronger focus on regional planning, greater investment in infrastructure, increased access to education and greater support for the young people, tree-changers and rural women and men in agriculture and small businesses who are the future of our rural industries and regions.

Notes

[1]The report is referred to as the Corish report because the President of the National Farmers' Association, Peter Corish, chaired the inquiry that led to the report.

[2]Peter McGauran, Minister for Agriculture Forests and Fisheries. *Government recognizes farmers' contribution to the environment*. Media release, 12 April 2006, accessed 19 April 2006. http://www.maff.gov.au/releases/06/06037pm.htm

[3]The writer is a member of a working party on rural mental health established by the NSW Farmers' Association following significant concerns of the executive consequent on drought.

[4]Social capital is the term typically used to describe the degree of social trust, reciprocity and networks within a community that may enhance civic cooperation and community well-being. Greater levels of social capital have been argued to enhance economic performance through increased investment in human and physical capital, reduced transaction costs and greater levels of research and innovation (BTRE, 2003).

References

ABARE (Australian Bureau of Resource and Agricultural Economics) (2006) Online interactive statistics, www.abare.gov.au (accessed 19 April 2006)

ABS (Australian Bureau of Statistics) (2005) *Agricultural Commodities, Australia, 2003–04*. ABS, Canberra.

ACSWC (Australian Catholic Social Welfare Commission) (2000) A litany of disadvantage: rural communities of Australia. http://www.catholicwelfare.com.au/publications/Discussion_Papers.htm (accessed 15 April 2006)

AIHW (Australian Institute of Health and Welfare) (1998) *Health in Rural and Remote Australia*. AIHW, Canberra.

Agriculture and Food Policy Reference Group, chaired by Peter Cornish (2006) *Creating our Future: Agriculture and Food Policy for the Next Generation. Report by the Agriculture and Food Policy Reference Group to the Minister for Agriculture, Fisheries and Forests*. ABARE, Canberra.

Alston, M. (2000) *Breaking through the Grass Ceiling: Women, Power and Leadership in Rural Australia*. Harwood Academic Publishers, Amsterdam.

Alston, M. (2006) 'I'd like to just walk out of here': women's experience of drought. *Sociologia Ruralis* 46(2), 154–170.

Alston, M. and Kent, J. (2001) *Generation X-pendable: Young, rural and looking for work: a report to Woolworths Ltd on funded research*. Centre for Rural Social Research, Charles Sturt University, Wagga Wagga.

Alston, M. and Kent, J. (2006) *Impact of Drought on Rural and Remote Education Access: a report to DEST and Rural Education Fund of FRRR*. Centre for Rural Social Research, Wagga Wagga.

Alston, M., Kent, J. and Kent, A. (2004) *Social Impacts of Drought: report to NSW Agriculture*. Centre for Rural Social Research, Charles Sturt University, Wagga Wagga.

Alston, M., Allan, J., Dietsch, E., Wilkinson, J., Shankar, J., Osburn, L., Bell, K., Muenstermann, I., Georgias, D., Moore, E., Jennett, C., Wallace, J., Bacchus, R., Quinn, H., Ritter, L., Gibson, J., Harris, J., Gibson, R. and

Grantley, J. (2006) Brutal neglect: Australian rural women's access to health services. *Journal of Rural and Remote Health* 6, 475.

Anderson, J. (1999) One nation or two? Securing a future for rural and regional Australia: address to the National Press Club, Canberra, 19 February, http://www.ministers.dotars.gov.au/ja/speeches/1999/as01_99.htm (accessed 19 April 2006).

Barr, N (2004) *The Micro-Dynamics of Occupational and Demographic Change in Australian Agriculture: 1976–2001*. ABS, Canberra.

Board, H. (1997) Rural women. In: *Women in Agriculture and Resource Management Conference Proceedings*. DAFF, Canberra, pp. 1–3.

Bourke, L. (2001) One big happy family? Social problems in rural communities. In: Lockie, S. and Bourke, L. (eds) *Rurality Bites: the Social and Environmental Transformation of Rural Australia*. Pluto Press, Annandale, VA, pp. 89–102.

Bryden, J. (2000) Is there a 'new rural policy'? In: *European Rural Policy at the Crossroads Conference*. Arkleton Centre for Rural Development Research, University of Aberdeen, Scotland, pp. 1–17.

BTRE (Bureau of Transport and Regional Economics) (2003) *Government Interventions in Pursuit of Regional Development: Learning from Experience*. BTRE, Canberra.

Burnley, I. and Murphy, P. (2004) *Sea Change: Movement from Metropolitan to Arcadian Australia*. UNSW Press, Sydney.

Cheers, B. (1998) *Welfare Bushed: Social Care in Rural Australia*. Ashgate, Sydney.

Cheers, B. and Taylor, J. (2005) Social Work in Rural and Remote Australia. In: Alston, M. and McKinnon, J. (eds) *Social Work Fields of Practice*. 2nd edn, Oxford University Press, Melbourne.

Dibden, J. and Cocklin, C. (2005) Introduction. In: Cocklin, C. and Dibden, J. (eds) *Sustainability and Change in Rural Australia*. UNSW Press, Sydney, pp. 1–18.

Garnaut, J. and Lim-Applegate, H. (1998) *People in Farming*. ABARE, Canberra.

Garnaut, J., Robinson, J. and Lubulwa, M. (1997) *Issues in the Delivery of Commonwealth Social Support Programs to Farm Families*. ABARE, Canberra.

Gray, I. and Lawrence, G. (2001) *A Future for Regional Australia: Escaping Global Misfortune*. Cambridge University Press, Cambridge.

Hamilton, C. and Mail, E. (2003) Downshifting in Australia: a seachange in the pursuit of happiness. http://www.growthfetish.com/DP50sum.pdf (accessed 16 April 2006)

Hugo, G. (2005) The state of rural populations. In: Cocklin, C. and Dibden, J. (eds) *Sustainability and Change in Rural Australia*. UNSW Press, Sydney, pp. 56–79.

HREOC (Human Rights and Equal Opportunities Commission) (2000) *Emerging Themes: National Inquiry into Rural and Remote Education*. HREOC, Sydney.

Lawrence, G. (1987) *Capitalism in the Countryside*. Pluto Press, Sydney.

Lawrence, G. (2005) Globalization, agricultural production systems and rural restructuring. In: Cocklin, C. and Dibden, J. (eds) *Sustainability and Change in Rural Australia*. UNSW Press, Sydney, pp. 104–120.

Losch, B. (2004) Debating the multi-functionality of agriculture: from trade negotiations to development policies by the South. *Journal of Agrarian Change* 4, 336–360.

McCarthy, J. (2005) Rural geography: multi-functional rural geographies – reactionary or radical? *Progress in Human Geography* 29, 773–782.

National Rural Health Policy Forum (NRHPF) and the National Rural Health Alliance (NRHA) (1999) *Healthy Horizons 1999–2003: A Framework for Improving the Health of Rural, Regional and Remote Australians*. NRHPF, Canberra.

Potter, C. and Tilzey, M. (2005) Agricultural policy discourses in the European post-Fordist transition: neoliberalism, neomercantilism and multi-functionality. *Progress in Human Geography* 29, 581–600.

Productivity Commission (2005) *Trends in Australian Agriculture*. Productivity Commission, Canberra.

Rural Doctors' Association (2005) Governments must heed the evidence: small maternity units are safer. Media release December 7. http://www.rdaa.com.au/uploaded_documents/Governments%20Must%20Heed%20the%20Evidence%20–%20Small%20Maternity%20Units%20are%20Safer%20–%20December%202005.htm (accessed 19 April 2006)

Salt, B. (1992) *Population Movements in Non-Metropolitan Australia*. KPMG, Melbourne.

Salt, B. (2005) Australian life, work and relationships in the 21st century: what it might mean for the built environment. In: *Built Environment Conference*. PIPE Communications, Sydney.

Shortall, S. (2006) Gender and farming: an overview. In: Bock, B.B. and Shortall, S. (eds) *Rural Gender Relations: Issues and Case Studies*. CAB International, Wallingford, Oxfordshire, pp. 21–30.

Stehlik, D., Lawrence, G. and Gray, I. (1999) *Drought in the 1990s: Australian Farm Families' Experiences.* RIRDC, Kingston, ACT.

Stephens, M. and Laughton, K. (2003) Government interventions in pursuit of regional development: what have we learnt? *Sustaining Regions* 3, 5–12.

Tonts, M. (2000) The restructuring of Australia's rural communities. In: Pritchard, B. and McManus, P. (eds) *Land of Discontent: the Dynamics of Change in Rural and Regional Australia.* UNSW Press, Sydney, pp. 52–72.

3 Ethics on the Frontiers of Livestock Science

Paul B. Thompson
Michigan State University, USA

Abstract
Philosophers have tended to emphasize reductive approaches to rights, virtues and consequences, resulting in the 20th century in two dominant theories, utilitarianism and rights theory. These approaches are well represented in contemporary animal ethics, but are not easily adaptable for livestock science. In particular, the key problems associated with rDNA-based animal transformation are not well articulated by either of these approaches. The most natural way to articulate these issues is through an ethical approach that stresses virtue, but neither utilitarian nor rights theories are capable of accommodating a virtues approach. A more pragmatic style in ethics is introduced as a way to accommodate a pluralism of values in animal ethics, and a pragmatic approach is discussed for topics at the cutting edge of livestock science.

Introduction

This chapter provides a whirlwind introduction to the way that philosophers approach the subject of ethics, with special attention to non-human animals and even more pointed application to issues at the cutting edge of animal biotechnology. In moving quickly through a range of topics that could easily be the focus of much more detailed discussion, simplifications and illustrations are introduced that are potentially misleading. Readers are advised to adopt a generous reading of these generalizations, but also to recognize that more careful and exhaustive discussions would yield more plausible and insightful interpretations. The chapter begins by discussing ethics in the contemporary animal sciences, and moves rapidly to a very simplified model of human activity and decision making that is used to characterize three logically independent domains in which people attribute ethical significance to the things that human beings do. From there, main philosophical approaches in animal ethics are discussed with an emphasis on the way that these approaches have adopted a strategy that mimics reductionism in the biophysical sciences. The main goal in the chapter is to illustrate the inadequacy of these reductionist approaches as we move to the frontiers of livestock science.

A brief discussion of some recent critical philosophical responses to the leading theoretical approaches in animal ethics is followed by a discussion of livestock applications for biotechnology. These applications at the frontier of animal science pose additional challenges to leading theoretical views in philosophical animal ethics. Genetic approaches to production disease are central to some of the work currently being done in animal biotechnology. The ethical issues can be illustrated with the case of the blind chicken, an animal breed developed through classical animal breeding. In place of ethical reductionism, the chapter concludes by advocating a pragmatist approach. However, the sketch of pragmatism offered in the concluding section requires further development and amplification that cannot be undertaken within the bounds of this chapter.

Ethics in the Animal Sciences and in Human Action

The word 'ethics' has had something of an uncertain status in livestock science. Bernard Rollin has written that ethical norms have always been implicit within animal husbandry. Animal producers have never doubted that they had ethical responsibilities to provide humane care for their animals, though they have understood these responsibilities in largely personal terms. But Rollin notes that university and government-laboratory based programmes of animal husbandry were reconfigured as departments of animal science during the 1950s and 1960s, a transition that had been completed worldwide by the mid-1970s. Rollin believes that during this transition a scientific ethic of positivism became widespread in research programmes (though not among livestock producers) (Rollin, 1995a). According to positivism, scientists should constrain themselves to reporting empirical data and analytically derived relations among data. Because ethical norms go beyond data or analytically discoverable relations among data, ethics was not seen as appropriate for scientific discourse.

Ironically, the positivist norm for what scientists should do cannot itself be part of science. Scientists who are positivists can limit themselves to reporting data and relationships among data, but in following this limitation they cannot even state or advocate the positivist norm, much less engage in professional discussions of how this norm affects science. To do so would violate the positivist norm. Within philosophy of science this curious feature of positivism, that the very statement of positivist doctrine was inconsistent with what the doctrine advocated, was the undoing of positivism as an influential philosophical movement. As an institutional practice in the sciences, however, positivism has been more robust. As a guideline for how scientists should conduct their professional activity, positivism becomes reinforced through the peer review process for publications and research proposals, where efforts that exceed the limitation to reporting data and relations among data are simply rejected as 'unscientific'. Positivism may also be reinforced through hiring and promotion in that it may be possible to stifle the careers of those who fail to follow this norm. But within a science guided by this practice, positivist doctrines are themselves beyond criticism or debate by virtue of the fact that those who would either defend or criticize positivism are precisely those whose voices are stifled by others who follow the positivist norm. This kind of blind and autocratic adherence to any principle is deeply contrary to the spirit of science; as such it is high time for the influence of positivism to recede (Thompson, 2004a).

After several decades of adherence to positivism, the animal science profession is arguably lacking a vocabulary with which to discuss ethical norms. The traditional vocabulary of husbandry is not fully adequate for this task because of its focus on personal, as opposed to institutional, responsibility. That is, traditional husbandry portrays ethical responsibility in terms of actions undertaken by individuals who serve as caregivers to particular animals. Institutional responsibilities, in contrast, fall upon an industry, an organization, or a profession, and may be discharged by committee-based policy-making activities, as well as by personal conduct. Ethical questions about what kinds of science and technology should be done are not matters of personal choice for the individual researcher. They are questions that refer to social priorities, industry needs and consumer demands.

As such there is a critical need to develop a vocabulary with which such questions can be discussed and debated within the livestock sciences.

We may begin to recover that vocabulary by considering how we characterize human activities as guided by ethics in the most general sense. Human action involves making choices and acting under constraints. These constraints might be technological, that is, what is technically feasible (which is, of course, partly determined by the laws of physics and chemistry). Within philosophical ethics, the nostrum, 'ought implies can', indicates that one cannot be ethically required to do what is impossible to do. But livestock sciences alter these technical constraints, and thus sometimes the question about whether a technical constraint should be removed can take a form that is more familiar within the relatively formal and socially dictated constraints of law and policy. In law and policy a change in constraints (as well as enforcement of constraints) is done through processes of legislation, administration and judicial review in the case of government, or through corporate or organizational procedures in the case of private organizations. The kind of debate, review and decision process that would be used to alter a formal law or policy reflects a form of reasoning and communication characteristic of an institutional approach to ethics. So one way to understand ethics in livestock science is to ask the question, 'How can the forms of debate, review and decision making that are typical in forming law or policy be applied to technological innovation and research?'

However, it is also important to note that constraints may also take the shape of customs and norms, which are not as formal as law and policy but are nevertheless very influential on a person's behaviour. The norm of forming a queue to obtain service is informal, yet exceedingly robust in Western cultures. These informal norms change through processes that are not necessarily deliberative, and may even be quite mysterious. Nevertheless, informal norms are often associated with a very strong sense of propriety. People who accept informal norms also implicitly accept a presumption that following them is what one ought to do. In the present context, what is critical is simply to recognize that there are at least three general ways in which human activity operates under constraint: technical, formal and customary. The latter two, at least, are strongly associated with an understanding that they indicate what is *proper*, rather than what is *possible*. Thus, a collateral question for the ethics of livestock science is, 'How should informal norms and cultural expectations enter the debate over technological innovation and research?'

Given this initial orientation to ethics in livestock science, it is important to understand how the terminology in which ethical propriety is articulated reflects distinct elements of the decision-making situation. A very simple model of human activity looks something like this: someone considers alternative courses of action to take under these *constraints*, and then eventually chooses, taking one course of action or another. The action actually taken is his or her *conduct*. Although this characterization refers implicitly to individual human beings, it is important to note that it can be generalized to actions considered and taken by organizations or groups. The last key element in our simple model of activity is the outcome of action. Any action that individuals, organizations or societies take will have certain *consequences*, that is, your action is going to have impact upon the health and well-being of yourself and others. *Constraints*, *conduct* and *consequences* are thus three key elements in a very simple model of human activity.

These three elements are important because they link up with three distinct vocabularies or forms of discourse that we use to characterize human activity as ethical or unethical. Specifically, constraints become understood as ethically significant when they are characterized as *rights and duties*. The consequences of our actions may be considered to have a *beneficial or harmful* impact upon ourselves and others. Finally, certain types of conduct are described as *virtuous* while other types are *vicious*. In each case (but especially in the case of virtues) there are specific terms for distinct types of moral valuation. Thus conduct is honest or dishonest, courageous or cowardly, loyal or disloyal, etc. Such terms are more frequent in ordinary language than the terms 'virtue' or 'vice'. The connection between this elementary moral vocabulary and the simple model of human action provides a starting point for more systematic ethical enquiry.

From Ordinary Language Ethics to Philosophical Ethics

Philosophical ethics is the study of moral judgements, which is to say that it is a study of how human beings use ideas such as rights and duties, benefits and harms or virtue and vice to shape their own decision making as well as to evaluate the activity of others. Positivist-inclined scientists may well be impatient with the simplistic and unexceptional characterization of ordinary moral language given above, thinking that it only shows how chaotic and subjective ethical talk tends to be. For a long time, academic philosophers have been inclined to agree. For about the last 300 years in the discipline of philosophy, most philosophers working in ethics have been engaged in a project of reductionism, not unlike the sciences themselves. Briefly, the reductionist approach is to focus on one of these elements – rights and duties, virtues and vices, benefits and harms – in an attempt to show through philosophical argument and logical demonstration that one particular element is the most basic in determining ethical action. Reductionist philosophers have believed that if we focus on one of these elements, it will be possible to develop systematic and consistent ethical theories of the decision-making process.

For 300 years, philosophers have been refining theories in which one of these domains (usually rights or consequences) is seen as the touchstone for all the others. This historical observation is relevant here because reductionism has influenced the ways in which philosophers have approached animal ethics. What follows is a stripped down overview of some reductionist strategies that have influenced the way that ethical obligations to livestock are currently being discussed. The goal is to suggest that, as powerful and influential as these approaches have been, they do not do justice to the range of ethical problems at the frontiers of livestock science. I acknowledge that the characterization I give of these views below is itself over-simplified, and beg the forbearance of readers who are familiar with the philosophical details.

For example, the philosophy of utilitarianism focuses on the idea that it is impacts on health and well-being – the consequences of our actions – that are ethically basic. Here the word 'consequences' is used specifically as defined above, to indicate the outcome or impact that conduct has on the welfare of oneself and others. If a utilitarian talks about rights and duties, or virtues and vices, ultimately this has to be reducible to an underlying language relating to the way that our conduct affects health and well-being. Utilitarians reduce complexity in ethics by interpreting all ethical claims as referring ultimately to impact on welfare. Other elements typical of utilitarianism include the assumption that impacts on welfare can be ranked and summed. This allows the expected outcome from each possible course of action to be placed in a rank order, which (in combination with probability theory) allows one to assign an expected value to each option (each possible course of action). Utilitarians are known for the decision rule that mandates choosing the option that has the greatest expected value, once the welfare of all affected parties has been considered.

If animals experience the same kinds of impacts on their health and well-being as humans, then it would be logically inconsistent to ignore this when calculating the costs and benefits of our actions. Therefore, according to the principles of utilitarianism, the impact of human behaviour on the health and well-being of animals (for good or ill) should be reflected in our understanding of which actions are ethically justified. Impacts on the welfare of animals should be weighed along with impacts on the welfare of humans. An ethically responsible person or organization consistently chooses the course of action that produces maximal welfare, all affected parties (human or animal) considered. Peter Singer articulates these concepts in his famous book *Animal Liberation* (Singer, 2002).

A second reductionist strategy in philosophy has been a focus on moral rights or duties, sometimes called rights theory. The history of the rights tradition is rather complex, but Kant's philosophy has been especially influential in the debate over animals. Immanuel Kant (1724–1804) argued that for each of us, our own freedom or autonomy is basic to the way that we understand ourselves as capable of acting ethically. From this platform, we can derive a sense of what our most fundamental and basic rights and duties are, not only for ourselves but also for others. Kant's view was that a coherent picture of responsibility flows from a principle he called the Categorical Imperative: never treat others solely as a means to an end, but always as ends in themselves.

Kant was known for thinking animals could not have rights, or, more accurately, that humans could not owe moral duties directly to animals. He thought autonomy required much more advanced rational capacities than he believed animals possessed. Animal rights theorists, however, believe that, like humans, each animal has its own subjective identity. They argue that Kant was wrong to place so much emphasis on advanced rationality, noting that infants and mentally disabled human beings lack this capacity, as well. This pattern of reasoning, known as the Argument from Marginal Cases (AMC), provides the basis for including many animals in the sphere of rights-holding beings. If infants and mentally disabled human beings have rights, we must move to a more general criterion for our basic ethical principle. Once we do this, it is logically inconsistent to ignore the animal's subjectivity and rights when weighing our actions. Our duty is to respect this subjectivity and never reduce animals to a means to meet our ends. Tom Regan makes these points in *Animal Rights, Human Wrongs* (Regan, 2003).

A utilitarian emphasis on health and welfare outcomes and an alternative emphasis on the inherent rights of animals to live lives free from human interference are the two ethical philosophies that dominate the animal ethics debate. These two philosophies represent opposing strategies for reducing ethical complexity. Utilitarianism and rights theory represent the dominate alternatives within ethical theory for the 20th century, irrespective of whether responsibilities to animals are discussed. Many philosophical debates of the 20th century consisted in contrasts between utilitarian and rights-based positions. Furthermore, using Singer and Regan to illustrate these two theoretical approaches has proven to be an attractive pedagogical approach within college curricula. When many philosophers mention 'the animal rights debate' they are referring to the debate between Singer and Regan, as if these were the only valid philosophical alternatives.

Nevertheless, there are a number of challenges that might be made to utilitarian and rights philosophers in considering the ethics of how human beings relate to animals. Some recent challenges in the philosophical literature are discussed in the next section. With respect to livestock production systems, David Fraser has argued that reductionist approaches to animal ethics lead to three critical failures in insight: because they emphasize similarities between humans and other animals they neglect species-specific characteristics that are critical to good husbandry; they presume that the critical ethical problem lies in showing that humans *have* ethical responsibilities to animals, which the husbandry ethic has already recognized; and they neglect much of what is important about the agricultural context (Fraser, 1999). But the philosophical strategies utilized by Singer and Regan have not been immune from criticism by mainstream philosophers, and the next section reviews some recent debates.

Animal Ethics: Recent Philosophical Developments

If Fraser's 1999 paper is any indication, livestock researchers work with a representation of the philosophical approaches in animal ethics that is dominated by the work of Peter Singer, Tom Regan and Bernard Rollin. All three began to publish on animal issues in the late 1970s. The summary above links Singer to a reductionist strategy ending in a utilitarian theory of animal ethics, while Regan is linked to a reductionist strategy ending in a neo-Kantian ethic of rights. Rollin is better known perhaps among animal scientists than philosophers, and the work summarized above emphasizes the legacy of positivist philosophy of science. Rollin's ideas on animal ethics will be discussed below. Although an exhaustive discussion of the literature that has been generated by philosophers would be out of place in the present context, it may be useful to discuss a few long-standing debates as well as several recent developments. One goal is to familiarize readers with the breadth of approaches and viewpoints available in current literature. A second goal is to indicate how reductionist and pragmatist approaches in animal ethics offer alternative methodologies that may be useful for livestock researchers.

As noted above, philosophers have noticed a number of conceptual and logical problems with the theories of animal ethics developed by Peter Singer and Tom Regan. As early as 1978 Cora Diamond published an essay intended to show that the philosophical strategy being deployed by Singer and Regan would have far-reaching and probably unacceptable implications. Diamond notes that both utilitarian and rights-based approaches in animal ethics single out particular traits for which the difference between humans and other animals can be shown to be arbitrary. She finds fault with their claim that practices such as laboratory experimentation and meat eating are 'speciesist', that is, that these practices are deemed acceptable on moral grounds only as the result of a morally arbitrary bias favouring human interests. In rebuttal she notes numerous ways in which commonly accepted moral attitudes would be rendered incoherent if the reductionist argument against 'speciesism' is allowed to stand (Diamond, 1978).

Many of the deficiencies in the Singer/Regan approach become evident if one attempts to apply it in the domain of wildlife and biodiversity. The emphasis that both utilitarian and rights-based philosophies in animal ethics place upon the life and well-being appears to provide no philosophical basis for conservation of species in general, and is particularly deficient as a basis for the conservation of genetic diversity within species. This philosophical deficiency becomes particularly problematic in cases where it becomes necessary to control populations of wild or feral animals in order to preserve habitat. Ben Minteer (2004) notes in his review paper on the debate between animal protectionists and environmental preservationists that the philosophical approaches represented by Singer and Regan appear to provide little basis for hunting or population control efforts that cause suffering to individual animals in order to achieve this end. Because only individual animals (and not species) have welfare or rights, the approaches developed by Singer and Regan appear to rule out efforts that harm individual animals in service to species conservation.

There have been a number of philosophical responses to this problem. Gary Varner has argued that it is possible to augment both utilitarian and rights-based philosophies with a more sophisticated account of biological interests. Once this has been done, Varner believes that efforts to preserve habitat can be reconciled with these philosophies in view of the fact that depletion of habitat is itself a primary cause of suffering and the frustration of animal interests. Varner focuses specifically on hunting as a way to control the population of animals (such as deer) who tend to overshoot the carrying capacity of their habitat once predators have been reduced or removed (Varner, 1998).

Varner's main contribution to a more philosophically complex treatment of animal ethics consists in the way that he modifies the conceptions of animal welfare and animal rights as they have been developed by Singer and Regan, respectively. Varner moves away from the utilitarian claim that suffering or the rights view's claim that deprivation of the animal's control over its life (limited though that may be in the wild) are single factors on which an ethically acceptable human relationship to animals can be based. In place of these criteria, Varner offers a more complex account of biological interests, which include not only desire-based interests that might be articulated in terms of avoiding suffering or maintaining control, but also the satisfaction of functional needs that do not depend on conscious experience. Such needs reflect the physiology, behaviour and ecological adaptation of any given population of organisms. Indeed, Varner believes that plants, as well as animals, have functional needs that deserve some level of human recognition and moral accountability. This view allows Varner to argue that animals such as deer have an interest in having their population controlled (either through predation or what Varner calls 'therapeutic hunting') even though neither emphasis on suffering or autonomy provides a clear basis to recognize this interest (Varner, 1998).

Martha Nussbuam makes a move that is similar to Varner's with her recent attempt to apply 'the capabilities approach' to animal ethics. The key philosophical claim is that effective capacities, rather than rights or welfare, are what is morally basic. The capabilities approach had been developed by Nussbaum and by Amartya Sen as a response to deficiencies of utilitarian and rights approaches in the domain of development ethics. The view that human beings could be characterized in terms of suffering or happiness (a welfare criterion) or freedom and autonomy (a rights criterion) was criticized as overly reductive and also as not allowing the human beings who are the intended beneficiaries of development sufficient latitude to pursue goals that they themselves have articulated and conceptualized in language with which they are comfortable. 'Capabilities' are substituted and characterized as more complex, open-ended and situationally determined.

In the domain of human development, Nussbaum has compared the capabilities approach to the philosophy of Aristotle (384–322 BCE), who has been regarded as the leading exemplar of a 'virtues approach' in ethics (MacIntyre, 1982). Nussbaum lays particular stress on Aristotle's view that the overarching aim of ethical enquiry is to promote and realize human flourishing. The capabilities approach is intended to encourage the promotion of human flourishing through a non-reductive understanding of ethical enquiry as open-ended and even partially constitutive of the flourishing ideal. Thus, the particular specification of flourishing that is taken to be controlling in a given case can only be determined through ethical enquiry, and not by any criterion that can be known in advance of that. Again in the domain of human development, the capabilities approach requires the participation of all parties affected by the development programmes in question.

Nussbaum's suggestion that the capabilities approach can also be applied to animal ethics is based largely on her view that deficiencies in utilitarian and rights-based ethical theories have limited the creativity of philosophers working in this domain. Referring again in particular to problems with wildlife conservation, she notes that a capabilities approach would demand that human beings resist the temptation to specify what is good for animals in terms of what would be good for them. She suggests that animals themselves should be able to determine their own constitutive good without interference from human beings (Nussbaum, 2004). These arguments are advanced in support of the view that simply preserving habitat best articulates the basis of human responsibility to wild animals, rather than intervening in predator–prey interactions when a utilitarian or rights criterion suggests that a given animal should be entitled to such action on its behalf. It is not at all clear from Nussbaum's work on animals how a capabilities approach would be applied to food animals and livestock production. Nussbaum is important in the present context because she is among the most influential of contemporary philosophers and because her challenge to utilitarianism and rights theory emphasizes a non-reductive and open-ended philosophical enquiry.

One final example of recent philosophical criticisms of Singer and Regan comes from Elizabeth Anderson. Anderson's critique involves a detailed analysis of the argument structure developed by both Singer and Regan. Anderson emphasizes the use that both make of the aforementioned Argument from Marginal Cases (AMC). This argument notes that children and mentally disabled humans, who are regarded as members of the moral community, lack characteristics (such as complex rational capacity) that are put forward as a higher standard for moral standing. Recall that experiential welfare (in the case of utilitarianism) or subjective identity (in the case of rights) is advanced as the characteristic required for moral standing by Singer and Regan, respectively. Singer and Regan use the AMC to show that an alternative reductive approach requiring a complex ability such as the ability to use language is logically incoherent.

Anderson's criticism highlights the implicit social background required for making any sort of moral claim coherent. In place of reductive arguments for animal ethics, she substitutes accounts of the reciprocity that in fact does and has existed between humans and some species of animals. She suggests that these relationships provide a more promising way to conceptualize ethical responsibilities toward animals. Anderson (like Nussbaum) also emphasizes a view of moral norms that stresses how human beings endorse or respond to norms through 'affective, deliberative and agentic capacities' (Anderson, 2004, p. 291), rather than seeking a single feature or criterion that can be the foundation for all moral judgements. Like Nussbaum, Anderson uses this reasoning to support less invasive conceptions of human responsibility to intervene in the behaviour of wild animals, but Anderson also argues in favour of human efforts to control damage brought about by vermin or large predators. When animals threaten the stability of human communities, whether through destruction of life and property or as disease vectors, humans are fully justified in taking action to limit those threats. Also like Nussbaum, Anderson does not offer examples of how her view would apply to livestock.

Diamond, Varner, Nussbaum and Anderson each rebut Singer and Regan-style thinking in animal ethics by showing how the argument forms adopted by each lead to incoherence in our moral thinking, and respond to them by suggesting more complex approaches. It may be significant that three of these philosophers are women. Nussbaum and especially Anderson stress an active and deliberative conception of ethics in which the moral validity of a given judgement or norm derives from the circumstances in which the judgement or norm is developed or becomes operative. Both suggest that ethics cannot succeed by ignoring the procedural or deliberative circumstances in which normative judgements are made. Both reject a theoretical approach in which a single or at least very limited specification of characteristics or principles are held to be determinative of ethical validity in a comprehensive or totalizing sense.

For the most part, these commentaries are notable for their lack of attention to the context and practices of livestock production or research. Interestingly, Bernard Rollin's work is entirely omitted from the bibliographic essay that was prepared for the volume in which Nussbaum and Anderson's essays appear. This suggests that while animal ethics has become a central topic for mainstream philosophy, contributions by scholars known especially for their work with or on livestock are largely overlooked by researchers working in the leading research departments in philosophy. While these new trends in philosophical ethics may provide opportunities to move beyond the utilitarianism/rights theory dichotomy, they do no necessarily represent approaches that would be readily applicable in livestock research.

The Case of the Blind Hen

In 1999 a group of Danish researchers led by philosopher Peter Sandøe published a discussion of ethical issues raised by research on a strain of congenitally blind chickens. Previous research had established that these blind chickens were less likely to exhibit signs of stress or agitation under crowded conditions. This suggested that blind chickens might be a response to some of the animal welfare problems in poultry production, notably the aggressive behaviour of hens crowded together in the battery cage system of egg production that is, at this writing, still the most widely used approach in North America. Is it ethical to develop a strain of congenitally blind chickens? Would it be ethical to use them for egg production? The Danish group's aim was simply to establish the legitimacy of such ethical questions (Sandøe *et al.*, 1999).

The reaction to a summary of the Danish study that I made for a radio programme provides an anecdotal illustration of an important philosophical problem. The National Public Radio story on animal biotechnology was widely broadcast throughout the United States and Canada. My remarks follow:

> There's a strain of chickens that are blind, and this was not produced through biotechnology. It was actually an accident that got developed into a particular strain of chickens. Now blind chickens, it turns out, don't mind being crowded together so much as normal chickens do. And so one suggestion is that, 'Well, we ought to shift over to all blind chickens as a solution to our animal welfare problems that are associated with crowding in the poultry industry.' Is this permissible on animal welfare grounds?
>
> Here, we have what I think is a real philosophical conundrum. If you think that it's the welfare of the individual animal that really matters here, how the animals are doing, then it would be more humane to have these blind chickens. On the other hand, almost everybody that you ask thinks that this is an absolutely horrendous thing to do.
>
> (Kastenbaum, 2001)

The broadcast elicited numerous enquiries and an initial enraged response by the US poultry industry that was quieted only when they saw that the remarks were indeed based on peer-reviewed science. In addition, an animal protection group launched a campaign claiming that the story had advocated blinding chickens, urging their membership to write NPR in protest.

The outrage occurred in response to the very idea of blind chickens. The philosophical conundrum lies in the fact that leading theories of animal ethics tell us that this would be a good thing to do, while the moral intuitions reflected in that outrage tell us that it is an absolutely horrendous thing to do. Philosophers use the word 'intuition' to indicate a large class of seemingly immediate and involuntary cognitive experiences. *Perceptual intuitions* are raw sensations, like the cylindrical white shape I now see against a dull grey background. The white shape is my coffee cup and the background is my desk. However, seeing them *as* my coffee cup and desk may involve additional processing that I could defer at will, but it is hard to imagine how I could not see these shapes, so long as I can see at all. *Linguistic intuitions* are the 'sense' that we make of words and sentences when they are spoken or visually presented to us. Here, too, there is an involuntariness, a compulsive character that cannot be resisted. If someone says 'Move over, loser', I can *pretend* that I have not heard, but I cannot actually choose whether or not I want to understand (though I must, of course, understand English idioms to have this linguistic intuition). *Moral intuitions* are similar in that they are immediate, seemingly involuntary, and do not involve any conscious or thoughtful judgement. When confronted with a given situation (either in practice or, as above, by description), we just react to it as 'wrong'. It is quite possible that, as in the case of language, we are culturally educated into our moral intuitions, but this does not alter the fact that we seem unable to choose whether or not we will have them.

The normal case is that intuitions blend seamlessly into more carefully considered judgements. In the case of moral intuitions, we typically experience no dissonance between our immediate reaction and the judgement we reach when we thoughtfully review a situation in light of moral principles. But despite their immediacy, intuitions are not always reliable. Sometimes we realize that what we thought we saw or heard was not in fact what was there, what was actually said was not at all what we thought. The same is true for moral intuitions. In many cases where our first reaction is to think that something is morally wrong, we may be brought around to the idea that it is not wrong after all by reasoning carefully about the situation and considering all of the relevant details. But some moral intuitions are quite robust, and our sense of rightness or wrongness about them may remain even when thinking more carefully about them fails to support the initial reaction. Such intuitions produce conundrums.

Blind Hens and Reductionist Ethics

The thought of blind chickens producing our table eggs was apparently repulsive to many listeners; it just struck them as wrong. But reductionist theories of animal ethics do not support this judgement. Peter Singer's approach to animal welfare, for example, tells us that we should give equal consideration to interests, without regard to the animal that has these interests. We should take the suffering of animals into account in making our decisions and should not favour choices that produce trivial human benefits simply because the harm or suffering these choices cause happen to occur in non-human animals (Singer, 2002). Relevant in the present case are interests in avoiding the suffering that is associated with production disease. Conventional animals have these interests, and experience the suffering. Modified animals lack the interests and do not experience the suffering. If our goal is to minimize the unnecessary suffering in the world, as utilitarian philosophers have advocated for over 200 years, the choice seems direct. Organisms that lack the capacity to suffer cannot be harmed, so taking steps to create such organisms seems to be what a utilitarian would have us do.

Perhaps, one might think, a stronger animal rights view would not support this. The position advocated by Tom Regan, for example, might not support the use of blind chickens, for example, because even blind chickens still have an internal life experience, a sense of present and past, and a capacity to live their lives in a manner conducive to their own individual proclivities and interests. They are, as Regan would have it, *subjects-of-a-life*, and it would be wrong to treat them solely as tools or instruments for our own purposes (Regan, 1983, 2003). But more radical approaches in biotechnology might eliminate this capability altogether. By Regan's own reasoning, animals (such as insects or protozoa) that lack any conscious capability altogether are *not* subjects-of-a-life. If we can develop an animal that produces meat, milk or eggs and is not a subject-of-a-life, there is nothing or no one to be harmed by doing so. Further, if doing that is a step toward removing ordinary pigs, cattle and chickens from the production circumstances where their rights are, in Regan's view, currently being violated, it would seem that his ethic of 'empty cages' weighs in on the side of developing such literally mindless animals.

I use the case of blind chickens to suggest that it is not the impact on the chicken's health and well-being that motivates people to turn against the use of blind hens in poultry production systems. A much better way to account for this moral intuition notes that many people share the sense that blinding hens (even through these genetic means that do not deprive – and hence harm – a previously sighted hen) is not how virtuous people should act. That is, modifying the physical characteristics of animals to this level in order that they better fit modern production systems is not justifiable. Here is a case where eliminating the language of virtue from animal ethics tends to divide scientists and philosophers from the great majority of people, for whom the need to act virtuously is important. This perspective should not be overlooked in the animal ethics debate, nor should scientists hoping to apply ethical thinking in development of their research programmes neglect it.

Ethics and Animal Biotechnology: a Review

Bernard Rollin introduced the basic conceptual problem illustrated by the blind chicken problem in one of the first papers written on the ethics of using rDNA-based techniques for animal transformation over 20 years ago (Rollin, 1986). His 1995 book *The Frankenstein Syndrome* gave a lengthy discussion of the ethical issues that might be raised by transformation, and his analysis has been the subject of several subsequent discussions by philosophers. Rollin's analysis utilizes the notion of *telos*, a concept he introduced into animal ethics as a way to characterize the genetically based capacities, drives and functional needs that would be associated with animals of a given species or breed. He frames the ethical questions of genetic engineering by asking the question, 'Under what circumstances is it ethically permissible to alter an animal's *telos*?'

The answer to this question is summarized by the *Principle of Conservation of Welfare* (PCW), which holds that transformations are acceptable so long as the general health and propensity to endure pain or other forms of deprivation are comparable to animals of the species or breed that have not been developed through genetic engineering. Rollin is clear in stating that the main ethical concerns with genetic engineering lie in transformations that create dysfunctional animals. He cites the early experiments with swine transformed for faster growth (the so-called 'Beltsville pigs') as a case in point. The PCW rules out the intentional use of such animals in pursuit of more profitable production. The exception to this rule is an animal developed as a disease model, which would need to exhibit the symptoms and suffering characteristic of the disease in order for human or veterinary medical research to proceed. Rollin argues that such animals can be developed only on strict applications of laboratory research guidelines intended to limit suffering to the absolute minimum. In this connection, he endorses further types of genetic engineering that might address suffering by rendering research animals 'decerebrate' (Rollin, 1995b).

Other authors recognized fairly quickly that decerebrate animals might in principle be developed for routine food and fibre production as a response to welfare problems in concentrated animal feeding operations (CAFOs). Like blind hens, this possibility runs deeply afoul of moral intuitions. As such, fairly extensive literature has developed in response to Rollin's work. Many of the authors who have been moved to write in criticism of Rollin's analysis articulate a position that is actually just a restatement of Rollin's PCW rule. Their disagreement with Rollin derives from uncertainties they associate with the risk of causing harm through transgenic modification, which leads them to conclude that genetic engineering of animals is intrinsically wrong (see Fox, 1990; Sapontzis, 1991; Warkentin, 2006). Other authors note simply that there is disagreement about whether alteration of *telos* is morally acceptable (Appleby, 1999; Gifford, 2002). Rollin, for his part, has laid stress on the view that this conflation between animal welfare risks and the claim that genetic engineering is intrinsically wrong is just an example of faulty ethical reasoning (Rollin, 1996).

Another group of philosophical authors have attempted to articulate principled reasons that would provide a rationale to support the intuitive reaction that many people have to those genetic manipulations that, like the blind chicken problem, resolve a welfare problem by reducing or eliminating the organism's ability to experience well-being (or suffering) at all. Phillipp Balzer, Klaus Peter Rippe and Peter Schaber have argued that this kind of genetic engineering is an affront to the animal's dignity (Balzer *et al.*, 2000). Similarly, Bernice Bovenkirk, Frans Brom, and Babs van den Bergh analyse the trouble with genetic engineering as an ethically problematic challenge to animal integrity (Bovenkirk *et al.*, 2001). Each of these articles offers treatments of dignity or integrity respectively that would indeed be convincing if it were the case that one were robbing an animal that had the capacity in question, that is, if one were surgically or chemically blinding a sighted hen, to utilize the blind chicken example again. But animals that are produced through genetic modification have never had the capacities in question, as congenitally blind hens have never had sight.

These two groups of authors argue that genetic engineering is wrong because it does not show sufficient respect to animal dignity or animal integrity. In fact it does not matter whether it is dignity or integrity that is the object of respect. By tying the analysis to the norm of ethical respect, these authors wind up with a result that is very much like the argument already discussed in connection with animal rights views, which require that one show respect for animals' subjectivity, for their being the subject of a life. We may concede that the act of depriving a being that has subjectivity of its ability to experience a state of well-being fails to show respect, but how can the use of an organism that has no subjectivity be understood as a lack of respect. It seems that we would show greater recognition of and respect for the dignity or integrity of creatures that *have* dignity or integrity by leaving them alone and using an organism that lacks (and has always lacked) the basis for dignity or integrity, which must certainly have something to do with ability to experience its own life.

Yet another philosophical strategy has been deployed by Henk Verhoog and by Allan Holland, both of whom argue that the problem with altering *telos* cannot be analysed as a form of harm or lack of respect either to individual animals or to the species as a whole. Rather each sees the wrongness in these practices residing in how they affect human beings. Verhoog argues that these practices threaten the intellectual coherence of the way that we understand human relationships with other animals (Verhoog, 1993). Holland argues that this type of transgenic work adopts a totally instrumental attitude toward animal life, an attitude that runs contrary to the respect that human beings should show towards living beings, conscious or not (Holland, 1995). Both of these approaches could be reconciled with the previous suggestion that intuitions in the blind chicken case are actually relying on a virtues approach in ethics, one that operates somewhat independently of the way that human actions are understood as being unacceptable when they violate moral constraints (e.g. rights) or produce harmful consequences.

For his part, Rollin has responded to these criticisms mainly by reiterating his original view. He finds the idea that altering *telos* itself is immoral, inconsistent with centuries of human domestication of animals, and hence simply a case of poor ethical reasoning. He has analysed the types of argument offered by Verhoog and Holland as making 'aesthetic' rather than ethical objections to biotechnology. Here, Rollin accepts that consumers may resist such products, but argues that they are not doing so on moral grounds (Rollin, 1998). Only in a very brief paper responding to the suggestion that challenging species in biomedical research can be troubling to people does Rollin admit the possibility that such practices might be ethically problematic by virtue of their potential to damage the way that people think and speak to one another about the idea of humanity and the place of human beings in the broader world (Rollin, 2003a). Here, again, the moral logic of this concern resides in the domain of virtues, rather than that of rights or consequences.

Toward a more Pragmatic Ethic for Livestock Research

All elements of the ethical debate – including rights, virtues, benefits and harms – should be considered when determining practices in modern agricultural systems. We need an approach *opposite* to the reductionist approach. The animal welfare philosopher needs to ensure that all elements are included in the course of deliberation. Deliberation should avoid a situation in which one element dominates the others. Developing policy and making choices mindful of all these elements is an extremely complex process – there are no easy reductionist formulas that encapsulate all elements of the debate.

There are, for example, other ways in which virtue-thinking influences public attitudes toward contemporary agriculture. Many people have become suspicious that the use of animals in modern production systems is driven by profit and efficiency and not by a consideration of the wider ethical framework (McNaughton, 2004). My general sense is that the majority of people feel that general agriculture, as practised over thousands of years, is an acceptable way of relating to and using animals. While the animals may not be leading lives free from predators, disease and suffering, the gap between the domesticated livestock and their human caretakers is not so great. There was a degree of reciprocity between humans and animals in the pre-industrial agricultural setting. This was, of course, Rollin's point in noting the turn from husbandry to science as a worldview for understanding human–animal relationships.

As Rollin notes, the move to an industrial setting for agriculture has changed this dynamic. Technology has given humans too much of an upper hand (Rollin, 2003b). Rollin's writings on agricultural biotechnology have suggested that a very limited ethic focused on animal suffering and genetically based drives is sufficient to guide research. He has described concerns about animal genetics being altered as examples of poorly reasoned ethics (Rollin, 1996). The case of the blind hen suggests that Rollin himself may have viewed the transformation of animals too narrowly at the outset, though in subsequent writings he has admitted that what he calls 'aesthetic' considerations may lead people to prefer traditional forms of animal production to those that apply biotechnology as a means to eliminate welfare problems by reducing the animal's capacity to experience a welfare in the first place (Rollin, 1998).

The question is, how does one develop a philosophical approach in animal ethics that brings all of these elements together? Bovenkirk, Brom and van den Bergh's discussion of animal integrity demonstrates cognizance of the possibility that framing the problem in such terms can mislead people to look for either a rights violation or a form of consequential harm. In reply, they argue that the term 'integrity' has the advantage of directing people to look at the broader context of action, the way that animals and human beings interact in a social and ecological environment. They also call attention to the 'whole life' of the animal, rather than simply to its suffering or welfare. Finally, they conclude that the language of integrity is 'flawed but workable', arguing that the main philosophical reason for adopting it is that doing so facilitates a public debate over the ethics of animal biotechnology. In this respect, these authors, like Nussbaum and Anderson, are advocating an approach in ethics that stresses the processes and norms of debate and deliberation, rather than the specific conceptual content of an idea like integrity.

While no one thinks that animal production systems of the future will look like the traditional agrarian systems of the past, we need to reconsider this traditional relationship of humans and animals within the context of more scientifically based and industrial settings, to meet the goal of improving animal welfare outcomes. I am *not* saying that a virtues approach should *replace* consideration of benefits and harms, on the one hand, or rights and duties, on the other. Neither am I saying that profit has no place in animal production, as those who look at my other writings on farm animal welfare will understand. Rather, my point is that we need negotiated standards when talking about production systems and research practices. These standards need to emerge from open discussion and deliberation which is not driven by a purely strategic desire to ensure that 'the public' accepts what science deems to be an effective solution. These deliberations, however, must be informed by both science and philosophical ethics, and the philosophical ethics that informs them needs to encompass the full range of ethical vocabularies available to us. I have described how animal scientists should think of their responsibility to participate in such deliberations elsewhere (Thompson, 1999).

In closing, it may be useful to emphasize the shift in philosophical perspective that takes place when one stops thinking of ethics as a reductionist does and starts thinking as a pragmatist does. In one sense, many of the philosophers discussed in this review represent moves in the pragmatist direction, though it is unclear that any of them would be willing to identify themselves as such. Diamond, Nussbaum and Anderson each lay some degree of emphasis on the way that ethical norms and ethical vocabularies are interwoven with ordinary practices in very complex ways. They suggest that people have very sophisticated abilities to make ethical discriminations independent of any exposure to philosophical ethics. As such, it is wise to be cautious in developing theoretical approaches in ethics that imply radical departures from ordinary common sense. On the other hand, it is clear from history that entire societies and cultures have tolerated or even embraced practices that do not appear defensible given contemporary understandings of ethical norms. As such, any philosophy must address the question of how and under what conditions common-sense understandings should be challenged. One key element of pragmatist doctrine, in this respect, is that no normative understanding or practice is exempt from critique or revision; hence it is indeed possible to radically rethink responsibilities to non-human animals. However, pragmatists would reject wholesale upheavals in the entire framework of common, everyday ethical understandings as leading to conceptual incoherence, rather than insight. As such, proposals for ethical reform or change are viewed as hypotheses that must be subjected to a number of philosophical and also common-sense, public tests.

Bryan Norton (2005) has produced a careful philosophical treatment of how pragmatism might reshape the way that ethics is integrated into environmental decision making, and what he has written provides the philosophical backdrop for what is being said here in summary form. Pragmatism has its philosophical roots in observations on scientific method made by Charles Sanders Peirce in the 19th century. Peirce argued that science is a particular form of rule-governed social enquiry. The warrant for scientific results does not consist in their consistency with some predetermined metaphysical standard for what is real, or in their meeting epistemologically derived standards for truth. Rather, scientific results derive the authority that they have from the way that scientists establish networks of researchers who not only utilize publicly reproducible (and hence testable) methods to generate their results, but who actually undertake the testing and reproduction of each other's results. Given infinite time and resources, Peirce argued, the results that such a community of enquiry would produce are indeed true and factual, but it is not truth or factuality that is most important in the real world of limited time and resources. Instead, it is the collective result of scientists' compliance with these public methods of enquiry that provides the justification for regarding science as providing the best warranted basis for human action, even while most scientists expect that their results will eventually be overturned by future studies (Norton, 2005).

Beginning with John Dewey and subsequently with the German philosopher Jürgen Habermas, pragmatism in ethics has similarly moved toward a more procedural approach, one in which establishing the parameters for good discourse is seen as equally important as the specific norms or claims generated by moral discourse. Unfortunately, the positivist legacy in the sciences has made scientific researchers among the most resistant to actually participating in open, public argumentation on ethical norms, values and responsibilities. Ironically, scientists' reluctance to engage in continuing ethical debates over activities such as animal biotechnology has almost certainly contributed to the public's suspicion of genetically engineered crops (Thompson, 1997). But it is not enough simply to make public statements about biotechnology's benefits. It is also necessary to engage one's critics on a level playing field, to be willing to respond to criticisms in the same kind of ethical language in which they are advanced. Thus, a key problem of the sciences is that when they do engage in ethical discussions they are too narrowly focused on risks and benefits, that is, on the consequentialist dimension illustrated above. It is equally important that scientists learn to speak the language of rights and virtues (Burkhardt, 1998).

Interestingly, it seems likely that a concerted, institutionally organized and publicly visible effort to engage in serious practical deliberation about the ethics of animal transformation might be precisely the sort of thing that is needed to address the virtue arguments noted in connection with the blind chicken problem. There, it seemed as if scientists' willingness to find technical solutions to production disease and their disinterest in socially generated responses was at the heart of the concern that this is an ethically horrendous thing to do. Here, devoting time and energy to a serious and publicly open dialogue on ethicality of animal transformation is precisely what we would expect a virtuous person to do. If a public dialogue were to establish that a genetically engineered animal truly was the most ethical response, one may hypothesize that the virtue-argument levied against this kind of work would be seriously weakened. Yet such a dialogue would clearly have to be undertaken in a genuinely open-ended way, so that other types of change in production systems might stand a chance of being endorsed as the most ethical response. And it is important to acknowledge that this is only a hypothesis; there is no way to know what a public dialogue would produce in advance.

It is, perhaps, worth stressing exactly *how* this approach is being put forward as a response to ethical challenges raised by biotechnology. The argument above concludes that the ethical problem with blind chickens and biotechnology *cannot* be understood in terms of harm to (or violation of the rights of) animals produced using genetic alteration. Instead the ethical problem arises in connection with the view that this kind of activity is unvirtuous because it appears calculating and insensitive to deeply held intuitions. Conducting a public dialogue must be undertaken primarily as a redress of that insensitivity, and in order to be effective, concern for those insensitivities must be sincere. The hypothesis is that in *demonstrating* concern for these sensitivities, the primary basis for finding scientists' conduct unvirtuous is removed by the very act of conducting the dialogue. If scientists are understood to be making the best possible response to current problems in livestock production, given economic and political constraints, the objection from virtue should be removed, and perhaps these applications of biotechnology can go forward with public support. But it is also possible that a public dialogue will also result in the political will to change economic and political constraints so that more traditional production systems can meet animal welfare needs, obviating the need for animal transformation. Which one of these outcomes will occur can be determined only by conducting the public dialogue.

The good news for scientists is that as one moves towards pragmatic ethics the rather narrow and dogmatic assertions criticized so ably by David Fraser are replaced by a philosophical approach that is much more sensitive to the context of agricultural production (Thompson, 2004b). The bad news may be that a pragmatist approach requires continuous work, and a long-term commitment to critical engagement with philosophers and members of the broader public alike. Yet I am hopeful that once positivism is truly dead in the animal sciences, it will be possible to incorporate ethics at the very frontiers of research in a manner that is subtle, wise and comprehensive of the full range of values that guide human conduct.

References

Anderson, E. (2004) Animal rights and the values of nonhuman life. In: Sunstein, C. and Nussbaum, M. (eds) *Animal Rights: Current Debates and New Directions*. Oxford University Press, Oxford, pp. 277–298.

Appleby, M.C. (1999) *What Should We Do about Animal Welfare?* Blackwell, Oxford.

Balzer, P., Rippe, K.P. and Schaber, P. (2000) Two concepts of dignity for humans and non-human organisms in the context of genetic engineering. *Journal of Agricultural and Environmental Ethics* 13, 7–27.

Bovenkirk, B., Brom, F.W. and van den Bergh, B.J. (2001) Brave new birds: the use of integrity in animal ethics. *Hastings Center Report* 32, 16–22.

Burkhardt, J. (1998) The inevitability of animal biotechnology? Ethics and the scientific attitude. In: Holland, A. and Johnson, A. (eds) *Animal Biotechnology and Ethics*. Chapman and Hall, London, pp. 114–132.

Diamond, C. (1978) Eating meat and eating people. *Philosophy* 53, 465–479.

Fox, M.W. (1990) Transgenic animals: ethical and animal welfare concerns. In: Wheale, P. and McNally, R. (eds) *The Bio-Revolution: Cornucopia or Pandora's Box*. Pluto Press, London, pp. 31–54.

Fraser, D. (1999) Animal ethics and animal welfare science: bridging the two cultures. *Applied Animal Behaviour Science* 65, 171–189.

Gifford, F. (2002) Biotechnology. In: Comstock, G. (ed.) *Life Science Ethics*. Iowa State Press, Ames, Iowa, pp. 191–224.

Holland, A. (1995) Artificial lives: philosophical dimensions of farm animal biotechnology. In: Mepham, T.B., Tucker, G.A. and Wiseman, J. (eds) *Issues in Agricultural Bioethics*. University of Nottingham Press, Nottingham, pp. 293–306.

Kastenbaum, D. (2001) Analysis: debate over genetically altered fish and meat. National Public Radio News, Morning Edition, December 4, 2001. Transcript www.npr.org (accessed April 2006).

MacIntyre, A. (1982) *After Virtue*. University of Notre Dame Press, South Bend, Indiana.

McNaughton, P. (2004) Animals in their nature: a case study on public attitudes to animals genetic modification and 'nature'. *Sociology* 38, 533–551.

Minteer, B.A. (2004) Beyond considerability: a Deweyan view of the animal rights–environmental ethics debate. In: McKenna, E. and Light, A. (eds) *Animal Pragmatism: Rethinking Human–Nonhuman Relationships*. Indiana University Press, Bloomington, Indiana, pp. 97–118.

Norton, B. (2005) *Sustainability: A Philosophy of Adaptive Ecosystem Management*. University of Chicago Press, Chicago.

Nussbaum, M.C. (2004) Beyond 'compassion and humanity': justice for nonhuman animals. In: Sunstein, C. and Nussbaum, M. (eds) *Animal Rights: Current Debates and New Directions*. Oxford University Press, Oxford, pp. 299–320.

Regan, T. (1983) *The Case for Animal Rights*. University of California Press, Berkeley, California.

Regan, T. (2003) *Animal Rights, Human Wrongs: An Introduction to Moral Philosophy*. Rowman and Littlefield, Lanham, Massachusetts.

Rollin, B.E. (1986) The Frankenstein thing. In: Evans, J.W. and Hollaender, A. (eds) *Genetic Engineering of Animals: An Agricultural Perspective*. Plenum Press, New York, pp. 285–298.

Rollin, B.E. (1995a) *Farm Animal Welfare*. Iowa State University Press, Ames, Iowa.

Rollin, B.E. (1995b) *The Frankenstein Syndrome: Ethical and Social Issues in the Genetic Engineering of Animals*. Cambridge University Press, New York.

Rollin, B.E. (1996) Bad ethics, good ethics and the genetic engineering of animals in agriculture. *Journal of Animal Science* 74, 535–541.

Rollin, B.E. (1998) On *telos* and genetic engineering. In: Holland, A. and Johnson, A. (eds) *Animal Biotechnology and Ethics*. Chapman and Hall, London, pp. 156–187.

Rollin, B.E. (2003a) Ethics and species integrity. *American Journal of Bioethics* 3, 15–17.

Rollin, B.E. (2003b) Toxicology and new social ethics for animals. *Toxicologic Pathology* 31 (Suppl. 1), 128–131.

Sandøe, P., Nielsen, B.L., Christensen, L.G. and Sørensen, P. (1999) Staying good while playing God – the ethics of breeding farm animals. *Animal Welfare* 8, 313–328.

Sapontzis, S.F. (1991) We should not manipulate the genome of domestic hogs. *Journal of Agricultural and Environmental Ethics* 4, 177–185.

Singer, P. (2002) *Animal Liberation*, rev. edn. HarperCollins, New York.

Thompson, P.B. (1997) *Food Biotechnology in Ethical Perspective*. Chapman and Hall, London.

Thompson, P.B. (1999) From a philosopher's perspective, how should animal scientists meet the challenge of contentious issues? *Journal of Animal Science* 77, 372–377.

Thompson, P.B. (2004a) The legacy of positivism and the role of ethics in the agricultural sciences. In: Scanes, S.G. and Miranowski, J.A. (eds) *Perspectives in World Food and Agriculture 2004*. Iowa State University Press, Ames, Iowa, pp. 335–351.

Thompson, P.B. (2004b) Getting pragmatic about farm animal welfare. In: McKenna, E. and Light, A. (eds) *Animal Pragmatism: Rethinking Human–Nonhuman Relationships*. Indiana University Press, Bloomington, Indiana, pp. 140–159.

Varner, G. (1998) *In Nature's Interests? Interests, Animal Rights and Environmental Ethics*. Oxford University Press, Oxford.

Verhoog, H. (1993) Biotechnology and ethics. In: Brante, T., Fuller, S. and Lynch, W. (eds) *Controversial Science*. SUNY Press, New York, pp. 83–106.

Warkentin, T. (2006) Dis/integrating animals: ethical dimensions of the genetic engineering of animals for human consumption. *AI and Society* 20, 82–102.

4 The Impact of Genomics on Livestock Production

Ross Tellam
CSIRO Livestock Industries, Queensland Bioscience Precinct, Australia

Abstract

The unique genome of a livestock animal represents the potential of that individual in a particular production environment. In contrast, the different genomes of individuals of a livestock species reflect the genetic plasticity of the species as a whole and the diversity of genetically determined phenotypes that can be produced by the species. Exploitation of this individual variation within a species is at the heart of livestock breeding programmes. The bovine genome sequence in 'draft' form will be completed in 2007. The availability of the sequence and very large numbers of single nucleotide polymorphisms will have profound effects on beef cattle, dairy cattle and sheep production by enabling acceleration of the selection of animals with desirable production traits. Many livestock industries are poised to capture the benefits from this enormous wealth of genome sequence information because they have comprehensive databases containing phenotypic and pedigree data for large numbers of animals, intensively use genetics in breeding programmes and efficiently manage reproductive performance. The immediate challenge facing the livestock industry is the integration of new technological capabilities into existing breeding programmes and production systems.

The Fidelity and Plasticity of the Genome

The genome of a living organism consists of DNA sequence information that codes for the protein building blocks of life as well as instructions on their assembly and control. Within the context of a few generations the genome is largely unchanging, transmitting this information from parent to offspring with remarkable fidelity. The combination of this genetic information with environmental influences defines the production potential of a livestock animal. At the molecular level this potential reflects the interaction between environmental factors and gene activities. Contrasting in some ways with this process is the enormous phenotypic variation of individuals within a species and the remarkable form and physiological differences, even between highly related species that share great similarity in their DNA sequences. These observations are rationalized in the context of differing timescales, in which the generational fidelity applies over relatively brief periods of time, whereas genomic variation between individuals is a function of evolution of the species and occurs over many generations and long timescales. In the latter instance, advantageous mutations are propagated within the species and these both define the species and lead to individual variation within the species. The diversity of phenotypes generated by this process reflects the plasticity of the genome and the potential of a species to adapt to new challenges to its existence. The current genomics revolution is attempting to unmask these fundamentals of life in exquisite detail and in the process will provide tools that enable greatly improved management of livestock animals.

Since the dawn of domestication of plants and animals, perhaps some 8000 to 9000 years ago, humans have been practising geneticists, enriching for desirable attributes in these highly selected species (Wendorf and Schild, 1994). This practice provided enormous advantage to humans as food became relatively reliable and abundant and allowed specialization of humans for many different tasks within a society. This in turn fostered the growth of larger communities and the development of new knowledge, which reinforced this process. However, even today there are very few plants and animals that have been domesticated compared with the full repertoire of living organisms. Some do not provide adequate nourishment compared with the human effort involved in their production, others are unsuitable because they contain toxins or nutritional components not suitable for human consumption and still others 'resist' domestication because of a variety of factors including animal social structures and their reproductive strategies. In the context of Darwinian natural selection it can be argued that domesticated animals and plants 'chose' humans to facilitate the survival and selective advantage of their species. The success of this process is highlighted by the growing biomass of domesticated species, often at the direct expense of non-domesticated species. The phenotypes and presumably genomes of domesticated plants and animals of today are generally far removed from those of their wild ancestors due to intense selection pressures imposed by humans. The potential for additional changes that enhance the productivities of the industries associated with these domesticated plants and animals, while minimizing their environmental footprint, is clear.

Just as the 1970s was the decade of electronics, the first decade of the new century heralds the era of genomics and the promise of major advances in biology. In the livestock industry the ultimate benefits of genomics technology, which largely exploits the natural genetic variation in livestock animals, will be increased food production and quality for lower cost, enhanced food safety, a reduced environmental footprint and the development of novel foods with enhanced health, nutritional and lifestyle benefits. Importantly, these gains will be achieved in a manner that does not overly burden the animal itself.

Genomics

What is genomics? It means different things to different people but it is largely the study of the totality of the activities of all genes within an organism and how they express themselves in the different physical traits of a species, breed or individual. Operationally, genomics is defined as the science of simultaneous investigations of the structure and function of very large numbers of genes in the context of the entire DNA sequence of an organism. Animal breeders have been practitioners in the field for many thousands of years and this has markedly increased the economic value of many agriculturally important species. In this traditional activity, selection is based on observable characteristics of an animal but without knowledge of the genes that underlie the selected trait. There are some limitations to this breeding strategy, however, especially for selection of traits that are difficult or expensive to measure, have low heritability, are negatively correlated with other economically important traits or appear late in development and cannot be measured until the animal has contributed to the next generation. The difference now is that molecular tools based on DNA sequence information are available to understand, direct and speed the process of selection of animals and plants with desirable traits.

The new genomics capabilities are the result of the confluence of new technologies – molecular biology, chemistry, electronics, computing and robotics – applied to an old science, genetics. This is best exemplified by the remarkable achievement, announced at the beginning of this decade, of the determination of the human genome sequence, which encompasses about 3 billion pieces of genetic information or nucleotides. Since then the genomes of additional mammalian species such as the mouse, rat and dog have been determined. The genome sequence of the first production mammal, a Hereford cow named *L1 Dominette,* is nearing completion at a total cost of US$53 million. The cost of the human genome sequence has been reported at US$2.7 billion (NHGRI, 2005a).

The difference in these numbers reflects not only the differing extents of coverage and accuracies of these sequences (the bovine sequence is a 'draft' sequence*) but also the rapidly decreasing cost of sequencing, which is attributable to significant technological innovations and factory-based production systems. The majority of the current genome sequencing efforts involve large consortia and are tributes to international scientific cooperation and the mastery of technological challenges. The cost of DNA sequencing is halving on average every 22 months (Guttmacher and Collins, 2005). Moreover, new technologies are being developed that promise radical developments in this area and perhaps will be able to sequence a mammalian genome at low coverage for as little as US$1000 (NHGRI, 2006a). Thus, the public databases are increasingly awash with genome sequence information.

So Why Invest Large Sums of Money in Genome Sequences?

The culmination in February 2001 of the deciphering of the human genome sequence represents an historical milestone in human endeavours that will have far-reaching implications for our understanding of biology. The repercussions of this achievement, which will be felt far into the future, are impacting on studies of all species and have enabled the development of new approaches to the understanding of the typically complex relationship between DNA sequence and phenotype. One of the primary motivations driving the determination of the human genome sequence was the promise of new medical treatments especially for complex multi-factorial and chronic diseases such as cancer and type II diabetes. However, of equal importance was the recognition that DNA sequence can aid the understanding of the origins and evolution of humans as well as revealing the mechanisms underlying change in a species.

For humans, the genome sequence and its accompanying molecular tools provide detailed insight into the molecular mechanisms causing disease and thus will generate new opportunities for disease treatment and prevention. The genome sequence of individuals or associated DNA markers also provides the opportunity to tailor disease treatments to the specific needs of an individual thereby achieving better outcomes for the individual. The negative aspect of this strategy will be the increased cost of medical care associated with both the genotyping of individuals and their individualized treatment regimes. Offsetting this, there will be increased focus on preventative health approaches that may entail early dietary intervention in individuals identified as predisposed to the development of specific diseases later in life, an approach termed nutrigenomics. Here the interaction between genotype and environment is harnessed to produce a desirable outcome.

The second and perhaps more compelling motivation is the drive to understand the fundamental mechanisms underpinning inheritance, human development and the response of humans to their environment. The fidelity of inheritance patterns and the programmed nature of development argue that these biological processes are highly regulated but, as yet, we know little about how these systems function at the molecular level. There are likely to be fundamental design rules, as yet unknown, that govern and regulate the enormous complexities of cellular biology. Identification of these rules will also enable better health care through an ability to specifically target a component of the cellular machinery in a manner that has a highly predictable and beneficial outcome. Potentially, these fundamental rules of biology may also provide new understandings of unrelated highly complex processes such as weather patterns, stock market volatility and the ultimate example of complexity, traffic flows at peak hour. What is also being revealed is the responsiveness of genes to external stimuli, which allows adaptation of an organism in real time to its changing environment. The molecular mechanisms governing the detection of environmental cues, the transmission of this information into a cell and the subsequent molecular and cellular responses are only beginning to be elucidated. Again, the common feature of these processes is their complexity, which probably underlies their robustness, accuracy and reproducibility.

The motivations for sequencing livestock genomes, such as the bovine genome, are somewhat different. Here the primary motivation relates to economic benefit to humans, although human health care interests such as food quality and disease control are significant and growing in importance. The livestock genome sequence provides molecular tools for producers and processors to optimize the outcomes of the genetic potential of an animal and to better match product with the market. The nexus between food quality and human health, albeit recognized as important, is yet to be fully realized. The obesity epidemic in humans and its associated medical and economic consequences is indicative of this disjunction. Diseases of the food supply chain, such as bovine spongiform encephalopathy and foot and mouth disease, have taken economic and social prominence and consequently there is strong interest in minimizing and eliminating their potential future impacts using our increased understanding of biology, genetics and gene function. Related to this is the increasing prevalence of infectious diseases, such as severe acquired respiratory distress syndrome and the H5N1 strain of bird flu, which use wild animal populations, and in some cases livestock animals, as reservoirs for infection of humans. Two additional motivations for livestock genome sequences are to provide tools that address environmental and animal welfare concerns.

What Information is Contained Within a Genome Sequence?

The human genome sequence has been referred to as the book of life but it is actually four books (J. Shine, Garvan Institute Sydney, 2004, personal communication). The first is a history book that contains information about the distant past and the challenges for survival of the species. It is said that history is written by the victors and, in that sense, so is the genome sequence as it only records information contained in the animal lines that have survived the rigours of natural selection. The second book contains a list of the component parts or genes and an instruction manual for their use. While we now have a good catalogue of the approximate 25,000 genes in a mammal, the instruction manual for their use remains elusive. The third book is a medical text book that contains information about the specific adaptations of a species to disease challenges. Finally, there is a 'Who's Who' book that records the unique genetic characteristics of each individual in the species. The uniqueness of the individual is the strength of the species as it ensures survival of some individuals during times of strong and adverse natural selection, e.g. during disease challenges. Livestock genome sequences have similar information content.

We are still far from understanding all that is contained within a genome sequence. The present status is like having correctly reassembled a shredded white pages telephone directory for a large city, where phone numbers represent genes and the addresses correspond with chromosomal positions. We have the list of about 25,000 genes and their addresses but we don't have their functions or how they are controlled, i.e. a corresponding yellow pages phone directory is yet to be fully constructed. Intriguingly, the actual protein encoding genes in a mammalian genome only account for about 1.5% of the genome sequence. The remaining 98.5% has not been thoroughly explored but this non-coding DNA is likely to contain complex gene expression regulatory information and perhaps quite a few surprises as well.

The Bovine Genome Sequencing Project

The bovine genome consists of 29 pairs of autosomal chromosomes and an X and Y chromosome collectively containing approximately 2.7 billion base pairs of DNA. The plasticity of this genome is evident in the phenotypic diversity of the many bovine breeds which have become specialized for milk and meat production, ability to work and carry heavy loads, and to prosper in tropical environments in the presence of continuous parasite challenges. This specialization is relatively recent in the timescale of evolution of the species and reflects intense selection pressures on the breeding populations.

The Bovine Genome Sequence Project (BGSP) was born out of the Human Genome Project (HGP) and capitalized on its massive infrastructure investments and know-how at a time when the costs of DNA sequencing were dramatically decreasing due to technological innovations and industrial-scale management. It was also widely recognized that mammals have very similar complements of genes and thus the bovine genome sequence would allow capture of the enormous wealth of functional information associated with human and rodent genes. A discussion paper, authored by Richard Gibbs, George Weinstock, Steven Kappes, Lawrence Schook, Loren Skow and Jim Womack, proposing the sequencing of the bovine genome was submitted to the National Human Genome Research Institute (NHGRI) in 2002 and it was favourably received, but commencement of the project was dependent on raising the necessary finances (NHGRI, 2005b).

The BGSP was formally initiated in December 2003 by an international consortium of research organizations and funding agencies at a total cost of US$53 million. The consortium includes: the National Human Genome Research Institute (USA); CSIRO (Australia); the US Department of Agriculture; the State of Texas; Genome Canada; New Zealand's Agritech Investments Ltd, Dairy Insight Inc and AgResearch Ltd; the Kleberg Foundation (USA); and the National, Texas and South Dakota Beef Check-off Funds (USA). The sequencing and assembly of the bovine genome is being undertaken at the Baylor College of Medicine, Human Genome Sequencing Center (USA) and the Michael Smith Genome Science Centre (Vancouver, Canada). Both the HGP and BGSP were undertaken by public consortia with a strong commitment to early release of sequence data as significant scientific progress can be made even with partial and fragmentary sequence information. The project is driven by a sense of scientific discovery and involves a large group of scientists with an enormous wealth of technical expertize and biological knowledge.

The specific motivations for the BGSP at its commencement were multifaceted. It was reasoned that the cow sequence could be used to better identify functional regions in the human genome sequence through comparison of the human, rodent (murine and rat) and bovine sequences (Collins *et al.*, 2003). It was also recognized that the cow could be used directly as a model for some human diseases and in some instances to discover disease-causing genes. One example is a mutation in the limbin gene in Japanese Brown cattle that was shown to cause chondrodysplastic dwarfism (Takeda *et al.*, 2002). This discovery facilitated the identification of a mutation in the human orthologue that results in a similar genetic disease in humans (Ruiz-Perez *et al.*, 2000; Galdzicka *et al.*, 2002). Perhaps the power of this type of approach will be more fully apparent for complex cattle traits such as muscling, disease resistance and feed conversion efficiency, which may better inform human disorders such as sarcopena, susceptibility to disease, obesity and type II diabetes. The cohabitation of humans and livestock for thousands of years has resulted in disease interchange that has helped to mould the evolution of immune response genes in both species. Thus, the structure and evolution of livestock immune genes may help us understand the functions of the corresponding human genes. In the comparative genomics approach that underlies this approach, the primary DNA sequence is of greatest importance. In the interim period since the initiation of the BGSP, additional genome sequences for chimpanzee, opossum, rhesus macaque, dog, fugu/Japanese blowfish and chicken have been determined and several more are underway, at least at low sequence coverage (NHGRI, 2006b). The chicken and bovine sequences represent the only livestock animals although porcine and ovine sequences are likely in the near future. These additional genome sequences will enable discovery of functional regions of the bovine and human genomes.

One of the major scientific aims of genomics research being undertaken with cattle is to understand the reasons why ruminants are so successful as an animal group. One component of this research will undoubtedly reflect the special adaptations that allow digestion of forage and the generation of energy from volatile fatty acids released in the rumen. Analysis of the bovine genome and the expression of the genes contained within it may also allow an understanding of the adaptations of the species to intensive selection for enhanced growth, meat and milk production.

Another motivation for the BGSP was to understand the linkage between genetic variation of individuals and their resultant phenotypic variation. In particular, it was proposed that the BGSP would facilitate the more rapid development of DNA marker-assisted selection (MAS) capabilities to help breed cattle that fit better with different segments of the livestock market. MAS has the potential to increase the intensity, efficiency and accuracy of selective breeding programmes and to minimize the environmental footprint of the industry. The discovery of large numbers of single nucleotide polymorphisms (SNPs, or small DNA differences between individuals of the same species), their physical placement on the assembled bovine DNA sequence and, ultimately, their chromosomal positions are of prime interest to this objective.

The bovine genome sequence can also be used to identify the historical relationship between the many breeds and allow focus on those genes under selective pressures associated with the domestication of the species. These genes are likely to be pivotal contributors to many of the major production traits of interest to the livestock industry and thus their identification is of considerable importance. Comparison of genome sequences between different mammalian groups will allow identification of conserved and non-conserved regions. The former group will be dominated by protein encoding sequences and these are thought to represent about 1.5% of the genome sequence of most mammals. Recent studies in humans and mice, however, have demonstrated that as much as 5% of the mammalian genome is undergoing purifying selection i.e. there is selective pressure to maintain a nucleotide at a specific position (Human Genome Sequencing Consortium, 2001; Rat Genome Sequencing Consortium, 2004). The additional conserved sequence probably represents maintenance of common gene regulatory sequences and non-protein encoding genes (Kapranov *et al.*, 2002; Mattick, 2003; Ravasi *et al.*, 2006). The regions of bovine sequence that are not conserved compared with human, rodent and other mammalian genome sequences may provide valuable insight into what makes a cow a cow and comment on one of its major characteristics, the presence of a rumen, i.e. how does this species efficiently convert low-energy plant material into high-energy animal muscle and milk?

A single bovine genome from a Hereford animal has been sequenced on average seven times to ensure a high degree of accuracy in the sequence information. Genome sequences for six additional breeds have also been determined at much lower coverage. These breeds are Holstein, Angus, Jersey, Limousin, Norwegian Red and Brahman. This latter effort has discovered large numbers of SNPs that are present in the different breeds and different animals within those breeds. Some of these genetic differences either directly contribute to or are associated with breed-specific characteristics as well as differences between individuals within a breed. SNPs are scientifically and commercially valuable as they provide the practical linkage between genetic differences and desirable production traits such as muscling, feed conversion efficiency, milk production, product quality traits, reproductive efficiency and disease resistance. Simple genetic markers for easily characterized inherited diseases have been available for some time and more recently the first commercial markers for marbling and tenderness in beef cattle have been developed (van Eenennaam, 2006a). However, most production traits are complex, resulting from interactions between many genes and the environment. It is likely that specific combinations of SNP will be ultimately used to accurately link the unique genetics of an individual animal with desirable production traits.

Another major benefit of the bovine genome sequence is the ability to simultaneously measure the activities of all of the genes expressed in a given tissue. Just a few years ago, the tools that were available only allowed investigation of the activity of one gene at a time. Now the activities of tens of thousands of genes in a tissue can be simultaneously measured on a gene microarray, the size of an ordinary microscope slide. This has revolutionized biological research, as it is clear that most visible traits or phenotypes are the result of very complex interactions between the activities of large numbers of genes and these can now be identified. This area of research is called *functional genomics* and can be used, for example, to identify all of the genes that have enhanced activity in mammary tissue as a result of mastitis. Many of these genes are involved in defending mammary tissue from bacterial invasion and, once identified, various strategies can be implemented to enhance their activity and thus promote resistance to infection. This technology is now being used to discover genes involved in many complex traits of economic importance and to identify novel strategies for enhancement of desirable production traits.

Specific Aims and Current Status of the BGSP

Production of a sixfold 'draft' sequence anchored on to bovine chromosomes through the use of an integrated bovine map

The DNA sequence from a partially inbred Hereford female, *L1 Dominette 01449*, has been sequenced (Fig. 4.1). The L1 line of animals represented by *L1 Dominette 01449* has been inbred since the 1920s. The choice of this inbred animal was influenced by its decreased genome complexity, which in turn allows easier assembly of the DNA sequence information. The primary sequencing strategy has used whole genome 'shotgun' (WGS) and targeted sequencing of large fragments of bovine DNA cloned into bacterial artificial chromosomes (BACs). The sequence coverage on average was planned to be sixfold, i.e. on average the sequence in any region was sixfold redundant (i.e. 6X, a 'draft' sequence*). This approach is required to minimize gaps, provide high quality sequence and to facilitate the sequence assembly. Overlaps of WGS sequences are first assembled into 'clusters', and then 'paired-end' sequence information is used to generate higher-order 'scaffolds'. In addition low-sequence coverage of the minimal genome tiling path of BAC clones is being used to enlarge the scaffolds and minimize gaps. The scaffolds are then ordered on to bovine chromosomes using common markers derived from an integrated genetic and physical chromosomal map, which was independently constructed. At this time (December, 2006) a 7.1X draft assembly (Btau_3.1) has been publicly released and another more refined assembly is due in the next few months. The current status of the BGSP is available from a Baylor College of Medicine web site (www.hgsc.bcm.tmc.edu). The assembly also includes an additional 1% (30 Mb) of the total genome sequence derived from the independent ENCODE project (NHGRI, 2006c). This project is producing 'finished' quality sequence on 44 separate regions from several species including the cow. Its aim is to intensively

Fig. 4.1. *L1 Dominette*. Photo courtesy of Mike MacNeil, USDA/ARS Fort Keogh LARRL.

annotate these regions. The Hereford animal used for primary sequencing is female. Hence, an independent effort is required to sequence the Y chromosome. This will probably involve physical isolation of the Y chromosome and 'shotgun' as well as clone-by-clone sequencing. This is not an easy task as the Y chromosome DNA is technically difficult to sequence due to the prevalence of repetitive sequence elements.

Sequencing of 10,000 non-redundant full-length mRNAs

The Michael Smith Genome Science Centre in Vancouver is generating full-length mRNA sequences for 10,000 non-redundant bovine genes. These will be used to denote the boundaries of gene transcripts and may aid in assembly of the genome sequence as the mRNAs represent ordered exonic sequences which link long segments of genomic DNA. The combination of this transcript information with expressed sequence tags (ESTs; fragments of mRNA sequences) will also facilitate the construction of higher density microarrays, particularly oligonucleotide-based microarrays, and the identification of transcript profiling sequence tags derived from SAGE (serial analysis of gene expression) and MPSS (massive parallel signature sequencing) technologies. The latter two technologies have an open architectural design and therefore, unlike microarrays, have the capacity to discover novel transcripts, particularly non-coding transcripts. These techniques allow measurement of the activities of all genes in a particular tissue and the responses of those genes to environmental stimuli.

Discovery of 100,000 SNPs

SNPs provide very useful tools for genome-wide association studies that link genes with heritable traits. Complementing the primary WGS information is another 1.5X coverage of sequence derived from samples taken from individuals representing the Holstein, Angus, Jersey, Limousine, Brahman and Norwegian Red breeds. Comparison of these sequences with the Hereford sequence and among themselves has identified ~117,000 potential SNPs. It is likely that most of these SNPs will be present in the majority of these breeds but their allelic frequencies will be quite different. In addition, examination of heterozygosity in the 6X Hereford sequence, despite the inbreeding of this animal, may have potentially identified as many as 2.2 million SNPs, i.e. a SNP every 1250 bp (Richard Gibbs, George Weinstock and David Wheeler, Baylor College of Medicine, Texas, 2006, personal communication).

How many SNPs are required to cover the bovine genome to enable genome-wide SNP association studies to be undertaken? The latter approach links individual SNPs with production traits of interest and thus these can be used in selective breeding programmes. The number of SNPs required depends on the extent of linkage disequilibrium (LD), a measure of sequence diversity in the population, which is poorly defined at present. The level of LD in cattle is expected to extend well beyond the range that is characteristic of the human genome (Kruglyak, 1999; International HapMap Project, 2005). The human HapMap project has demonstrated that LD in humans varies in scale and is often discontinuous. These characteristics may also be present in cattle populations and make global estimates of LD very imprecise. Hence, it is difficult to identify the number of SNPs that will be required for genome-wide association studies in cattle, although speculations of 30,000–60,000 are typical. This number reflects the view that LD in livestock extends over tens of centimorgans compared with less than a centimorgan in humans. Interestingly, the human HapMap Project has demonstrated correlations between regional LD and several structural and positional aspects of the genome and, surprisingly, some functional classification of genes. In the latter case it was speculated that the extent of LD may reflect the influence of natural selection. For example, immune response genes were located in regions of low LD, suggesting increased diversity was associated with genes interfacing with the environment.

Utility of the Bovine Genome Sequence

One of the major applications of genomics technology in the livestock industry will be the use of genetic markers to select elite performing animals for breeding purposes. In the dairy industry it takes about 6 years to progeny-test a bull to determine whether it has elite breeding characteristics. Marker-assisted selection in combination with advanced reproductive technologies has the potential to short-circuit this expensive and time-consuming testing cycle. The primary tool will be a gene chip that defines the specific genotypes that impact on the production performance of an individual. The chip will be about the size of a thumbnail and will report the genetic variation found at hundreds, if not thousands, of positions across the genome of an individual. The genetic information is linked to production performance data, and subsequently a prediction can then be made about the future productivity of an animal sired by the characterized bull. In another application of this technology, animals will be identified that are best suited to a specific production environment or market segment.

Parentage assignment

Perhaps one of the best direct examples of the application of DNA technology in livestock animals is its use in defining parentage. This capability is already commercially available. In the beef industry as much as 9.8% of recorded parentage assignments and 11.4% of individual identity assignments are erroneous (Pollak, 2005). The magnitude of these numbers and the importance of parentage assignment in progeny testing of bulls suggest that there is a need for increased DNA testing for parentage assignment. Moreover, the greater application of marker-assisted breeding programmes to industry herds will require resolution of parentage issues that will inevitably be presented as discordant genotyping information.

Traceback to individual production animals

There is increasing need to have the capability to trace agricultural products back to their production source. This capability is required to identify the source of disease and as a quality assurance system for consumers. In addition, market feedback can be given directly to the primary producer. Genetic tests can be used to trace the origin of a carcass. Heaton *et al.* (2002) concluded that as few as 30 appropriately selected SNP loci could be used to identify 900,000 individual cattle.

Infectious disease intervention

The bovine genome sequence provides the complete repertoire of bovine immune response genes; some of the encoded proteins may potentially be used as therapeutics and prophylactics to selectively direct the immune system and combat some of the more intransient and chronic diseases of cattle. In addition, it will be possible to select for animals that are genetically more resistant to disease. The potential for this is evident in the differing disease susceptibilities of different breeds and perhaps most clearly exemplified in different strains of inbred mice. Removal of 'at risk' animals from the breeding herds could be a long-term disease preventative strategy. For example, in sheep, mice and humans a particular group of genetic variations in a gene named PrP is associated with increased susceptibility to some prion diseases including scrapie and variant Creutzfeldt-Jakob disease (variant CJD). In cattle, structural changes in the protein expressed from this gene are thought to contribute to bovine spongiform encephalopathy (BSE or mad cow disease). If susceptibility genotypes in the PrP gene or other genes can be identified in cattle, then these animals could be removed from the breeding cycle thereby reducing the likelihood of spread of the disease should it gain entry into a population.

Genetically modified cattle

Genetically modified cattle can be very useful. The bovine genome sequence will provide greater opportunity for accurate insertion of transgenes into the cow genome. At the current time, however, standard livestock commodities produced from these animals are unlikely to be accepted by consumers. What may be accepted are livestock products that are demonstrably safe and have positive health and lifestyle benefits to humans whilst not compromising the health and well-being of the animal. In one general application, transgenic cattle could be used to produce a wealth of human therapeutics that could not be technically or economically produced by alternative processes. Genetically modified (GM) foods are highly controversial but GM therapeutics such as insulin to treat diabetes or the hepatitis B vaccine have been widely accepted and proven to be effective and safe. In another type of application, transgenic cows with increased resistance to mastitis have recently been produced thereby helping to alleviate a chronic animal welfare issue (Wall *et al.*, 2005). In the longer term, cloning and transgenic technologies will be coupled with genomics to deliver animals with particularly advantageous traits. However, it is clear that these technologies are neither technologically mature at present nor generally acceptable to the consumer of traditional foods.

Monogenic diseases and traits

There are now a considerable number of genetic diseases and physical defects of simple genetic origin that have been well documented for livestock animals. These are listed in the 'Online Mendelian Inheritance In Animals' (OMIA) database (OMIA, 2006). Many of these diseases are due to single locus genes with normal Mendelian inheritance patterns. The causal mutations of many of these traits have been identified, the majority of which affect coding gene sequences and therefore presumably alter protein structure and/or function (Womack, 2005). DNA testing can unambiguously identify animals carrying these gene mutations. Variants of a number of genes have also been significantly associated with a range of diseases (Kutzer *et al.*, 2003). The availability of the bovine genome sequence and its associated mapping resources will allow more rapid identification of additional causal genes or DNA markers physically associated with these genes. These tools can be used in marker-assisted breeding programmes to select against unwanted traits. Already, genetic tests are available for the single gene traits Polled/Horn and Black/Non-black genes. A number of bovine genetic diseases also have counterparts in humans and hence can provide animal models for medical research.

DNA marker-assisted selection of superior animals and animals better suited to different production environments

With few exceptions economically important production traits are the result of small contributions from a number of genes impacting with environmental factors. Variations in the DNA sequences of these genes, or alleles, can be expressed as variation in the trait. Moreover, variants in DNA sequence that are physically closely associated with a particular allelic variant of a gene directly contributing to a trait can be used as indirect markers for the trait, as they have a high probability of being co-inherited with the causal allelic variant. Thus, the inheritance of a trait can be predicted without knowing the specific causal gene contributing to the trait. The individual allelic variants of genes directly contributing to continuously variable traits are referred to as quantitative trait loci (QTL) and these can be statistically linked with relatively large chromosomal regions using highly structured populations of cattle. Polymorphic DNA markers within these regions can be used as indirect selection markers for the trait. In a new paradigm for livestock scientific research, the availability of very large numbers of SNPs enables genome-wide SNP association studies to be undertaken using linkage disequilibrium analysis in commercial populations of cattle. The SNPs in this instance are markers for haplotype blocks, i.e. blocks of DNA inherited from a common ancestor which are associated with the measured trait. In this approach the discovery *and* use of SNP markers for selection of desirable (or undesirable) production traits often occurs in the same population of animals.

The application of these technologies is primarily directed toward genetic markers that are associated with production traits that are difficult or expensive to measure (e.g. disease resistance), sex-limited traits (eg. milk production traits), traits that are only obtainable when the animal is an adult or, in the case of beef cattle, at slaughter (i.e. after it has contributed progeny to the next generation), and traits that have low heritabilities (i.e. traits, such as disease resistance, with poor predictor value). The beef and dairy traits that are most likely to benefit from MAS include disease resistance, carcass quality, fertility, reproductive efficiency, feed conversion efficiency, carcass quantity and yield, milk production and composition, and growth performance (Van Eenennaam 2006a). In the dairy industry longevity and temperament are also traits that may benefit from MAS. The aim is to increase the frequency of desirable alleles of genes contributing to a particular trait. This can be achieved by positive selection within a herd or by introgression of the desirable allele from another herd. MAS technology is likely to be applied primarily to breeding animals in the first instance. However, the technology can also be used on animals at birth or even as an embryo. The samples required from cattle can be hair, semen, blood or tissue. Validation of these markers is required in large commercial herds independent of the herds in which the marker was originally discovered. This will also yield information on cost/benefit ratios.

A large number of QTL have been identified particularly for dairy traits but also disease, reproduction and various beef traits (e.g. Hanotte *et al.*, 2003; Rupp and Boichard, 2003; Khatkar *et al.*, 2004). QTL experiments are inherently expensive due to costs associated with constructing a suitable pedigree of cows, and the phenotyping and genotyping of these animals. However, very few QTL have progressed to commercially viable genetic tests. One reason for this may be that some QTL studies, because of their expense, are statistically underpowered and result in QTL that cannot be independently validated. Another limitation has been the general inability to efficiently fine-map the identified QTL regions and find DNA markers more closely associated with the causal gene. Further, some gene markers for QTL may not be commercially viable for a variety of reasons. The availability of the bovine genome sequence and high densities of informative and ordered SNP markers, in combination with large numbers of phenotypically well-characterized animals, will markedly accelerate the success of this scientific strategy and lead to many more commercial markers for industry-relevant traits. SNPs are particularly useful not only because of their large numbers distributed across the genome at known locations, but also they are very amenable to industrial-scale genotyping, which has the attributes of high accuracy, high reproducibility and low cost per SNP assay. It is interesting to note that a human SNP chip containing approximately 500,000 SNP assays is now commercially available and currently a number of cattle studies using 10,000–20,000 SNP are in progress. These numbers represent a quantum leap in discovery capability. A major driver for these high-throughput studies has been technology developments that have decreased the cost of genotyping by about 25-fold in the space of only 2 to 3 years following the beginning of the human HapMap Project (Broudy, 2005).

Since 2003, commercial gene markers for the beef cattle traits of meat marbling, tenderness and commercial yield-grading have been marketed after independent validations. At present, there are five commercial genetic tests for complex production traits and these are marketed by seven companies (Paschal, 2005; Van Eenennaam, 2006b). This trend presents a significant challenge. Each major production trait may require several genetic tests to accurately predict economically significant phenotypic variation. This, coupled with the potential need to measure multiple traits and to purchase these tests from multiple suppliers, means that the breeder or producer will need to pay close attention to the benefit/cost ratio. Market forces will probably promote commercial mergers that minimize the number of future suppliers of genetic tests for cattle. Genotyping costs are relatively expensive (US$10–45 (2006)) but rapid technology developments in this area, driven by the concept of individualized medical care of humans, will simultaneously decrease costs and markedly increase capacity. This in turn is likely to promote multi-trait testing not only of breeding animals but also of production animals. A preview of future trends can also be seen in the commercialization of A2 milk (the A2 variant of β–casein) based on the promotion of its purported direct health benefits to humans (Truswell, 2005).

Several factors influence the rate at which desirable traits are introduced into a herd by MAS. The effects of MAS in a population of animals and therefore the rate of genetic gain are limited by the selection intensity, the size of the differences in phenotype between animals of different genotypes and the rate of replacement of a herd with the genetically superior animals. The genetic gain must also be made in a marketing environment where there are competing selection objectives. One of the early impacts of MAS will be to more rapidly select elite breeding bulls. At present, progeny testing of bulls takes as long as 5 to 6 years. Therefore, there are significant economic gains to be had from acceleration of this testing programme. Perhaps one of the most intriguing possibilities is the combination of MAS with advanced reproductive technologies to select desirable animals as early as the embryo stage, which could markedly accelerate genetic gain in a herd (Raadsma and Tammen, 2005).

Challenges

We are now faced with a situation where rapid and profound technology advancements have produced very large quantities of data but our ability to efficiently convert that data into knowledge benefiting the livestock industry is currently a major limitation. A similar situation in human medical sciences will ensure that rapid progress is made in this area over the next few years.

Marker-assisted selection

Significant challenges still remain in the commercial exploitation of MAS technology (Dekkers, 2004; Pollak, 2005). MAS will need to be integrated into existing selection indices and appropriate weighting given to the marker. The existing selection indices, largely based on progeny testing of bulls, are good indicators of the total genetic merit of an animal and an individual gene marker should not adversely influence that predictive ability. Consideration will also need to be given to the possibility that a desirable allele of one gene contributing to a complex trait may have detrimental effects on another trait. Gene by gene interactions may also mean that expression of the trait could be dependent on the genetic background of an individual. This means that specific gene markers may not work in some breeds or even families and consequently there will be a need for validation in the animal population of interest.

Some traits may be more amenable to MAS and others more recalcitrant. It is interesting to note that disease resistance probably falls within the latter category. The genetic variation underlying this trait is complex, probably reflecting a much greater extent of gene by gene and gene by environment interactions. Indeed, immune responsiveness is characterized by intrinsic flexibility designed to cope with a broad range of foreign challenges and hence it might be expected that it involves a multitude of gene interactions superimposed on previous disease, and even nutritional, history. Thus, identification of gene markers for disease resistance may be considerably more difficult than other production traits. Selection for disease-resistant animals may select for tolerance rather than resistance, in which case the selected animals may represent a disease reservoir. Hence, the definition of the 'resistance' phenotype is critical in any selection programme.

Identification of causal polymorphisms

It has been argued that indirect gene markers are sufficient for use in the commercial cattle industry and the identification of causal polymorphisms is unnecessary. However, an intellectual property position based on one or even a few gene markers is not strong as there may be a large number of equally informative but unprotected gene markers immediately adjacent to the former markers in the same haplotype block. Alternatively, definition of a complete haplotype block may be useful, but this is still not as powerful as the identification of the causal polymorphism and the former's definition may involve a similar research investment. Thus, identification of causal genetic variants underpins a strong commercial position, although much greater effort is required to achieve this end.

Although many bovine QTL and gene associations have been discovered for a diversity of cattle traits, particularly in dairy cattle, very few causal genes have been identified. Even in the mouse, where genetically highly defined inbred populations are readily available, there has been a similar difficulty (Flaherty *et al.*, 2005). Two successful examples of positional cloning in dairy cattle are the discovery of *DGAT1*, a gene contributing to milk fat composition, and *ABCG2*, a gene contributing to protein concentration and milk fat composition (Grisart *et al.*, 2002, 2004; Cohen-Zinder *et al.*, 2005). While QTL and LD association studies are useful in their own right as they can provide DNA markers for desirable production traits, they do have intrinsic limitations. The genetic variation underlying most complex production traits is the result of multiple genes each contributing a small effect. Thus, analyses involving large populations of cattle are required to confidently identify a gene association or QTL. Moreover, while most analyses dissect the variation primarily in terms of additive effects of the contributing genes, it is clear that gene interactions can have substantial impact on phenotype. Often QTL define a large chromosomal segment, typically a physical size of about 30 Mb, that may contain a large number of candidate genes (Barendse, 2005). Sorting through these genes to identify the causal gene is a demanding task. Thus, the identification of causal polymorphisms in genes is difficult. In reality, causal polymorphisms are likely to be identified by a combination of the following: chromosomal positional information; biochemical function linked to the phenotype; location of the protein encoded by the gene in the tissue of interest and within a cell; and clues derived from orthologue gene function in other mammalian species. The difficulty in the identification of causal polymorphisms is further confounded by restricting analyses to protein coding genes (~1.5% of the genome), as recent studies suggest that as much as 45% of the non-repetitive mammalian genome may be transcribed (Cheng *et al.*, 2005). Some causal genetic variations may lie in non-obvious DNA segments including non-coding RNA and long-distance gene regulatory elements. For example, the muscle hypertrophy phenotype characteristic of *Callipyge* sheep is due to a single point mutation in an intergenic region which resides within a long-range regulatory element controlling the expression of six imprinted genes (Charlier *et al.*, 2001; Freking *et al.*, 2002; Cockett *et al.*, 2005). Finally, genome-wide association studies will need to be rigorously performed and analysed as the statistical use of tens to hundreds of thousands of SNPs may produce a confounding level of false positive associations (Ioannidis, 2005). Validation of associations in independent cattle populations and additional independent evidence will be required to offset this potential outcome.

Genome complexity

Mammalian genomes contain similar repertories of protein encoding genes and for humans it is estimated that there are only approximately 25,000 genes, markedly fewer than estimates as high as 100,000 made before the human genome sequence was determined (Human Genome Sequencing Consortium, 2001). Moreover, less complex organisms like *Drosophila* and *Caenorhabditis elegans* do not contain proportionally fewer genes compared with a mammalian genome. So where does the additional genetic information reside in mammals to account for their increased functional and structural complexity? Clearly, the answer resides with the relatively large size of the mammalian genomes (2700–3000 Mb) compared with other species (*Drosophila*, 150 Mb; *C. elegans*, 108 Mb). Mammalian genomes are complex, as is the link between physiological function and gene. Factors contributing to this structural complexity include extensive alternative mRNA splicing involving as many as 40–60% of mammalian genes (Modrek and Lee, 2003), large numbers of non-protein encoding transcripts (Mattick, 2003; Ravasi *et al.*, 2006), a multitude of combinatorial factors regulating gene expression, widespread non-equivalence of allelic expression (Lo *et al.*, 2003; Pastinen *et al.*, 2004; Lee, 2005), large-scale copy-number variations (Carter, 2004), epigenetic status in general (Fazzari and Greally, 2004) and more specifically imprinted genes which show monoallelic expression in a parent-of-origin-dependent manner (Reik and Walter, 2001), and potential regulatory roles of the large number of repetitive sequence elements present in the genome (Whitelaw and Martin 2001). Thus, the current concept of a gene as largely a protein encoding DNA sequence needs to revert to its original meaning and include any stably inherited modification or allele that is associated with a specific trait. Despite the complexity of the mammalian genome, it is interesting to note that most monogenic traits and the few complex traits that have been genetically defined overwhelmingly show that the causal mutation alters protein encoding genes. Whether this is a reflection of bias in the ease of discovery of these genes or an intrinsic commentary on the importance of protein encoding genes is not yet clear.

Epigenetics and gene interactions

Perhaps the greatest challenges are to understand the effects of environment on gene expression and the role of gene interactions contributing to phenotype. In the former case for example, there is substantial evidence that maternal nutritional impositions on a fetus can have profound and lifelong influence on gene expression and phenotype especially as an adult (Waterland and Garza, 2002). Epigenetic modifications of DNA may underlie many of these responses (epigenetics is defined as 'the study of mitotically and/or meiotically heritable changes in gene function that cannot be explained by changes in DNA sequence' (Riggs *et al.*, 1996)). Moreover, there is now a hint that the effects of some nutritional and behavioural impositions can be stably transmitted to the next generation, raising the possibility of metastable epialleles (Morgan *et al.*, 1999; Whitelaw and Martin, 2001; Weaver *et al.*, 2004). Normally the epigenome is reset in each generation; however, if this process is incomplete or imprecise then it could lead to heritable epigenetic effects. These could particularly confound QTL analyses. Poor epigenetic reprogramming may underlie some of the structural and functional abnormalities of cloned animals (Dean *et al.*, 2001).

The genetic component of a trait associated with gene interactions is known as epistasis. Knowledge of these gene interactions will provide a better predictive capability of the performance of an animal particularly when it is produced from very different genetic backgrounds. Gene interactions probably lie at the heart of many complex phenotypes but as yet the tools necessary to understand and utilize these interactions in terms of livestock production are not sufficiently developed. While there are many examples of the combinatorial complexity of gene product interactions within a cell, perhaps one of the more recent discoveries is particularly pertinent. MicroRNAs are small ~21 bp non-coding RNAs that result in the post-transcriptional silencing of genes, typically by translational repression but also cleavage of target mRNA. The specificity of targeting is defined by the complementarity between the microRNA and its target mRNA sequence. However, the majority of the several hundred known microRNAs do not have perfect complementarity and, as a consequence, each targets a considerable number of mRNA. Indeed, it is thought that about one-third of the coding mRNAs are targeted by miRNAs (Lewis *et al.*, 2005). In addition, many mRNAs are targeted by multiple microRNA (Farh *et al.*, 2005). The combinatorial complexity of these interactions is immense and exemplifies just one of many systems using similar strategies, e.g. transcription factors controlling gene expression. It is probable that gene interactions such as these represent an autonomous information system in their own right, which contains within it memory of the past and flexibility to respond to a spectrum of future environmental stimuli.

The Future

The link between gene (and protein) expression and QTL/association studies will be much more strongly developed in the near future. Microarray-based gene expression analysis can measure the activities of tens of thousands of genes in a rapid, high-throughput manner. Gene expression itself can be treated as a continuously variable trait and hence, using a specifically structured population of animals, it is possible to identify genetic determinants that control the variation in transcription of each gene expressed in a tissue (expression QTL or eQTL) (Cheung *et al.*, 2005; Flaherty *et al.*, 2005). Alternatively, it is possible to perform a genome-wide association study with large numbers of SNP to identify associations with genetic loci based on variation in gene expression. This combination of genetic and genomic approaches is a powerful strategy to identify causal variants of genes that are contributing to production traits of interest to the livestock industry.

The impressive genetic technologies that have been developed since the year 2000, in terms of both new capabilities and scale, will enable development of novel approaches to understanding the relationship between gene and phenotype. In particular, the discovery of as many as 2.2 million bovine SNPs in combination with the bovine genome sequence and large populations of animals with well-recorded phenotypes will have a major impact on the livestock industry. Rapid technological advances driven by the HGP will markedly reduce genotyping costs, allowing assessment of the genetic potential of not only elite breeding animals but also primary production animals. The genotyping information will allow a prediction of the production potential of an animal at birth, or perhaps even as a fetus, also the selection of animals best suited to a specific production environment. In particular, the combination of new reproductive technologies and MAS has the potential to make rapid genetic gains for the livestock industry.

Genetics will increasingly make contributions to the productivity of the livestock industry. However, the immediate challenges with the new genetic technologies are to integrate the enhanced capability into the current production system, understand their limitations and prove their usefulness in the industry.

In a scientific sense the genome sequences are a voyage of discovery, there is much that is unknown but tremendous rewards for those who can navigate through these waters and exploit this knowledge. As recently as the early 1990s, the thought of a mammalian genome sequence was stretching scientific credibility. What exciting developments await us in the next decade?

Acknowledgment

I thank a long list of colleagues, particularly members of the BGSP and its advisory groups, for stimulating thoughts and their enormous breadth of knowledge.

Note

*'Finished' sequence is defined as 'the clone insert is contiguously sequenced with high quality standard error rate of 0.01% – there are usually no gaps in the sequence'. A 'draft' sequence can have different definitions depending on the specific project, but a typical definition is: 'at least 3–4X of the estimated clone insert is covered in Phred Q20 bases in the WGS stage. The clone sequence may contain several pieces of sequence, separated by gaps. The true order and orientation of these pieces may not be known' (NCBI).

References

Barendse, B. (2005) The transition from quantitative trait loci to diagnostic test in cattle and other livestock. *Australian Journal of Experimental Agriculture* 45, 831–836.

Broudy, T. (2005) HapMap-tested genotyping assays and microarrays used for association studies. *Microarray Bulletin* 1, 1–5.

Carter, N. (2004) As normal as normal can be? *Nature Genetics* 36, 931–932.

Charlier, C., Segers, K., Karim, L., Shay, T., Gyapay, G., Cockett, N. and Georges, M. (2001) The callipyge mutation enhances the expression of coregulated imprinted genes in *cis* without affecting their imprinting status. *Nature Genetics* 27, 367–369.

Cheng, J., Kapranov, P., Drenkow, J., Dike, S., Brubaker, S., Patel, S., Long, J., Stern, D., Tammana, H., Helt, G., Sementchenko, V., Piccolboni, A., Bekiranov, S., Bailey, D.K., Ganesh, M., Ghosh, S., Bell, I., Gerhard, D.S. and Gingeras, T.R. (2005) Transcriptional maps of 10 human chromosomes at 5-nucleotide resolution. *Science* 308, 1149–1154.

Cheung, V.G., Spielman, R.S., Ewens, K.G., Weber, T.M., Morley, M. and Burdick, J.T. (2005) Mapping determinants of human gene expression by regional and genome-wide association. *Nature* 437, 1365–1369.

Cockett, N.E., Smit, M.A., Bidwell, C.A., Segers, K., Hadfield, T.L., Snowder, G.D., Georges, M. and Charlier, C. (2005) The callipyge mutation and other genes that affect muscle hypertrophy in sheep. *Genetics Selection Evolution* 37, S65–S81.

Cohen-Zinder, M., Seroussi, E., Larkin, D.M., Loor, J.J., Everts-van der Wind, A., Lee, J.H., Drackley, J.K., Band, M.R., Hernandez, A.G., Shani. M., Lewin, H.A., Weller, J.I. and Ron, M. (2005) Identification of a missense mutation in the bovine ABCG2 gene with a major effect on the QTL on chromosome 6 affecting milk yield and composition in Holstein cattle. *Genome Research* 15, 936–944.

Collins, F.S., Green, E.D., Guttmacher, A.R. and Guyer, M.S. (2003) A vision for the future of genomics research. *Nature* 422, 835.

Dean, W., Santos, F., Stojkovic, M., Zakhartchenko, V., Walter, J., Wolf, E. and Reik, W. (2001) Conservation of methylation reprogramming in mammalian development: aberrant reprogramming in cloned embryos. *Proceedings of the National Academy of Sciences of the United States of America* 98, 13734–13738.

Dekkers, J.C.M. (2004) Commercial application of marker- and gene assisted selection in livestock: strategies and lessons. *Journal of Animal Science* 82, E313–E328.

Farh, K.K., Grimson, A., Jan, C., Lewis, B.P., Johnston, W.K., Lim, L.P., Burge, C.B. and Bartel, D.P. (2005) The widespread impact of mammalian MicroRNAs on mRNA repression and evolution. *Science* 310, 1817–1821.

Fazzari, M.J. and Greally, J.M. (2004) Epigenomics beyond CpG islands. *Nature Reviews Genetics* 5, 446–455.

Flaherty, L., Herron, B. and Symula, D. (2005) Genomics of the future: identification of quantitative trait loci in the mouse. *Genome Research* 15, 1741–1745.

Freking, B.A., Murphy, S.K., Wylie, A.A., Rhodes, S.J., Keele, J.W., Leymaster, K.A., Jirtle, R.L. and Smith, T.P.L. (2002) Identification of the single base change causing the *callipyge* muscle hypertrophy phenotype, the only known example of polar overdominance in mammals. *Genome Research* 12, 1496–1506.

Galdzicka, M., Patnala, S., Hirshman, M.G., Cai, J.F., Nitowsky, H., Egeland, J.A. and Ginns, E.I. (2002) A new gene, EVC2, is mutated in Ellis-van Creveld syndrome. *Molecular Genetics and Metabolism* 77, 291–295.

Grisart, B., Coppieters, W., Farnir, F., Karim, L., Ford, C., Berzi, P., Cambisano, N., Mni, M., Reid, S., Simon, P., Spelman, R., Georges, M. and Snell, R. (2002) Positional candidate cloning of a QTL in dairy cattle: identification of a missense mutation in the bovine DGAT1 gene with major effect on milk yield and composition. *Genome Research* 12, 222–231.

Grisart, B., Farnir, F., Karim, L., Cambisano, N., Kim, J.J., Kvasz, A., Mni, M., Simon, P., Frere, J.M., Coppieters, W. and Georges, M. (2004) Genetic and functional confirmation of the causality of the DGAT1 K232A quantitative trait nucleotide in affecting milk yield and composition. *Proceedings of the National Academy of Sciences of the United States of America* 101, 2398–2403.

Guttmacher, A.E. and Collins, F.S. (2005) Realizing the promise of genomics in biomedical research. *Journal of the American Medical Association* 294, 1399–1402.

Hanotte, O., Ronin, Y., Agaba, M., Nilsson, P., Gelhaus, A., Horstmann, R., Sugimoto, Y., Kemp, S., Gibson, J., Korol, A., Soller, M. and Teale, A. (2003) Mapping of quantitative trait loci controlling trypanotolerance in a cross of tolerant West African N'Dama and susceptible East African Boran cattle. *Proceedings of the National Academy of Sciences of the United States of America* 100, 7443–7448.

Heaton, M.P., Harhay, G.P., Bennett, G.L., Stone, R.T., Grosse, W.M., Casas, E., Keele, J.W., Smith, T.P., Chitko-McKown, C.G. and Laegreid, W.W. (2002) Selection and use of SNP markers for animal identification and paternity analysis in U.S. beef cattle. *Mammalian Genome* 13, 272–281.

Human Genome Sequencing Consortium (2001) Initial sequencing and analysis of the human genome. *Nature* 409, 860–921.

International HapMap Project (2005) HapMap homepage, www.hapmap.org (accessed June 2006).

Ioannidis, J.P. (2005) Why most published research findings are false. *PLoS Medicine* 2, e124. 0696–0701.

Kapranov, P., Cawley, S.E., Drenkow, J., Bekiranov, S., Strausberg, R.L., Fodor, S.P. and Gingeras, T.R. (2002) Large-scale transcriptional activity in chromosomes 21 and 22. *Science* 296, 916–919.

Khatkar, M.S., Thomson, P.C., Tammen, I. and Raadsma, H.W. (2004) Quantitative trait loci mapping in dairy cattle: review and meta-analysis. *Genetics Selection Evolution* 36, 163–190.

Kruglyak, L. (1999) Prospects for whole-genome linkage disequilibrium mapping of common disease genes. *Nature Genetics* 22, 139–144.

Kutzer, T., Leeb, T. and Brenig, B. (2003) Current state of development of genome analysis in livestock. *Current Genomics* 4, 487–525.

Lee, M.P. (2005) Genome-wide analysis of allele-specific gene expression using oligo microarrays. *Methods in Molecular Biology* 311, 39–47.

Lewis, B.P., Burge, C.B. and Bartel, D.P. (2005) Conserved seed pairing, often flanked by adenosines, indicates that thousands of human genes are microRNA targets. *Cell* 120, 15–20.

Lo, H.S., Wang, Z., Hu, Y., Yang, H.H., Gere, S., Buetow, K.H. and Lee, M.P. (2003) Allelic variation in gene expression is common in the human genome. *Genome Research* 13, 1855–1862.

Mattick, J.S. (2003) Challenging the dogma: the hidden layer of non-protein coding RNAs in complex organisms. *Bioessays* 25, 930–939.

Modrek, B. and Lee, C.J. (2003) Alternative splicing in the human, mouse and rat genomes is associated with an increased frequency of exon creation and/or loss. *Nature Genetics* 34, 177–180.

Morgan, H.D., Sutherland, H.G., Martin, D.I. and Whitelaw, E. (1999) Epigenetic inheritance at the agouti locus in the mouse. *Nature Genetics* 23, 314–318.

NHGRI (2005a) The Human Genome Project Completion: Frequently Asked Questions, www.genome.gov/11006943 (accessed June 2006).

NHGRI (2005b) Bovine Genomic Sequencing Initiative: Cattle-izing the Human Genome, www.genome.gov/Pages/research/Sequencing/SeqProposals/bovineSEQ.pdf (accessed June 2006).

NHGRI (2006a) NHGRI expands effort to revolutionize sequencing technologies, www.genome.gov/15015208 (accessed June 2006).

NHGRI (2006b) Genome sequencing proposals, www.genome.gov/10002154 (accessed June 2006).

NHGRI (2006c) The ENCODE Project: ENCyclopedia of DNA Elements, www.genome.gov/10005107 (accessed June 2006).

OMIA (2006) OMIA – Online Mendelian inheritance in animals homepage www.omia.angis.org.au (accessed June 2006).

Paschal, J.C. (2005) Marker assisted selection for genetic improvement, www.agfacts.tamu.edu/D11/Extension/animal%20science/E-352Genetics.pdf. (accessed June 2006).

Pastinen, T., Sladek, R., Gurd, S., Sammak, A., Ge, B., Lepage, P., Lavergne, K., Villeneuve, A., Gaudin, T., Brandstrom, H., Beck, A., Verner, A., Kingsley, J., Harmsen, E., Labuda, D., Morgan, K., Vohl, M.C., Naumova, A.K., Sinnett, D. and Hudson, T.J. (2004) A survey of genetic and epigenetic variation affecting human gene expression. *Physiological Genomics* 16, 184–193.

Pollak, E.J. (2005) Application and impact of new genetic technologies on beef cattle breeding: a 'real world' perspective. *Australian Journal of Experimental Agriculture* 45, 739–748.

Raadsma, H.W. and Tammen, I. (2005) Biotechnologies and their potential impact on animal breeding and production: a review. *Australian Journal of Experimental Agriculture* 45, 1021–1032.

Rat Genome Sequencing Consortium (2004) Genome sequence of the brown Norway rat yields insights into mammalian evolution. *Nature* 428, 493–521.

Ravasi, T., Suzuki, H., Pang, K.C., Katayama, S., Furuno, M., Okunishi, R., Fukuda, S., Ru, K., Frith, M.C., Gongora, M.M., Grimmond, S.M., Hume, D.A., Hayashizaki, Y. and Mattick, J.S. (2006) Experimental validation of the regulated expression of large numbers of non-coding RNAs from the mouse genome. *Genome Research* 16, 11–19.

Reik, W. and Walter, J. (2001) Genomic imprinting: parental influence on the genome. *Nature Reviews Genetics* 2, 21–32.

Riggs, A.D., Martienssen, R.A. and Russo, V.E.A. (1996) Introduction. In: Russo, V.E.A., Martienssen, R.A. and Riggs, A.D. (eds) *Epigenetic Mechanisms of Gene Regulation*. Cold Spring Harbor Laboratory Press, Woodbury, New York, p. 1.

Ruiz-Perez, V.L., Ide, S.E., Strom, T.M., Lorenz, B., Wilson, D., Woods, K., King, L., Francomano, C., Freisinger, P., Spranger, S., Marino, B., Dallapiccola, B., Wright, M., Meitinger, T., Polymeropoulos, M.H. and Goodship, J. (2000) Mutations in a new gene in Ellis-van Creveld syndrome and Weyers acrodental dysostosis. *Nature Genetics* 24, 283–286.

Rupp, R. and Boichard, D. (2003) Genetics of resistance to mastitis in dairy cattle. *Veterinary Research* 34, 671–688.

Takeda, H., Takami, M., Oguni, T., Tsuji, T., Yoneda, K., Sato, H., Ihara, N., Itoh, T., Kata, S.R., Mishina, Y., Womack, J.E., Moritomo, Y., Sugimoto, Y. and Kunieda, T. (2002) Positional cloning of the gene LIMBIN responsible for bovine chondrodysplastic dwarfism. *Proceedings of the National Academy of Sciences of the United States of America* 99, 10549–10554.

Truswell, A.S. (2005) The A2 milk case: a critical review. *European Journal of Clinical Nutrition* 59, 623–631.

Van Eenennaam, A. (2006a) Marker assisted selection in beef cattle, http://animalscience.ucdavis.edu/animalbiotech/Outreach/Marker_Assisted_Selection_in_Beef_Cattle.pdf (accessed June 2006).

Van Eenennaam, A. (2006b) Animal Biotechnology Marker Assisted Selection and Breeding homepage, http://animalscience.ucdavis.edu/animalbiotech/Biotechnology/MAS/index.htm (accessed June 2006).

Wall, R.J., Powell, A.M., Paape, M.J., Kerr, D.E., Bannerman, D.D., Pursel, V.G., Wells, K.D., Talbot, N. and Hawk, H.W. (2005) Genetically enhanced cows resist intramammary *Staphylococcus aureus* infection. *Nature Biotechnology* 23, 445–451.

Waterland, R.A. and Garza, C. (2002) Early postnatal nutrition determines adult pancreatic glucose-responsive insulin secretion and islet gene expression in rats. *Journal of Nutrition* 132, 357–364.

Weaver, I.C., Cervoni, N., Champagne, F.A., D'Alessio, A.C., Sharma, S., Seckl, J.R., Dymov, S., Szyf, M. and Meaney, M.J. (2004) Epigenetic programming by maternal behavior. *Nature Neuroscience* 7, 847–854.

Wendorf, F. and Schild, R. (1994) Are the early Holocene cattle in the Eastern Sahara domestic or wild? *Evolutionary Anthropology* 3, 4.

Whitelaw, E. and Martin, D.I. (2001) Retrotransposons as epigenetic mediators of phenotypic variation in mammals. *Nature Genetics* 27, 361–365.

Womack, J.E. (2005) Advances in livestock genomics: opening the barn door. *Genome Research* 15, 1699–1705.

5 Precision Animal Breeding

Kishore C. Prayaga[1] and Antonio Reverter[2]
[1]*CSIRO Livestock Industries, QLD, Australia;* [2]*CSIRO Livestock Industries, Queensland Bioscience Precinct, Australia*

Abstract
Precision in selecting breeders for improved production has increased over the years with the advances in the application of quantitative genetic theory to animal breeding practices. This was further assisted by the development of statistical tools designed to exploit performance and pedigree information to understand the genetic basis of improved performance. Recently, molecular genetic revolution has led to the identification of quantitative trait loci influencing traits of economic importance. In this review, we have summarized these advancements and presented some examples related to beef cattle breeding. We have also presented a futuristic vision of precision animal breeding through integrated approaches using a systems biology framework. While most of these developments in statistical methodology, molecular markers, and gene expression patterns are advancing independently, greater challenges lie ahead in drawing these together to derive greater benefits from efficient animal production systems.

Introduction

> It is the task of science, as a collective human undertaking, to describe from the external side, such statistical regularity as there is in a world in which every event has a unique aspect, and to indicate where possible the limits of such description. It is not a part of its task to make imaginative interpretations of the internal aspect of reality. The only qualification is in the field of introspective psychology in which each human being is both observer and observed, and regularities may be established by comparing notes.
>
> Sewall G. Wright (1889–1988)

Animal breeding, as a scientific discipline, refers to the application of biological and quantitative genetic knowledge to the improvement of the desirable phenotypes through planned breeding of animals. As genes are passed on from generation to generation, the study of inheritance laws has played a significant role in guiding animal breeding research. Significant advances in applied animal breeding research were possible because of the advances in quantitative genetics and its delivery through the breeding programmes with targeted breeding goals. These advances can be sustained and improved further through the integration of molecular genetic information into the traditional breeding tools. Over the last few years, this has been the objective of several research programmes in genetic improvement applied to animal breeding where the aim was to deliver precise genetic evaluation and parameter estimation techniques, hopefully leading to precision animal breeding.

Admittedly, such advanced techniques have experienced mixed results when it comes to their ability to generate more accurate predictions of future performance. Hence, the implementation and adoption of such techniques in real-case scenarios (e.g. in national and international genetic evaluation schemes) has been an ongoing challenge.

The present vignette discusses the state of affairs with quantitative and molecular genetics applied to animal breeding. It is not intended to be a comprehensive review of literature, rather to give the readers an overview of this field and with the possibly pretentious attempt to conclude with a futuristic outlook. Several examples related to beef cattle breeding are included because of the authors' affiliation and experience in that area.

Quantitative Genetics Applied to Animal Breeding

Sir Ronald A. Fisher (1890–1962; for a biography see http://www-groups.dcs.stand.ac.uk/~ history/Biographies/Fisher.html), Sewall G. Wright (1889–1988; http://www.harvardsquarelibrary.org/unitarians/wright-sewall.html) and Charles R. Henderson (1911–1989; http://newton.nap.edu/html/bio73h/henderson.html) laid the foundation for quantitative genetics by outlining main principles. Since then, quantitative genetics gave the theoretical basis for the advances in animal and plant breeding programmes. Understanding the nature of quantitative genetic variation, applying selection techniques, quantifying the consequences of inbreeding and outcrossing, and estimating breeding values have been the major areas of animal breeding research over the last few decades (Lynch and Walsh, 1998). The importance of genetic parameters in estimating the true genetic value of an animal for various production traits of economic importance led to the development of several statistical tools such as restricted maximum likelihood (REML; Patterson and Thompson, 1971) and best linear unbiased prediction (BLUP; Henderson, 1975). REML and BLUP in combination have revolutionized animal breeding. REML estimates the (co)variance components (additive genetic, maternal genetic, permanent environmental) and BLUP provides solutions for the random animal effects, known as estimated breeding values (EBVs) or the value of an individual as a parent. Following Mendelian inheritance laws, half the EBV of an individual is expected to be passed on from parent to offspring. Such value is denoted as expected progeny differences (EPDs), also common in genetic evaluation summaries.

The driving force behind all these developments in the field of quantitative genetics has been the desire to understand the genetic value based on the available phenotypic information and the structure of the pedigree that gave rise to the population under study. Notably, an animal's own phenotypic information alone cannot be used for estimating an accurate genetic value for the majority of traits because of the environmental influences on those traits. Even after accounting for all the possible environmental influences, it is still difficult to partition the non-additive genetic effects from additive genetic effects, which represents the true genetic value passed on from parents to progeny, based on the animal's own phenotype. Hence, pedigree and relatives' phenotypic information leads to a more accurate prediction of genetic value via BLUP (Henderson, 1975, 1977), the most widely used statistical tool developed for this purpose.

General mixed models are routinely used by animal breeders for variance component estimation and subsequent (or simultaneous) estimation of breeding values. Hence, it is helpful to understand the general mixed model as it is the basis for several other developments discussed in later sections. In matrix notation, the general mixed model can be written as:

$$y = \mathbf{X}\beta + \mathbf{Z}\mu + e,$$

where y is the vector of phenotypes for the trait(s) under study; **X** and **Z** are the incidence matrices (whose entries are 0s or 1s, indicating the absence/presence of an incidence) relating the fixed and random effects to the observations; β is the vector of fixed effects (e.g. contemporary groups, typically defined as the group of animals of the same sex, year of birth, raised in the same environment and slaughtered in the same abattoir); μ is the vector of random effects (e.g. animal effects, sire effects, maternal effects, common environmental effects); and e is the vector of residual deviations (e.g. temporary environmental effects).

Estimation of breeding values

According to Falconer and Mackay (1996), the breeding value of an animal is its genetic merit as judged by the mean performance of its progeny. Hence, the estimation of breeding value has been of primary interest because of its importance in improving the herd productivity through inheritance. Models for the estimation of breeding values (EBVs) are all variants of the general mixed model described above and can be briefly defined as follows:

1. Animal Model: generates EBVs for all the individuals in the data.
2. Gametic Model: estimates individual breeding values in terms of parental combinations.
3. Sire Model: generates EBVs for sires only, assuming dams are randomly chosen from the entire population.
4. Reduced Animal Model: first introduced by Quaas and Pollak (1980), it combines the features of the animal and the gametic models and yields EBVs for parents only. Computational efficiency is its advantage over the animal model.

Inclusion of more than one random effect is among the other variations or developments (e.g. a maternal effect on growth along with the direct additive effect). Repeated measures on the same individual for certain traits (e.g. tick counts) can be accommodated in the analysis by including the permanent environmental effect of the animal. The general form of the mixed model equations are:

$$\begin{bmatrix} \mathbf{X}'R^{-1}\mathbf{X} & \mathbf{Z}'R^{-1}\mathbf{X} \\ \mathbf{X}'R^{-1}\mathbf{X} & \mathbf{Z}'R^{-1}\mathbf{Z}+G^{-1} \end{bmatrix} \begin{bmatrix} \beta \\ \mu \end{bmatrix} = \begin{bmatrix} \mathbf{X}'R^{-1}y \\ \mathbf{Z}'R^{-1}y \end{bmatrix}$$

Where, var(u) = G; var(e) = R; X, β, and μ are as explained above. The vector μ could contain a greater number of random effects as explained above. The vector in y could contain more traits as well in multi-trait analyses.

The variance–covariance structure of the model including additive genetic, maternal genetic and permanent environmental effects is given below:

$$V\begin{bmatrix} a \\ m \\ c \\ e \end{bmatrix} = \begin{bmatrix} A\sigma^2_a & A\sigma_{am} & 0 & 0 \\ A\sigma_{am} & A\sigma^2_m & 0 & 0 \\ 0 & 0 & I_c\sigma^2_c & 0 \\ 0 & 0 & 0 & I_n\sigma^2_e \end{bmatrix}$$

where, A is the numerator relationship matrix, and I_c and I_n are identity matrices of order equal to number of animals and number of records, respectively; σ^2_a is the direct additive genetic variance; σ^2_m is the maternal additive genetic variance; σ^2_c is the variance due to permanent environmental effects of the animal; σ_{am} is the covariance between direct and maternal additive genetic effects; and σ^2_e is the residual variance.

Walsh (2001) discusses classical quantitative genetics and its evolution into a neoclassical theory of quantitative genetics. Under the new framework, Walsh argues that genotypes of animals at some loci are known and hence, the total genetic value can be partitioned into genotypic value associated with known genotypes and polygenic value. The known genotypes can be either actual genes of interest or the closely linked markers affecting the phenotype.

$$\mu = Gm + g + Gm * g$$

Hence, the total genetic value is presumedly composed of genetic value associated with marker genotypes (Gm) plus the genetic value associated with genes with unknown genotypes (g) combined with their interaction. Hence it should be noted that this framework is 'marker-assisted selection', rather than 'marker-based selection'.

Depending on the magnitude of genetic variation explained by the molecular genetic markers, their importance in predicting the genetic value varies. Several models suggest that the information on genotypes will add incremental improvement to the traditional methods of estimating breeding values rather than act as a replacement selection method. This observation is more relevant in the case of significant epistatic interactions and genotype × environment interactions. As the complexity of gene actions and their interaction with the environment increases, the importance of more sophisticated quantitative genetic approaches increases. In reality, most of the production traits of interest in livestock are under the influence of multiple genes with differential performance across varying environments (see Olson *et al.*, 1991 for a beef cattle example). Hence, a combined approach integrating molecular information and quantitative methods is a more tenable solution. Walsh (2001) also argues that the phenotype should always be a part of selection criteria as doing so captures mutational variance more easily than a selection solely based on genotype. This is certainly the case for multifactorial traits, where there may always be a polygenic component needed to be included while partitioning variances to estimate breeding values. Over recent years, advances in cost-effective computational power allowed the routine implementation of more sophisticated quantitative genetic approaches through the increased use of likelihood-based models.

Inbreeding

Through the advances in scientific animal breeding involving the application of accurate genetic evaluation techniques, animal breeders have been able to successfully breed highly productive animals. Higher intensity of selection and preferential mating to increase production (e.g. dairy cattle selection for milk production) can lead to increased levels of inbreeding and as a consequence, fitness attributes of animals in the following generations have suffered. Soon, animal breeders realized that to achieve sustainable benefits of selection, inbreeding needs to be avoided because of its negative impact on traits connected with reproductive capacity or physiological efficiency. The phenomenon is often referred to as 'inbreeding depression' (Falconer and Mackay, 1996).

Recently, Kristenson and Sorensen (2005) reviewed the effects of inbreeding and outlined the lessons to be learnt from animal breeding, evolutionary biology and conservation genetics. Detrimental effects of inbreeding for traits under additive genetic control (e.g. growth, milk yield) is through reduced genetic variance and carry-over effect of potentially reduced response to selection. Inbreeding mainly affects traits that are under the influence of non-additive gene action through inbreeding depression. This can be easily understood by looking at the formula for change of population mean due to inbreeding i.e. $-2F\Sigma dpq$ (Falconer and Mackay, 1996). This implies that inbreeding depression is dependent on inbreeding coefficient (F), gene frequencies in the population (p, q), and most importantly the directional dominance (d); Σ indicates summation is over all loci. This formula supports the earlier observation that the negative effects of inbreeding are more pronounced in traits affected by non-additive gene action.

It has now been widely accepted that the traits that are mostly affected by inbreeding are the traits that respond well in crossbreeding because of hybrid vigour or heterosis. In other words, fitness lost due to inbreeding tends to be restored on crossing of inbred lines (Falconer and Mackay, 1996). Another important concept related to inbreeding is the effective population size (Ne). In simple terms, Ne refers to the effective number of breeding individuals. To avoid inbreeding depression, Franklin (1980) and Franklin and Frankham (1998) advocated an effective population size of 50 in the short term, and 500 to 5000 to retain genetic variation in the longer term.

Crossbreeding

Crossbreeding has evolved as a potent breeding tool to counter the negative effects of inbreeding. Crossbreeding results in hybrid vigour or heterosis emerging when two inbred lines or strains are bred, leading to an improved performance of the progeny over the mean performance of the parental populations. This boost in performance is generally attributed to dominance gene action, which involves non-additive interaction between the alleles at a locus. The amount of heterosis is equal to 'dy^2' (Falconer and Mackay, 1996), where d is the directional dominance and y is the difference in allele frequencies between two distinct populations. Hence, loci that are inherited predominantly through additive gene action without any dominance expression are not affected by inbreeding and do not display heterosis.

From an animal breeding perspective, a majority of the research over the last few decades was concentrated on either selection (trying to exploit additive genetic variance) or crossbreeding (trying to exploit non-additive genetic variance). Besides the advantages to be gained through heterosis, another advantage with crossbreeding is its scope to utilize breed complementarity. For example, crossbreeding has been used in tropical beef cattle production systems to complement the tropical adaptability attributes of Zebu breeds with higher fertility attributes of British breeds (Prayaga, 2004) in the resulting crossbred progeny. Gregory *et al.* (1999) in temperate beef cattle production systems and Prayaga (2003a,b, 2004, and references therein) in tropical beef cattle production, systems have outlined the breed differences and heterosis estimates for various production traits. Crossbreeding parameters estimated through partitioning the overall breed genetic effects into breed additive and breed dominance effects aim at understanding the genetic basis of improved performance in certain genotypes. These crossbreeding parameters were used to optimize the breed proportions in a tropical beef composite (Prayaga *et al.*, 2003). Based on these experimental results, differential ranking of various crossbred combinations for various production traits is presented in Table 5.1.

Recently, Cundiff (2006) has emphasized the impact of quantitative genetics on improving beef cattle productivity while summarizing the major challenges faced by the beef industry. These include the need for reducing the costs of production, matching genotypes to environments, producing leaner meat, and improving palatability, tenderness and consistency. Understanding breed difference and the use of crossbreeding to exploit heterosis and breed complementarity can play a role in meeting these challenges. Burrow (2006) highlighted the fact that many important tropically adapted breeds still need to be fully evaluated. There is also a need to implement a conservation programme where appropriate to ensure the availability of varied genetic resources for beef cattle production in tropical environments.

Adaptation to Environment

The symbiotic relationship between evolutionary biology and animal breeding has been well recognized. This relationship is particularly apparent while studying adaptation of certain genotypes to certain environments. The experience and knowledge from animal and plant genetics emphasizes that genotypes vary in their performance and fitness levels across varying environments. Evolutionary biologists have been studying this genetic basis of adaptation and various adaptive evolution theories are widely accepted. Orr (2005) suggested that empirical studies such as QTL (quantitative trait loci-a chromosomal region that contains one or more genes influencing a polygenic trait) analyses and molecular population genetic studies generated renewed interest in adaptation. From an evolutionary point of view, alleles favourable to a particular environment may be fixed in the population. But then the question is how the adaptation in an evolutionary perspective deals with changing environments and changing production targets in domestic animals supplying to various markets?

Precision animal breeding is not just about increasing the accuracy of estimating genetic worth of the animal, but also to do with precisely matching the genotype to the environment. Highly valued breeds or animals in a particular environment may not perform well in a different environment due to the inxfluence of genotype–environment interactions. Hence, precision in breeding should come from identifying the optimal

Table 5.1. Differential indicative performance ranking of various combinations of beef cow and bull breed groups for various traits under tropical conditions[a] B – Tropically adapted British (Hereford–Shorthorn), S – Sanga derived (e.g. Belmont Red), Z – Zebu (e.g. Brahman), C – Continental (e.g. Charolais), * denotes poor performance for the trait of interest and increments up to **** indicating good performance. (Source: Prayaga, 2003c).

		Bull breed											
		B	S	Z	C	B	S	Z	C	B	S	Z	C
		Birth weight				Weaning weight				Final weight (18 months)			
Cow breed	B	**	**	****		*	**	**		*	**	***	
	S	****	****	****		***	***	***		**	***	***	
	Z	*	**	*	***	***	***	***	****	***	***	***	****
		Tick resistance				Worm resistance				Heat resistance			
	B	*	**	****		**	*	***		*	***	***	
	S	**	*	****		**	*	***		**	***	***	
	Z	****	***	****	*	****	***	****	****	****	****	****	***
		Scrotal circumference				Calving success (wet cows)[b]				Calving success (dry cows)[b]			
	B	****	****	***		***		***		**		***	
	S	****	***	***			**	****			***	**	
	Z	**	*	*	*	***	***	**	*	****	****	***	****
		Days to calving[b]				Calf survival				Tenderness			
	B	***		**		**	****	**		****	***	**	
	S		**	***		****	****	****		***	***	**	
	Z	***	****	**	**	****	****	****	****	**	**	*	***

[a]these results are based on the Belmont crossbreeding experiment conducted under tropical conditions
[b]recorded on the parental generation (dam genotypes)

genetic mix given the environment. This capability can be developed only through clear understanding of the environmental stressors under various environments.

Increased production levels brought about through selective breeding have often led to serious decline in adaptation ability. Examples include reduced fitness in dairy cattle, leg problems in pigs, and respiratory diseases in poultry. This has forced the community of animal breeders to consider adaptation and welfare issues along with the traditional goal of achieving higher production. Hence, sustainable animal breeding has emerged as a system where genetic modifications causing increased production do not have negative effects on welfare-related traits of the animals (Jensen and Andersson, 2005). An animal's adaptation to a particular environment can be measured through understanding its behaviour and quantifying its resistance to various environmental stressors. Though the progress in behavioural genetics has been slow because of the difficulties in measuring the traits especially under commercial farm conditions, there have been several studies on measuring resistance to environmental stressors under tropical conditions.

In the northern Australian beef industry, cattle are finished either in feedlots or on pastures in more benign environments or even shipped out overseas as live cattle. Hence, producers are interested in cattle that are not just tropically adapted, but capable of adapting to changing environments. It is important to understand the genetic basis of adaptation and the evolutionary basis for certain breeds to be better equipped to handle harsh tropical environments. Management systems and environments are changing more rapidly than animal populations can adapt to such changes through natural selection. Hence, artificial selection and breeding methods assume greater importance to match various genotypes to the changing environments. Tropical adaptation can be defined as an animal's ability to survive, grow and reproduce despite the environmental stressors such as ectoparasites (ticks, *Boophilus microplus*; buffalo flies, *Haematobia irritans*), endoparasites (gastro-intestinal helminths such as *Cooperia* spp., *Haemonchus* spp.), hot humid climates, seasonally poor nutrition, and tropical diseases. These stressors are quantified by measuring tick counts, faecal egg counts, rectal temperatures under heat stress conditions, coat scores, weight gain under dry conditions. In a recent review on genetics of tropical adaptation, Prayaga *et al.* (2006) reported low to moderate heritability estimates but high variability in these adaptive traits, highlighting the potential for selection.

From a quantitative genetics point of view, genetic correlations among these adaptive traits and their relation to other economically important traits is of paramount importance to implement breeding programmes without compromising adaptation. Reports from tropical adaptation studies (Burrow, 2001; Prayaga and Henshall, 2005) suggest that moderate favourable genetic correlations exist among resistance attributes. Hence, selection for improved resistance in any of these traits will have favourable correlated responses in other resistance attributes. However, genetic associations between resistance traits and growth traits are less conclusive, implying that associations are more complex and are affected by the breed genetic effects. Mackinnon *et al.* (1991) and Burrow (2001) reported positive genetic correlations between body weights and fly counts, leading to various hypotheses such as high testosterone levels or unknown metabolic products in heavier animals being attractive to flies. However, this correlation could also be due to bigger surface area in heavier animals leading to a greater number of flies. The lack of negative effect on growth led researchers to believe that flies are more of an animal welfare and hide damage issue, rather than a production issue, at least in the northern Australian production environment.

Genetic correlations between resistance to heat and growth traits are generally significantly negative, suggesting that as growth in tropics increases, body temperature decreases, implying increased resistance to heat. Because of the thermoregulatory nature of coat (sleek coat versus woolly coat), a significant genetic correlation, is expected between resistance to heat and coat scores. However, Prayaga and Henshall (2005) reported a low, insignificant correlation, indicating the complexity of thermoregulation with components such as sweating, respiratory cooling and lower metabolic heat production (Turner, 1984). Favourable genetic correlations were reported between heat resistance and measures of female fertility (Turner, 1982; Burrow, 2001). However, more research is required to understand female fertility in the tropics, in general, and its relationship with adaptation, in particular. Current research programmes undertaken by the Cooperative Research Centre for Beef Genetic Technologies (Beef CRC) address these issues in Australia. Prayaga and Henshall (2005) reported moderate favourable genetic correlations between parasite and heat resistance traits and temperament, indicating that docile animals are more resistant to parasites and heat. Significantly high genetic correlations between flight time and meat tenderness were reported by Reverter *et al.* (2003) in tropical beef cattle, suggesting the use of temperament as an indirect selection measure for meat tenderness.

Understanding the genetic correlations between productive and reproductive traits and adaptation attributes is important, but it is crucial to implement this knowledge in the breeding programmes to achieve any long-term progress. However, most of these adaptive traits are difficult to measure under field conditions and hence cannot be easily improved through selection. One of the alternatives to improve tropical adaptability has been exploiting breed complementarity through crossbreeding. With the advent of new molecular genetic technologies, efforts are being made to identify major genes and QTL associated with these traits. Olson *et al.* (2003) reported evidence of a major gene responsible for sleek hair coat in Senepol and Criollo cattle. Barendse (2005a) reported five genetic regions of the bovine genome for associations to tick resistance. These developments were possible because of the advances in molecular genetics.

Molecular Revolution

The molecular revolution has the potential to have a huge impact on domestic animal breeding. Simultaneously animal breeding, through its extensively recorded phenotypes, pedigrees, and resource populations in wide-ranging environmental conditions subjected to both natural and artificial selection, contributes to the advances in molecular genetic studies. This is no different to the symbiotic relationship between animal breeding and statistics.

Ever since the studies on molecular genetic markers were initiated, there have been several questions regarding the relevance of traditional animal breeding tools such as pedigree and performance-based selection and there were predictions about one or more DNA tests essentially replacing these traditional genetic evaluation systems. We are now into two decades of this debate and almost reached the completion of bovine genome sequencing and now, more than ever, we are convinced that the molecular information needs to be utilized in conjunction with the traditional selection tools for making breeding decisions. However, this does not undermine the importance of the knowledge gained through molecular genetic advances and there are several areas where the genetic markers or the discovery of functional mutations has directly led to characterizing Halothane (RYR1) locus and mutation causing increased glycogen content in meat (PRKAG3) in pig breeding (Andersson, 2001). Several of these mutations also cause simple monogenic disorders as catalogued in the Online Mendelian Inheritance in Animals (OMIA – http://omia.angis.org.au/).

Andersson and Georges (2004) and Womack (2005) have reviewed the recent advances in livestock genomics and their realized and potential contributions to both human biology and agricultural science. Some of the early developments that led to the current ability to sequence the bovine genome are:

1. Early mapping techniques (somatic cell genetics and *in situ* hybridization) leading to synteny and cytogenetic maps (Womack and Moll, 1986; Yerle *et al.*, 1995);
2. DNA-level markers for building maps and mapping traits (Beckman and Soller, 1983)
3. Microsatellite markers leading to linkage maps (Barendse *et al.*, 1994; Bishop *et al.*, 1994);
4. Further expansion of maps (Barendse *et al.*, 1997; Kappes *et al.*, 1997);
5. Radiation hybrid map for higher resolution comparative mapping (Womack *et al.*, 1997; Williams *et al.*, 2002; Itoh *et al.*, 2005).

A single partially inbred Hereford female was selected to contribute 6x whole genome shotgun reads and another 1.5x from individual animals of Holstein, Angus, Jersey, Limousin, Brahman, and Norwegian Red breeds for SNP (Single Nucleotide Polymorphism) detection. Current map status and genomic resources available in various livestock species are presented in Table 5.2. Recent reviews of the bovine genome sequence and its likely impact on genetics research of beef cattle include those of Dalrymple (2005) and Womack (2006).

The primary aim of any of the farm animal genome sequencing and/or development of SNPs is initially to locate the regions, and ultimately, the causal muta-

Table 5.2. Map status and genomic resource for livestock species (Source: Womack, 2005).

	Cow	Pig	Sheep	Horse	Buffalo	Goat
Chromosome	60	38	54	64	50	60
Mapped markers						
Synteny	1800	1000	250	500	60	–
Linkage	2500	2700	1500	480	–	350
Cytogenic	300	300	850	450	300	350
RH	3200	3000	300	800	30	–
Resources						
EST sequence	450k	300k	20k	30k	–	–
cDNA libraries	80	100	12	15	–	–
BAC libraries	3	5	3	3	–	–
Microarrays	3	2	–	–	–	–
Whole genome sequence	6×	*	–	–	–	–

Note:
*Whole genome sequencing of pig genome begun in 2005

tions underlying the desirable phenotypes of various production, reproduction and adaptation traits. Although hundreds of QTL are mapped in livestock species, very few causative mutations are identified. One notable exception is that of the DGAT1 locus contributing to fat composition in milk followed by functional confirmation of the effect of a missense mutation (Grisart *et al.*, 2002, 2004). These are termed as the quantitative trait nucleotides (QTN) signifying the actual nucleotide affecting the quantitative trait of interest.

Another tool that needs to be exploited is the field of comparative genomics, or comparative analyses of genomes. This is one of the ways of dissecting the genetic basis of phenotypic variation across species. The power of this tool is enhanced by the availability of whole genome sequences for the livestock species. These comparative studies can direct research to species-specific chromosomal regions that could not be revealed by focusing on studying individual genomes. Womack (2005) remarked that identification of QTL for disease resistance in livestock is the next big frontier, leading to the understanding of host–pathogen interaction and subsequent improvement of both animal and human health. In the tropics, however, resistance not only to endemic diseases but also to natural tropical stressors such as ticks, endo-parasites and heat is essential to achieve desired production and health targets of farm animals.

QTL mapping

While the information on molecular genetic markers has revolutionized the biological research in animals, there have been equally important developments in statistical methods, aided by the availability of increased computing power. Detection of QTL requires accurate phenotypes for traits of interest from preferably a structured population of animals, genotypic data on those animals followed by statistical analyses correlating phenotypic and genotypic data. Traditional animal breeding relied on the performance recording for selection decisions. Later developments emphasized the importance of pedigree recording. This enables animal breeders to exploit the recent advances in molecular genetics assisting them in the mapping of genes affecting traits of economic importance. The recent developments in polymorphic DNA markers (microsatelites) as well as SNPs led to the onus on statistical developments in analysing this information.

A comprehensive review of the development of statistical methods dealing with QTL analysis was given by Rocha *et al.* (2002). A summary of this review and important references are given in Table 5.3.

Table 5.3. Methods of QTL mapping.

Methods	Reference
Crosses between outbred lines	Beckman and Soller (1988)
Multiple-marker mapping – regression approach	Haley *et al.* (1994)
	Knott *et al.* (1998)
Half-sib design	Soller and Genizi (1978)
Regression approaches	Knott *et al.* (1994)
	Knott *et al.* (1996)
Maximum likelihood	Elsen *et al.* (1999)
Full sib design	Soller and Genizi (1978)
Regression approaches	Lander and Botstein (1989)
Maximum likelihood	Knott and Haley (1992)
Mixture of full-sib and half-sib design	LeRoy *et al.* (1998)
Granddaughter design	Weller *et al.* (1990)
Complex pedigree structure:	
a. Variance component approaches	Lander and Botstein (1989)
	Fernando and Grossman (1989)
	George *et al.* (2000)
b. Maximum likelihood	
i. Jansen's mixture model	Jansen *et al.* (1998)
ii. Complex segregation and linkage analyses	Meuwissen and Goddard (1997)

Choosing the genes to be investigated in mapping studies can be a difficult task. The approaches can be either based on known gene function (the candidate gene approach), or based on their location on the genome (the positional cloning approach). Currently, the method of choice involves a combination of these two approaches, known as positional candidate approach (Barendse, 2005b).

Linkage and linkage disequilibrium

Linkage is the tendency of closely located genetic loci on the same chromosome to be inherited together more often than expected and linkage disequilibrium (LD) is the non-random association among their respective alleles. While linkage refers to physical location of genes on the same chromosome, LD can result even when the genes are on different chromosomes due to selection, migration and mutation, although it is most notable in closely paired genes. LD can be thought of as statistical dependence and exists only when copies of a given molecular variant shared by two individuals are clonal copies of the same ancestral alleles, identical-by-descent (IBD). Similarly, markers nearby are shared as well because of a paucity of recombination between the loci historically, that is linkage (see Terwilliger, 2001).

Goddard (1991) proposed the use of linkage disequilibrium (LD) for mapping quantitative traits in livestock. The use of linkage analysis to map a QTL is widely known and practised. Meuwissen and Goddard (2000) contrast the use of linkage and linkage disequilibrium analysis in QTL mapping. To position the QTL, linkage mapping uses only the recombinations that occurred within the dataset (may be two to three generations), whereas the LD mapping uses all recombinations since the mutation occurred thereby increasing the precision of the estimate of the position. Hence, Meuwissen and Goddard (2000) proposed a practical multistage approach to QTL mapping outline as follows:

1. Genome-wide scan for QTL – through linkage analysis narrows the region to 20 cM.
2. Markers 1 cM apart in the identified regions – through LD analysis narrows the region to 3 cM.

3. Markers 0.25 cM apart – through LD analysis narrow the region to 0.75 cM.
4. Sequential steps of comparative mapping, evolutionary tree mapping, and positional cloning.

LD can be measured, allowing the mapping of QTL. However, false positives can occur as LD between two loci, even if they are unlinked, caused by variation in the relatedness of pairs of animals, which needs to be adjusted in the model. Goddard and Meuwissen (2005) suggested that methods based on coalescence and IBD need to be improved for practical analysis of large data sets.

However, since 2000, the availability of large numbers of high-density markers (SNP markers) available for whole genome scan at higher density made a significant difference to this approach. Currently, several QTL mapping projects in beef cattle are underway based on whole genome scans with more than 10,000 SNP markers. The biggest challenge lies in unravelling the genes that underlie QTL as the associated problems such as mapping precision, lack of direct relationship between genotype and phenotype, small phenotypic effect of most QTL, and numerous loosely linked QTL complicate the mapping techniques (Andersson and Georges, 2004). In addition, the importance of validation of identified QTL regions is paramount because of false positives. Validation of markers for traits that are not usually recorded under farm conditions is a key issue in this regard. For example, age at puberty in tropical beef cattle is perceived to be influencing lifetime reproduction performance. However, this trait is difficult to define and measure, and hence it is difficult to validate genetic markers that are developed based on research herd data.

As pointed out by Barendse (2005b), QTL mapping is at a maturing stage with reagent costs dropped by two orders of magnitude and throughput genotyping efficiency increased by five orders of magnitude over the previous 5 years. As this trend continues with the availability of the 6x coverage of bovine genome sequence, the next 5 to 10 years will possibly be a 'golden era' for gene discovery and commercialization. However, the development of suitable statistical frameworks to deal with complex pedigrees is still in developmental stages. One of the immediate challenges in this process is to understand and account for epistatic interactions among QTL. Research should also aim to focus on effectively data mining the vast amounts of information available and integration into practical breeding programmes.

Systems Biology – the Inferential Validity of Integrated Approaches

For the last century, a great deal of research in biological sciences has been undertaken with the principal aim of predicting and enhancing future performance. Examples include breeding programmes, diet treatments, management regimes, addressing environmental stressors or challenge from a particular pathogen, or a combination of these.

In recent years, gene expression profiling has become part of the suite of technologies used by animal geneticists investigating livestock traits (Nobis *et al.*, 2003; Suchyta *et al.*, 2003; Lehnert *et al.*, 2004; Donaldson *et al.*, 2005). While some progress has been made in the identification of differentially expressed genes between two (or more) experimental conditions, it is also thought that the analysis of large datasets generated from expression profiles may allow the development of predictive models of the systems underpinning the genetics of complex traits (Andersson and Georges, 2004; Womack, 2005). However, these new advances present formidable computational challenges. CSIRO Livestock Industries has initiated and, to some degree, pioneered a number of strategies with the objective to establish a framework for the understanding of complex gene interaction networks underlying traits of economic importance (Reverter *et al.*, 2004). Methods for data normalization had to be tested and anchored to large-scale mixed-model equations; we have identified and published the formal details of the optimal approach (Reverter *et al.*, 2005a). Based on this approach and applied to data from the Australian Beef CRC, Reverter *et al.* (2005b) engineered a gene network for bovine skeletal muscle with 102 genes, constructed by relating the genes surveyed in five Beef CRC studies with muscle-specific genes identified from the National Cancer Institute, Cancer Genome Anatomy Project, SAGE database. This method is currently being extended to the construction of gene networks in bovine skeletal muscle and adipose tissue, based on gene-expression profile data with 822 genes from the entire set of Beef CRC experiments comprising 147 hybridizations across nine experiments and 47 conditions. Furthermore, our approaches are being applied to gene expression data from other species and conditions as they become available. These include the identification of changes in the network topology in the presence of fleece rot in Merino sheep.

At the same time and with the advent of robust bioinformatics approaches, it is anticipated that biotechnology laboratories will not be inclined to undertake new *in vitro*, let alone *in vivo*, experiments unless their computational/mathematical models anticipate a high chance of success. This is further exacerbated by the raise of 'Reduce, Refine, Replace' – the Three Rs – as the basic tenets of research and other policies concerning the use of animals in scientific testing and animal experimentation.

The availability of genomics data on a large scale has led to the development of bioinformatics inferential algorithms for statistical prediction of causal molecular pathways. However, these potentially powerful algorithms are limited by our inability to evaluate their accuracy, as we do not know the true biological network with which to compare them and experimenters cannot physically perform in reasonable time the multiple gene knockouts (or other types of interventions) necessary to test the predicted networks systematically. Nevertheless, the study of the basic principles involved in the dynamics and topological changes in genomes as a result of a given perturbation is gaining momentum. Given: (i) the potentially large number of gene-to-gene interactions resulting from epistatic effects; (ii) the inherent stochastic nature of the individual main effects of each gene in the network; and (iii) the fact that most gene knockouts result in little or no phenotypic change; it follows that successful (i.e. directed to a desired biological outcome) testing of postulated hypotheses about desired biological outcomes becomes a non-trivial task and/or too complex to allow for an exact solution. Furthermore, given a well-defined genetic model, postulating biologically meaningful hypotheses could be a task prone to large Type III error rates (i.e. correctly rejecting the null hypothesis, but incorrectly attributing the cause).

The advent of new high-throughput genetic technologies has brought the potential to drive the prediction of future performance using vastly improved frameworks, most importantly via integrative approaches. In its broad sense, such a shift toward biological integration is the approach advocated in systems biology. The ultimate vision of our systems biology endeavours is to provide a comprehensive integration of data-driven knowledge of genetic and environmental architectures to develop functional models that provide qualitative description of the subject matter with predictive power (Hwang *et al.*, 2005a, b).

A Venn diagram representation of how this integration is taking place is presented in Fig. 5.1. Within its core, three distinct types of data are clearly distinguishable: (i) Phenotype and pedigree (the basis for genetic evaluation and parameter estimation); (ii) Phenotype and marker (the basis for the simplest of the marker association studies); and (iii) Gene (protein or metabolite) expression (the basis for differential expression and differential connectivity studies). While attempts exist to integrate any two of these three types of data with mixed results, a great deal of research is still required to successfully integrate the three domains.

Hence, although we are moving in the right direction in the pursuit of precision animal breeding, the target itself is highly dynamic in nature. As we are learning more about the biology of traits under polygenic inheritance, it is becoming very obvious that we can only make incremental advances and precision animal breeding has to include and integrate information from all possible sources such as phenotype, genotype, and gene expression.

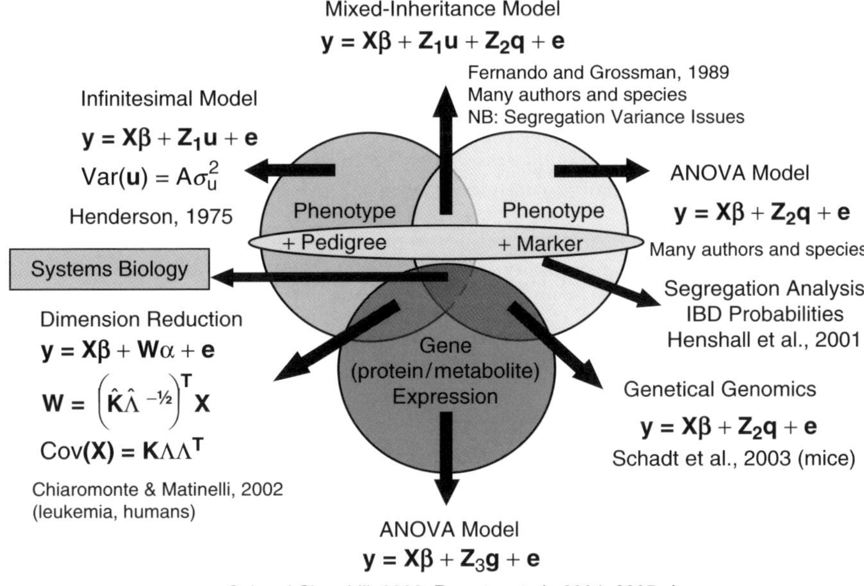

Fig. 5.1. Diagrammatic representation of data integration approaches.

References

Andersson, L. (2001) Genetic dissection of phenotypic diversity in farm animals. *Nature Reviews Genetics* 2, 130–138.

Andersson, L. and Georges, M. (2004) Domestic-animal genomics: deciphering the genetics of complex traits. *Nature Reviews Genetics* 5, 202–212.

Barendse, W. (2005a) DNA markers for tick resistance. *Provisional patent application*.

Barendse, W. (2005b) The transition from quantitative trait loci to diagnostic test in cattle and other livestock. *Australian Journal of Experimental Agriculture* 45, 831–836.

Barendse, W., Armitage, S.M., Kossarek, L.M., Shalom, A., Kirkpatrick, B.W., Ryan, A.M., Clayton, D., Li, L., Neiberg, H.L., Zhang, N., Grosse, W.M., Weiss, J., Creighton, P., McCarthy, F., Ron, M., Teale, A.J., Fries, R., McGraw, R.A., Moore, S.S., Georges, M., Womack, J.E. and Hetzel, D.J.S. (1994) A genetic linkage map of the bovine genome. *Nature Genetics* 6, 227–238.

Barendse, W., Vaiman, D., Kemp, S.J., Sugimoto, Y., Armitage, S.M., Williams, J.L., Sun, H.S., Eggen, A., Agaba, M., Aleyasin, S.A., Band, M., Bishop, M.D., Buitkamp, J., Byrne, K., Collins, F., Cooper, L., Coppettiers, W., Denys, B., Drinkwater, R.D., Easterday, K., Elduque, C., Ennis, S., Erhardt, G., Ferretti, L., Flavin, N., Gao, Q., Georges, M., Gurung, R., Harlizius, B., Hawkins, G., Hetzel, J., Hirano, T., Hulme, D., Jorgensen, C., Kessler, M., Kirkpatrick, B.W., Konfortov, B., Kostia, S., Kuhn, C., Martin Burriel, I., McGraw, R.A., Miller, J.R., Moody, D.E., Moore, S.S., Nakane, S., Nijman, I.J., Olsaker, I., Pomp, D., Rando, A., Ron, M., Shalom, A., Teale, A.J., Thieven, U., Urquhart, B.G.D., Vage, D.I., Van de Weghe, A., Varvio, S., Velmala, R., Vilkki, J., Weikard, R., Woodside, C., Womack, J.E., Zanotti, M. and Zaragoza, P. (1997) A medium density genetic linkage map of the bovine Genome. *Mammalian Genome* 8, 21–28.

Beckman, J.S. and Soller, M. (1983) Restriction fragment length polymorphisms in genetic improvement: methodologies, mapping and costs. *Theoretical and Applied Genetics* 67, 35–43.

Beckman, J.S. and Soller, M. (1988) Detection of linkage between marker loci and loci affecting quantitative traits in crosses between segregating populations. *Theoretical and Applied Genetics* 76, 228–236.

Bishop, M.D., Kappes, S.M., Keele, J.W., Stone, R.T., Sunden, S.L.F., Hawkins, G.A., Salinas Toldo, S., Fries, R., Grosz, M.D., Yoo, J. and Beattie, C.W. (1994) A genetic linkage map for cattle. *Genetics* 136, 619–639.

Burrow, H.M. (2001) Variances and covariances between productive and adaptive traits and temperament in a composite breed of tropical beef cattle. *Livestock Production Science* 70, 213–233.

Burrow, H.M. (2006) Utilization of diverse breed resources for tropical beef production. *Proceedings of the 8th World Congress on Genetics Applied to Livestock Production*. Ed: Organizing Committee WCGALP, Brazil, Communication No: 32–01.

Chiaromonte, F. and Martinelli, J. (2002) Dimension reduction strategies for analyzing global gene expression data with a response. *Mathematical Biosciences* 176, 123–144.

Cui, X. and Churchill, G.A. (2003) Statistical tests for differential expression in cDNA microarray experiments. *Genome Biology* 4, 210.

Cundiff, L.V. (2006) The impact of quantitative genetics on productive, reproductive and adaptive traits in beef cattle. *Proceedings of the Conference Australian Beef – the Leader!* Beef CRC, Armidale, NSW, pp. 29–46.

Dalrymple, B.P. (2005) Harnessing the bovine genome sequence for the Australian cattle and sheep industries. *Australian Journal of Experimental Agriculture* 45, 1011–1016.

Donaldson L., Vuocolo, T., Gray, C., Strandberg, Y., Reverter, A., McWilliam, S.M., Wang, Y.H., Byrne, K.A and Tellam, R. (2005) Construction and validation of a bovine innate immune microarray. *BMC Genomics* 6, 135.

Elsen, J.M., Mangin, B., Goffinet, B., Boichard, D. and Le Roy, P. (1999) Alternative models for QTL detection in livestock I. General introduction. *Genetics Selection Evolution* 31, 213–224.

Falconer, D.S. and Mackay, T.F.C. (1996) *Introduction to Quantitative Genetics*. Pearson Education, Harlow, Essex.

Fernando, R.L. and Grossman, M. (1989) Marker-assisted selection using best linear unbiased prediction. *Genetics Selection Evolution* 21, 467–477.

Franklin, I.R. (1980) Evolutionary change in small populations. In: Soule, M.E. and Wilcox, B.A. (eds) *Conservation Biology: an Evolutionary–Ecological Perspective*. Sinauer, Sunderland, pp. 135–150.

Franklin, I.R. and Frankham, R. (1998) How large must populations be to retain evolutionary potential? *Animal Conservation* 1, 69–70.

George, A.W., Visscher, P.M. and Haley, C.S. (2000) Mapping quantitative trait loci in complex pedigrees: a two step variance component approach. *Genetics* 156, 2081–2092.

Goddard, M.E. (1991) Mapping genes for quantitative traits using linkage disequilibrium. *Genetics Selection Evolution* 23 suppl. 1, 1315–1345.

Goddard, M.E. and Meuwissen, T.H.E. (2005) The use of linkage disequilibrium to map quantitative trait loci. *Australian Journal of Experimental Agriculture* 45, 837–845.

Gregory, K.E., Cundiff, L.V. and Koch, R.M. (1999) *Composite breeds to use heterosis and breed differences to improve efficiency of beef production*. USDA-ARS, Clay Center, USA.

Grisart, B., Coppieters, W., Farnir, F., Karim, L., Ford, C., Berzi, P., Cambisano, N., Mini, M., Reid, S., Simon, P., Spelman, R., Georges, M. and Snell, R. (2002) Positional candidate cloning of a QTL in dairy cattle: identification of a missense mutation in the bovine DGAT1 gene with major effect on milk yield and composition. *Genome Research* 12, 222–231.

Grisart, B., Farnir, F., Karim, L., Cambisano, N., Kim, J.J., Kvasv, A., Mini, M., Simon, P., Frere, J.M., Copieters, W. and Georges, M. (2004) Genetic and functional confirmation of the causality of the DGAT1 K232A quantitative trait nucleatide in affecting milk yield and composition. *Proceedings of the National Academy of Sciences of the United States of America* 101, 2398–2403.

Haley, C.S., Knott, S.A. and Elsen, J.M. (1994) Mapping quantitative trait loci in crosses between outbred lines using least squares. *Genetics* 136, 1195–1207.

Henderson, C.R. (1975) Best linear unbiased estimation and prediction under a selection model. *Biometrics* 31, 423–447.

Henderson, C.R. (1977) Best linear unbiased prediction of breeding values not in the model for records. *Journal of Dairy Science* 60, 783–787.

Henshall, J.M., Tier, B. and Kerr, R.J. (2001) Estimating genotypes with independently sampled descent graphs. *Genetical Research* 78, 281–288.

Hwang, D., Rust, A.G., Ramsey, S., Smith, J.J., Leslie, D.M., Weston, A.D., Atauri, P. de, Aitchison, J.D., Hood, L., Siegel, A.F. and Bolouri, H. (2005a) A data integration methodology for systems biology. *Proceedings of the National Academy of Sciences of the United States of America* 102, 17296–17301.

Hwang, D., Smith, J.J., Leslie, D.M., Weston, A.D., Rust, A.G., Ramsey, S., Atauri, P. de, Siegel, A.F., Bolouri, H., Aitchison, J.D. and Hood, L. (2005b) A data integration methodology for systems biology: experimental verification. *Proceedings of the National Academy of Sciences of the United States of America* 102, 17302–17307.

Itoh, T., Watanabe, T., Ihara, N., Mariani, P., Beattie, C.W., Sugimoto, Y. and Takasuga, A. (2005) A comprehensive rediation hybrid map of the bovine genome comprising 5593 loci. *Genomics* 85, 413–424.

Jansen, R.C., Johnson, D.L. and Van Arendonk, J.A. (1998) A mixture model approach to the mapping of quantitative trait loci in complex populations with an application to multiple cattle families. *Genetics* 148, 391–399.

Jensen, P. and Andersson, L. (2005) Genomics meets ethology: a new route to understanding domestication, behaviour, and sustainability in animal breeding. *Ambio* 34, 320–324.

Kappes, S.M., Keele, J.W., Stone, R.T., McGraw, R.A., Sonstegard, T.S., Smith, T.P., Lopez-Corrales, N.L. and Beattie, C.W. (1997) A second generation linkage map of the bovine genome. *Genome Research* 7, 235–249.

Knott, S.A. and Haley, C.S. (1992) Maximum likelihood mapping of quantitative trait loci using full-sib families. *Genetics* 132, 1211–1222.

Knott, S.A., Elsen, J.M. and Haley, C.S. (1994) Multiple marker mapping of quantitative trait loci in half-sib populations. *Proceedings of the 5th World Congress on Genetics Applied to Livestock Production.* WCGALP, Guelph, Canada, pp. 33–36.

Knott, S.A., Elsen, J.M. and Haley, C.S. (1996) Methods for multiple-marker mapping of quantitative trait loci in half-sib populations. *Theoretical and Applied Genetics* 93, 71–80.

Knott, S.A., Marklund, L., Haley, C.S., Andersson, K., Davies, W., Ellegren, H., Fredholm, M., Hannson, I., Hoyheim, B., Lundstrom, K., Moller, M. and Andersson, L. (1998) Multiple marker mapping of quantitative trait loci in a cross between outbred wild boar and large white pigs. *Genetics* 149, 1069–1080.

Kristenson, T.N. and Sorensen, A.C. (2005) Inbreeding – lessons from animal breeding, evolutionary biology and conservation genetics. *Animal Science* 80, 121–133.

Lander, E.S. and Botstein, D. (1989) Mapping Mendelian factors underlying quantitative traits using RFLP linkage maps. *Genetics* 121, 185–199.

Le Roy, P., Elsen, J.M., Boichard, D., Mangin, B., Bidanel, J.P. and Goffinet, B. (1998) An algorithm for QTL detection in mixture of full and half sib families. *Proceedings of the 6th World Congress on Genetics Applied to Livestock Production.* (Ed.) Organizing Committee WCGALP, Armidale, NSW, pp. 257–260.

Lehnert, S.A., Wang, Y.H. and Byrne, K.A. (2004) Development and application of a bovine cDNA microarray for expression profiling of muscle and adipose tissue. *Australian Journal of Experimental Agriculture* 44, 1127–1133.

Lynch, M. and Walsh, B. (1998) *Genetics and Analysis of Quantitative Traits.* Sinauer Associates Inc., Sunderland MA, USA.

Mackinnon, M.J., Meyer, K. and Hetzel, D.J.S. (1991) Genetic variation and covariation for growth, parasite resistance and heat tolerance in tropical cattle. *Livestock Production Science* 27, 105–122.

Meuwissen, T.H.E. and Goddard, M.E. (1997) Estimation of effects of quantitative trait loci in large complex pedigrees. *Genetics* 146, 409–416.

Meuwissen, T.H.E. and Goddard, M.E. (2000) Final mapping of quantitative trait loci using linkage disequilibria with closely linked marker loci. *Genetics* 155, 421–430.

Nobis, W., Ren, X., Suchyta, S.P., Suchyta, T.R., Zanella, A.J. and Coussens, P.M. (2003) Development of a porcine brain cDNA library, EST database, and microarray resource. *Physiological Genomics* 16, 153–159.

Olson, T.A., Filho, K.E., Cundiff, L.V., Koger, M., Butts, W.T. Jr and Gregory, K.E. (1991) Effect of breed group by location interaction on crossbred cattle in Nebraska and Florida. *Journal of Animal Science* 69, 104–114.

Olson, T.A., Lucena, C., Chase C.C. Jr, and Hammond, A.C. (2003) Evidence of a major gene influencing hair length and heat tolerance in *Bos taurus* cattle. *Journal of Animal Science* 81, 80–90.

Orr, H.A. (2005) Theories of adaptation: what they do and don't say. *Genetica* 123, 3–13.

Patterson, H.D. and Thompson, R. (1971) Recovery of interblock information when block sizes are unequal. *Biometrika* 63, 83–92.

Prayaga, K.C. (2003a) Evaluation of beef cattle genotypes and estimation of direct and maternal genetic effects in a tropical environment 1. Growth traits. *Australian Journal of Agricultural Research* 54, 1013–1025.

Prayaga, K.C. (2003b) Evaluation of beef cattle genotypes and estimation of direct and maternal genetic effects in a tropical environment 2. Adaptive and temperament traits. *Australian Journal of Agricultural Research* 54, 1027–1038.

Prayaga, K.C. (2003c) *Final Report on Belmont Crossbreeding Project.* Meat and Livestock Australia, Sydney.

Prayaga, K.C. (2004) Evaluation of beef cattle genotypes and estimation of direct and maternal genetic effects in a tropical environment 3. Fertility and calf survival traits. *Australian Journal of Agricultural Research* 55, 811–824.

Prayaga, K.C. and Henshall, J.M. (2005) Adaptability in tropical beef cattle: genetic parameters of growth, adaptive and temperament traits in a crossbred population. *Australian Journal of Experimental Agriculture* 45, 971–983.

Prayaga, K.C., Henshall, J.M. and Burrow, H.M. (2003) Optimization of breed proportions in tropically adapted beef composites based on growth and resistance traits. *Proceedings of the Association for the Advancement of Animal Breeding and Genetics*. AAABG, Melbourne, pp. 302–305.

Prayaga, K.C., Barendse, W. and Burrow, H.M. (2006) Genetics of tropical adaptation. *Proceedings of 8th World Congress on Genetics Applied to Livestock Production*. WCGALP, Brazil, CDROM; Communication no. 16–01.

Quaas, R.L. and Pollak, E.J. (1980) Mixed model methodology for farm and ranch beef cattle testing programs. *Journal of Animal Science* 51, 1277–1287.

Reverter, A., Johnston, D.J., Ferguson, D.M., Perry, D., Goddard, M.E., Burrow, H.M., Oddy, V.H., Thompson, J.M. and Bindon, B.M. (2003) Genetic and phenotypic characterization of animal, carcass, and meat quality traits from temperate and tropically adapted beef breeds. 4. Correlations among animal, carcass, and meat quality traits. *Australian Journal of Agricultural Research* 54, 149–158.

Reverter, A., Wang, Y.H., Byrne, K.A., Tan, S-H., Harper, G.S. and Lehnert, S.A. (2004) Joint analysis of multiple cDNA microarray studies via multivariate mixed models applied to genetic improvement of beef cattle. *Journal of Animal Science* 82, 3430–3439.

Reverter, A., Barris, W., McWilliam, S.M., Byrne, K.A., Wang, Y.H., Tan, S-H., Hudson, N.J. and Dalrymple, B.P. (2005a) Validation of alternative methods of data normalization in gene co-expression studies. *Bioinformatics* 21, 1112–1120.

Reverter, A., Barris, W., Moreno-Sánchez, N., McWilliam, S.M., Wang, Y.H., Harper, G.S., Lehnert, S.A. and Dalrymple, B.P. (2005b) Construction of gene interaction and regulatory networks in bovine skeletal muscle from expression data. *Australian Journal of Experimental Agriculture* 45, 821–829.

Rocha, J.L., Pomp, D. and Van Vleck, L.D. (2002) QTL analysis in livestock. In: Camp, N.J. and Cox, A. (eds) *Quantitative Trait Loci: Methods and Protocols*. Humana Press Inc., Totowa, New Jersey, pp. 311–346.

Schadt, E.E., Monks, S.A., Drake, T.A., Lusis, A.J., Che, N., Colinayo, V., Ruff, T.G., Milligan, S.B., Lamb, J.R., Cavet, G., Linsley, P.S., Mao, M., Stoughton, R.B. and Friend, S.H. (2003) Genetics of gene expression surveyed in maize, mouse and man. *Nature* 422, 297–302.

Soller, M. and Genizi, A. (1978) The efficiency of experimental designs for the detection of linkage between a marker locus and a locus affecting a quantitative trait in segregating populations. *Biometrics* 34, 47–55.

Suchyta, S.P., Sipkovsky, S., Kruska, R., Jeffers, A., McNulty, A., Coussens, M.J., Tempelman, R.J., Halgren, R.G., Saama, P.M., Bauman, D.E., Boisclair, Y.R., Burton, J.L., Collier, R.J., DePeters, E.J., Ferris, T.A., Lucy, M.C., McGuire, M.A., Medrano, J.F., Overton, T.R., Smith, T.P., Smith, G.W., Sonstegard, T.S., Spain, J.N., Spiers, D.E., Yao, J. and Coussens, P.M. (2003) Development and testing of a high-density cDNA microarray resource for cattle. *Physiological Genomics* 15, 158–164.

Terwilliger, J.D. (2001) On the resolution and feasibility of genome scanning approaches. *Advances in Genetics* 42, 351–391.

Turner, H.G. (1982) Genetic variation of rectal temperature in cows and its relationship to fertility. *Animal Production* 35, 401–412.

Turner, H.G. (1984) Variation in rectal temperature of cattle in a tropical environment and its relation to growth rate. *Animal Production* 38, 417–427.

Walsh, B. (2001) Quantitative genetics in the age of genomics. *Theoretical Population Biology* 59, 175–184.

Weller, J.I., Kashi, Y. and Soller, M. (1990) Power of daughter and granddaughter designs for determining linkage between marker loci and quantitative trait loci in dairy cattle. *Journal of Dairy Science* 73, 2525–2537.

Williams, J.L., Eggen, A., Ferretti, L., Farr, C.J., Gautier, M., Amati, G., Ball, G., Caramorr, T., Critcher, R., Costa, S., Hextall, P., Hills, D., Jeulin, A., Kiguwa, S.L., Ross, O., Smith, A.L., Saunier, K., Urquhart, B. and Waddington, D. (2002) A bovine whole genome radiation hybrid panel and outline map. *Mammalian Genome* 13, 469–474.

Womack, J.E. (2005) Advances in livestock genomics: opening the barn door. *Genome Research* 15, 1699–1705.

Womack, J.E. (2006) The impact of sequencing the bovine genome. *Australian Journal of Experimental Agriculture* 46, 151–153.

Womack, J.E. and Moll, Y.D. (1986) Gene map of the cow: conservation of linkage with mouse and man. *Journal of Heredity* 77, 2–7.

Womack, J.E., Johnson, J.S., Owen, E.K., Rexroad, C.E., Schlapfer, J. and Yang, Y.P. (1997) A whole-genome radiation hybrid panel for bovine gene mapping. *Mammalian Genome* 8, 854–856.

Yerle, M., Lahbib-Mansais, Y., Mellink, C., Goureau, A., Pinton, P., Echard, G., Gellin, J., Zijlstra, C., De Haan, N., Bosma, A.A., Chowdhary, B., Gu, F., Gustavsson, I., Thomsen, P.D., Christensen, K., Rettenberger, G., Hameister, H., Schmitz, A., Chaput, B. and Frelat, G. (1995) The PiGMap Consortium cytogenetic map of the domestic pig (*Sus scrofa domestica*). *Mammalian Genome* 6, 176–186.

6 Germ Cell Transplantation – a Novel Approach to 'Designer' Animals?

Ina Dobrinski[1] and Jonathan R. Hill[2]
[1]*Center for Animal Transgenesis and Germ Cell Research, School of Veterinary Medicine, University of Pennsylvania;* [2]*CSIRO Livestock Industries, Australia*

Abstract
Transplantation of male germ line stem cells from fertile donor mice to the testes of infertile recipient mice results in donor-derived spermatogenesis and transmission of the donor's genetic material to the offspring of recipient animals. Although most widely used in rodents, germ cell transplantation has been applied to pigs, goats and cattle, as well as primates. In large mammals, germ cells can be transplanted to a recipient testis using ultrasound guidance. Importantly, germ cell transplantation was successful between unrelated, immuno-competent donors and recipients in livestock species, whereas transplantation in rodents requires genetically matched or immuno-compromized recipients. Efficiency of colonization of the recipient testis by donor-derived germ cells can be improved by pretreatment of the recipient animal to deplete endogenous germ cells. Genetic manipulation of germ line stem cells before transplantation will result in the production of transgenic sperm. Transgenesis through the male germ line has tremendous potential in livestock species where embryonic stem cell technology is not available and current options to generate transgenic animals are inefficient. Introduction of a genetic change prior to fertilization will circumvent problems associated with manipulation of early embryos and developmental abnormalities associated with somatic cell nuclear transfer and reprogramming. Current research is directed toward improving isolation and culture of male germ line stem cells from livestock species to increase efficiency of transgene transmission and to allow for gene targeting prior to germ cell transplantation. Germ cell transplantation then represents an approach to germ line manipulation through use of transgenic sperm for natural breeding that is potentially more efficient than currently existing strategies.

Introduction

Male reproductive efficiency relies on continuous, efficient production of spermatozoa. Throughout the adult life of the male, the process of spermatogenesis is highly organized into sequential steps of cell proliferation and differentiation, resulting in the production of virtually unlimited numbers of spermatozoa (Russell *et al.*, 1990). The foundation of this system is the male germ line stem cell, which has the unique potential for both self-renewal and production of differentiated daughter cells which will ultimately form spermatozoa (Huckins, 1971; Clermont, 1972; Meistrich and van Beek, 1993). Among the stem cells in a male individual, the spermatogonial stem cell is unique in that it is the only cell in an adult body that divides and can contribute genes to subsequent generations, making it an obvious target for genetic manipulations. Because stem cells are ultimately defined by

 Redesigning Animal Agriculture (eds D. Swain, E. Charmley, J. Steel and S. Coffey)

function, unequivocal identification depends on an assay to demonstrate the potential to reconstitute the appropriate body system.

This assay became available for spermatogonial stem cells when, in 1994, Dr Ralph Brinster and colleagues at the University of Pennsylvania reported that transplantation of germ cells from fertile donor mice to the testes of infertile recipient mice results in donor-derived spermatogenesis and sperm production by the recipient animal (Brinster and Zimmermann, 1994). The use of donor males carrying the bacterial β-galactosidase gene allowed for identification of donor-derived spermatogenesis in the recipient mouse testis and established the fact that the donor haplotype is passed on to the offspring by recipient animals (Brinster and Avarbock, 1994).

Germ Cell Transplantation – a Brief History

In the years following the initial report of the technique, several important advances were accomplished. Jiang and Short (1995) applied the technique to germ cell transplantation between rats, which was subsequently also reported by Ogawa *et al.* (1999a) and Zhang *et al.* (2003). In 1996, Brinster's group showed that mouse spermatogonial stem cells can be cryopreserved for prolonged periods of time before transplantation and still establish spermatogenesis in the recipient testis (Avarbock *et al.*, 1996; Kanatsu-Shinohara *et al.*, 2003c). In the following year, a detailed technical analysis of the technique was published (Ogawa *et al.*, 1997) and in 1999, we established an image analysis approach that allows quantification of colonization of recipient testes by donor stem cells (Dobrinski *et al.*, 1999a).

With these fundamental techniques in place, it became possible to study the stem cell niche in the testis and to characterize putative spermatogonial stem cells. The pattern and kinetics of colonization after transplantation were described in detail (Parreira *et al.*, 1998; Nagano *et al.*, 1999; Ventela *et al.*, 2002), and cross-species transplantation showed that the cell cycle during spermatogenesis is controlled by the germ cell, not the Sertoli cell (Franca *et al.*, 1998). Sperm arising from transplanted donor germ cells are capable of fertilization *in vivo* and *in vitro* (Brinster and Avarbock, 1994; Goossens *et al.*, 2003; Honaramooz *et al.*, 2003a). Recently, experiments demonstrated the developmental potential of mouse primordial germ cells to initiate spermatogenesis when transplanted into a post-natal testis (Chuma *et al.*, 2005).

Cross-species Transplantation of Germ Cells

Using immunocompromised mice as recipient animals, rat sperm developed in mouse testes following cross-species spermatogonial transplantation from rats to mice (Clouthier *et al.*, 1996) and transplantation was subsequently also successful from mice to rats (Ogawa *et al.*, 1999a; Zhang *et al.*, 2003). For its obvious practical potential, cross-species germ cell transplantation was then explored further. Hamster spermatogenesis could also occur in the mouse testis (Ogawa *et al.*, 1999b); however, with increasing phylogenetic distance between donor and recipient species, complete spermatogenesis could no longer be achieved in the mouse. Transplantation of germ cells from non-rodent donors, ranging from rabbits, dogs, pigs, bulls, and horses to non-human primates and humans, resulted in colonization of the mouse testis but spermatogenesis became arrested at the stage of spermatogonial expansion (Dobrinski *et al.*, 1999b, 2000; Nagano *et al.*, 2001a, 2002a). Therefore, the initial steps of germ cell recognition by the Sertoli cells, localization to the basement membrane and initiation of cell proliferation are conserved between evolutionary divergent species. However, when donor and recipient species are phylogenetically more distant than rodents, the recipient testicular environment appears to become unable to support spermatogenic differentiation and meiosis. This incompatibility of donor germ cells and recipient testicular environment could be overcome by co-transplantation of germ cells and Sertoli cells (Shinohara *et al.*, 2003) or by testis tissue transplantation (Honaramooz *et al.*, 2002a). Although cross-species spermatogonial transplantation did not have the envisioned immediate practical application, it none the less provides a bioassay for stem cell potential of germ cells isolated from other species (Dobrinski *et al.*, 1999b, 2000; Izadyar *et al.*, 2002).

Germ Cell Transplantation in Livestock Animals

While the majority of studies are performed in rodent models, germ cell transplantation is also applied to non-rodent species like pigs, goats, cattle, primates and recently fish (Honaramooz *et al.*, 2002b, 2003a,b; Schlatt *et al.*, 2002; Izadyar *et al.*, 2003; Takeuchi *et al.*, 2003; Yoshizaki *et al.*, 2005; Okutsu *et al.*, 2006).

The application of germ cell transplantation technology to non-rodent mammalian species was initially difficult. Due to differences in testicular anatomy and physiology, germ cells cannot be delivered by the same technique as in rodents. Instead, we developed a technique combining ultrasound-guided cannulation of the centrally located rete testis with delivery of germ cells by gravity flow. Ultrasound-guided placement of the injection catheter into the rete testis with the recipient animal under general anaesthesia provides access into the network of seminiferous tubules, which can be confirmed by visualization of the spread of fluid into the tubules. A successful injection has a characteristic ultrasonographic appearance and can be seen when a small amount of air is present in the needle tip, which then flows quickly into the tubules, causing a characteristic echogenic appearance. Alternatively, injection of a commercial ultrasound-opaque fluid can highlight the path of the fluid through the tubules (Izadyar *et al.*, 2003). Thus, the 'quality' of a cell transfer injection can be assessed immediately and the most promising candidate recipient animals can be identified early on in the process, thereby minimizing cost associated with maintenance of recipient animals through to breeding age and mating. With this technique, we succeeded in transplanting donor cells from pigs and goats to the testes of prepubertal recipient animals (Honaramooz *et al.*, 2002b, 2003a,b). Transplantation of donor cells from transgenic goats provided proof-of-principle that germ cell transplantation results in donor-derived sperm production and fertility in a large animal species (Fig. 6.1; Honaramooz *et al.*, 2003b). In cattle, Izadyar *et al.* (2003) first reported the technique of germ cell transfer and showed that transplanted autologous germ cells can initiate spermatogenesis in the recipient testis. The study did not succeed in efficient donor cell colonization using homologous germ cells. These results are in contrast to our findings in cattle, pigs and goats and have to be interpreted with caution as only a small number of relatively old recipient animals were analysed, as acknowledged by the authors. We have shown that heterologous transplantation of bovine germ cells can succeed between breeds of cattle (Hill and Dobrinski, 2006) by transplanting testis cells from prepubertal *Bos taurus* (Angus) bull calves labelled with a fluorescent dye to prepubertal *Bos indicus* cross (predominantly Brahman bloodline) bull calves. Each of the recipients retained the donor cells in the tubule epithelium for at least 8 weeks after transfer, which suggests that transfer between different animals, and indeed even between animals of different breeds of cattle, can be achieved. Further studies will aim to demonstrate whether donor cells are able to undergo spermatogenesis in the recipient animals.

Recipient Animal Preparation

In rodents, using young recipient mice and treating recipients with GnRH-agonists improved donor cell colonization (Ogawa *et al.*, 1998, 1999a; Dobrinski *et al.*, 2001; Shinohara *et al.*, 2001). Interestingly, germ

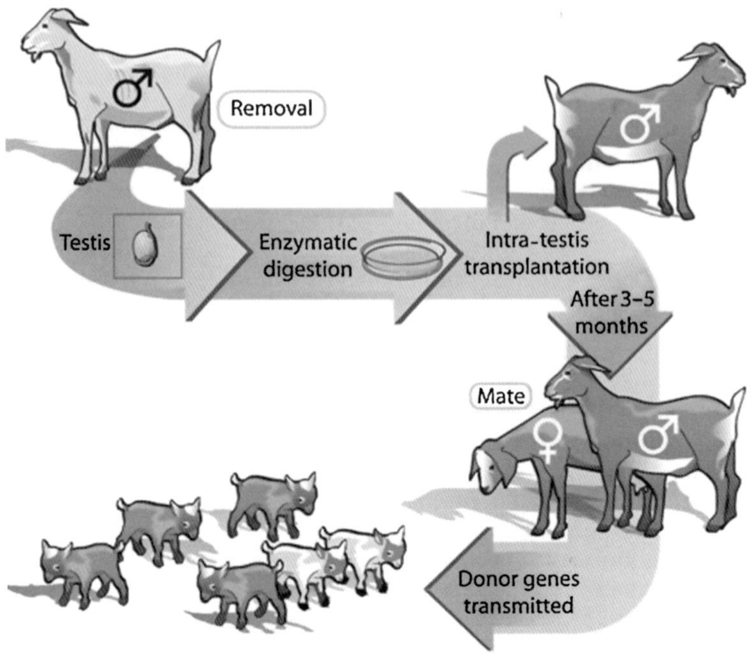

Fig. 6.1. Schematic overview of germ cell transplantation in goats. A single-cell suspension is prepared from the testes of a transgenic donor goat. The cells are infused into the seminiferous tubules of wild-type recipient goats. Donor-derived spermatogonial stem cells generate colonies of transgenic spermatogenesis. Mating the recipient goat to a wild-type doe produces progeny, some of which are transgenic for the donor transgene (modified from Honaramooz *et al.*, 2003b).

cell transplantation in rodents and perhaps also older cattle requires that donor and recipients are closely related or that recipient animals are immuno-suppressed (Kanatsu-Shinohara *et al.*, 2003a; Zhang *et al.*, 2003; Izadyar *et al.*, 2003), whereas germ cell transplantation in pigs, goats and younger cattle was successful also between unrelated individuals (Honaramooz *et al.*, 2002b; 2003a,b; Hill and Dobrinski, 2006). In most studies of successful germ cell transplantation reported so far in domestic animal species, recipient animals were pre-pubertal (Honaramooz *et al.*, 2002b; 2003a, b; Hill and Dobrinski, 2006). At this stage of testicular development, there is only a single layer of germ cells and somatic cells lining the seminiferous tubules and tight junctions between adjacent Sertoli cells are not fully formed, which will facilitate access of the donor cells to the putative stem cell niches in the recipient seminiferous epithelium.

The efficiency of colonization of seminiferous tubules by the transplanted germ cells can be improved if the recipient testes have little or no endogenous spermatogonia. Testicular irradiation has been explored to deplete endogenous germ cells in mice, cats, cattle, goats and monkeys (Creemers *et al.*, 2002; Schlatt *et al.*, 2002; Izadyar *et al.*, 2003; Honaramooz *et al.*, 2005; Kim *et al.*, 2006). This approach has the advantage of minimizing systemic effects but it requires specialized radiotherapy equipment which in its present form is not very practical for field application in livestock species.

Busulfan, a DNA alkylating agent that destroys proliferating cells, is frequently used in rodents to deplete recipient germ cells prior to germ cell transplantation. However, the stem cell depleting dose of busulfan is species- and strain-specific and treatment can be lethal due to severe bone marrow depression (Ogawa *et al.*, 1999a;

Brinster *et al.*, 2003). Treatment of pigs and rams with busulfan between 5 and 17 weeks of age resulted in a significant reduction of testis weight and germ cell number but also in systemic toxicity. In litter-bearing animals such as pigs, *in utero* treatment is effective as previously described in the mouse (Brinster *et al.*, 2003; Honaramooz *et al.*, 2005). Administration of busulfan to pregnant sows at d 98 and 108 of gestation (a period of high proliferation of fetal gonocytes) resulted in depletion of germ cells with no observed adverse effects on the piglets or on sow health or fertility (Honaramooz *et al.*, 2005). It was shown previously in rodents that complete suppression of spermatogenesis might not be desirable in recipient animals as some residual endogenous spermatogenesis is an indication that the testicular environment has not been permanently damaged by the cytotoxic treatment (Brinster *et al.*, 2003).

Donor Germ Cell Isolation

It has been estimated that there are only about 2×10^4 stem cells in 10^8 cells of a mouse testis (Meistrich and van Beek, 1993; Tegelenbosch and de Rooij, 1993). To improve efficiency of donor-derived spermatogenesis after transplantation and to allow manipulation of donor cells prior to transplantation, it is desirable to obtain a cell population enriched in germ cells. Isolation and enrichment of spermatogonia from pre-pubertal animals where significant testis growth has already occurred but germ cell differentiation has not yet started should yield the largest number of germ cells for subsequent experimentation. Therefore, enrichment of germ cells from immature testes is more efficient than from adult testes. In the mouse, selection of germ cells for expression of α_6- and β_1-integrin, the absence of c-kit receptor, as well as collection of cells from experimentally induced cryptorchid testes, resulted in a significant enrichment for spermatogonial stem cells (Shinohara *et al.*, 1999, 2000a,b). More recently, expression of Thy-1, CD9 or Egr3 was utilized as a marker for enrichment of mouse germ line stem cell populations (Kubota *et al.*, 2003, 2004; Kanatsu-Shinohara *et al.*, 2004a; Hamra *et al.*, 2004). Fluorescence- or magnetic-activated cell sorting (FACS or MACS) are frequently employed to isolate spermatogonia in mice (von Schonfeldt *et al.*, 1999; Shinohara *et al.*, 2000a). Alternatively, testicular cells subjected to velocity sedimentation and differential adhesion to extracellular matrix components like laminin or fibronectin were shown to be enriched in spermatogonia in mice and rats (Shinohara *et al.*, 1999; Hamra *et al.*, 2004). In large animal models, it is important that enrichment techniques yield large enough cell numbers for transplantation into a large animal recipient testis without compromising cell viability (Honaramooz *et al.*, 2002b; 2003a,b). Because the number of cells that can be isolated by FACS or MACS is limited, these approaches are not as suitable for enrichment of large numbers of spermatogonia from large animal testes, while differential plating and velocity sedimentation are potentially more practical for enriching spermatogonia from pre-pubertal testes for transplantation into large animals. A combination of differential plating and a discontinuous Percoll density gradient have been employed to enrich bovine type A spermatogonia (Izadyar *et al.*, 2002). We recently reported a modified differential plating approach to obtain enriched cell populations from pre-pubertal porcine testes that contained about 1 spermatogonial cell in 6 total cells (Luo *et al.*, 2006). While not all spermatgonial cells are necessarily stem cells, this degree of enrichment appears comparable to those reported for surface marker-based cell-sorting strategies for rodent germ line stem cells (Ryu *et al.*, 2005).

Germ Cell Culture

Nagano *et al.* (1998) showed first that mouse germ line stem cells could be maintained in culture for a long period of time. Improving culture conditions for male germ line stem cells still is under intense study as evidenced by several reports of improved culture systems for mouse germ cells (Kanatsu-Shinohara *et al.*, 2003b, 2005a,b; Jeong *et al.*, 2003; Kubota *et al.*, 2004; Hamra *et al.*, 2005). The majority of the work to date was performed with primary cultures of putative male germ line stem cells. While progress in this area has been significant, availability of immortalized cell lines would provide tremendous potential for the study and manipulation of male germ cells *in vitro*. To date, there are reports of immortalized germ cell lines from rat and mouse (Feng *et al.*, 2002; van Pelt *et al.*, 2002; Hofmann *et al.*, 2005). Recently, the amazing plasticity of male germ line stem cells was illustrated by reports of derivation of pluripotent stem cells from neonatal and adult mouse testes (Kanatsu-Shinohara *et al.*, 2004b; Guan *et al.*, 2006) and by the demonstration that adult male germ line stem cells can even give rise to fertilization of competent eggs when transplanted into an undifferentiated gonad in fish (Okutsu *et al.*, 2006).

Only a few studies to date have investigated culture of germ cells from pigs and cattle (Dirami *et al.*, 1999; Izadyar *et al.*, 2003; Oatley *et al.*, 2004; Luo *et al.*, 2006). Because germ cell transplantation in livestock species is still in its infancy and donor-derived spermatogenesis is not as easily quantifiable in the large animal testis as it is in the rodent system, studies in large animals mostly relied on morphological or immunological criteria to monitor maintenance of germ cells *in vitro*. Development of efficient, large-scale culture systems for enriched germ line stem cells from livestock animals will be a major milestone to enable widespread application of this technology.

Genetic Manipulation of Germ Cells

In recent years, transplantation of transfected germ cells has been investigated as an alternate means of generating transgenic animals through the manipulation of the male germ line. Although the lack of pure starting populations of germ line stem cells and optimized culture systems together with the low proliferating activity of stem cells have made this a difficult task, some success has been reported in generating transgenic mice and rats by retroviral or lentiviral transduction of germ cells prior to transplantation (Nagano *et al.*, 2001b, 2002b; Hamra *et al.*, 2002; Orwig *et al.*, 2002). An alternative vector that integrates recombinant adeno-associated virus into non-replicating cells showed promising results (Fig. 6.2; Honaramooz *et al.*, 2003c) and is a topic of current investigation.

Practical Application of Germ Cell Transplantation

Germ cell transplantation has a wide range of research opportunities and novel applications in livestock. For example, germ cell transplantation provides an opportunity to preserve genetic material of valuable males. Germ cell transplantation has an advantage over the only currently available approach, cryopreservation of sperm, in that it can be applied to pre-pubertal animals where sperm cannot be obtained or even to adults rendered azoospermic or teratozoospermic by disease. It has also been discussed as a tool to store genetic material from barrows (castrated male pigs) used in performance testing *in lieu* of using littermates of the top quality animals for breeding. Another potential application would be delivery of genetic material to a closed production herd through resident recipient animals. In cattle maintained under range conditions where artificial insemination is not practical, germ cell transplantation can be employed to deliver desired male genetics, e.g. bulls adapted to the environmental conditions (*Bos indicus*) delivering genetic material of breeds with higher carcass quality (*Bos taurus*).

Perhaps the most intriguing application of the technique will be the transplantation of transfected germ cells as an alternate means to generate transgenic – 'Designer' – animals. Genetic modification of livestock animals can result in improved production traits, resistance to specific diseases and improved adaptation to intensive management systems. In addition transgenic dairy animals are increas-

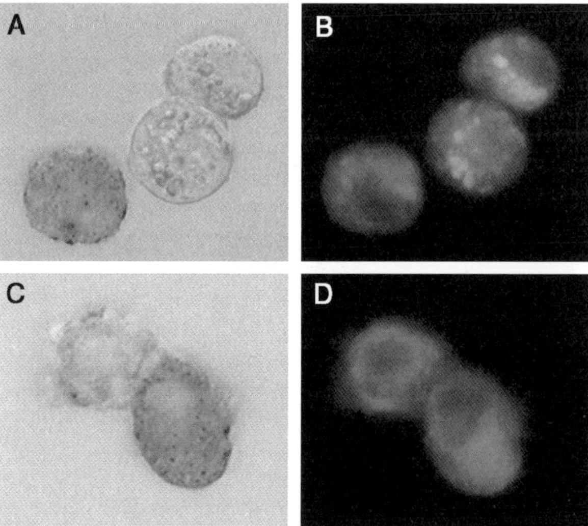

Fig. 6.2. Porcine testis cells exposed to rAAV-CMV/GFP and cultured for 5 days before analysis. A, C: light microscopic images showing germ cells stained for germ cell nuclear antigen-1 (GCNA; Enders and May, 1994) (A) or the mouse VASA-homologue (Toyooka *et al.*, 2000) (C) as well as unlabelled somatic cells. B, D: Corresponding fluorescence micrographs showing GFP expression in germ cells and somatic cells.

ingly sought after to produce biopharmaceutical proteins in their milk, and transgenic 'humanized' pigs are under investigation as a much needed source of organs for transplantation. Transgenesis through the male germ line using transplantation of genetically modified germ cells has tremendous potential in livestock species where embryonic stem cell technology is not available. Current options to generate genetically modified large animals include pronuclear microinjection of DNA (Hammer *et al.*, 1985) and nuclear transfer technology using modified donor cells (Lai *et al.*, 2002; Park *et al.*, 2002; Hyun *et al.*, 2003; Phelps *et al.*, 2003; Kolber-Simonds *et al.*, 2004), as well as sperm-mediated DNA transfer (Lazzereschi *et al.*, 2000; Lavitrano *et al.*, 2002, 2003). However, currently available technology is frequently fraught with low efficiency and developmental abnormalities in the few resulting offspring, making the approach of using germ cell transplantation a potentially very valuable alternative. Introduction of a genetic modification prior to fertilization will circumvent problems associated with manipulation of gametes and early embryos and developmental abnormalities associated with nuclear reprogramming. In addition, even if embryonic stem cell technology becomes available for domestic animals, the time required until transgenic sperm can be harvested will be significantly shorter using germ cell transplantation (Fig. 6.3). Introduction of a genetic modification before meiosis will allow for recombination to occur. A recipient animal therefore can produce sperm with different transgene integrations that may permit screening of offspring to select the most desirable genotype. Finally, genetic manipulation of male germ line stem cells may provide an approach to generate males that produce unisex sperm by introducing a sex-linked mutation that selectively prevents formation of haploid cells carrying either an X or a Y chromosome. Production of all male or all female offspring is of economic interest to different production systems. In addition, it might provide an avenue to avoid routine castration of males, a procedure that is coming under increasing scrutiny due to animal welfare concerns.

While germ cell transplantation as an alternative to artificial insemination for transfer of desired genetics into specific environments is

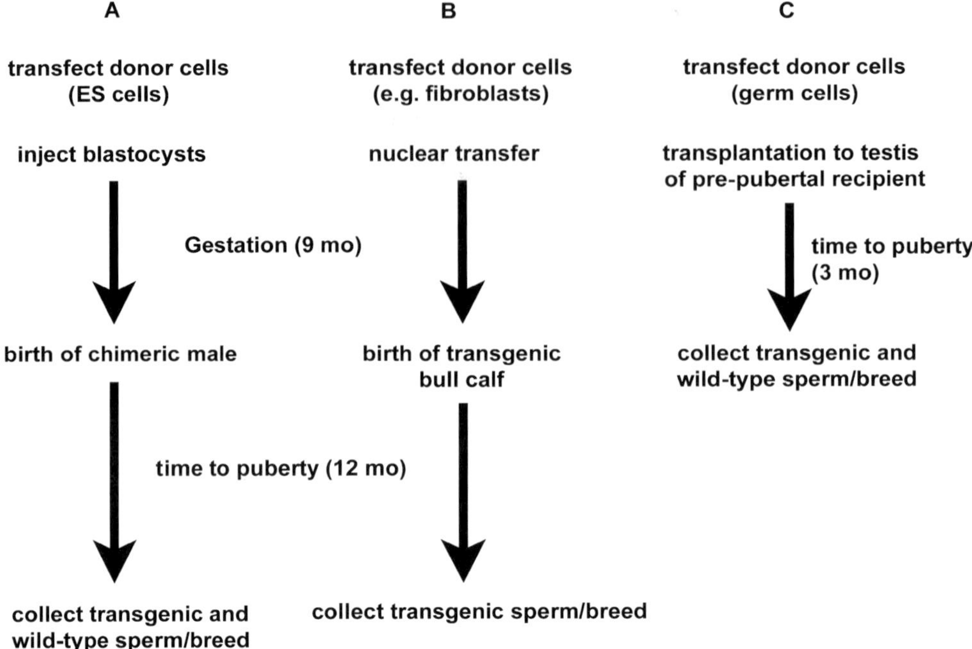

Fig. 6.3. Comparison of three potential approaches to transgenesis in cattle. A: Transgene introduction through embryonic stem cells (Note: This has been adapted in principle from work in rodents as the technology is currently not available in livestock species). B: Transgenesis through somatic cell nuclear transfer (cloning). C: Transgenesis through germ cell transplantation (work in progress).

practically possible at this point and currently under intense investigation, several steps still need to be addressed to make transgenesis through the male germ line commercially feasible. Currently, the main limitation of the approach is a lack of effici ent systems to maintain and propagate germ line stem cells from different livestock species in culture. Efficient *in vitro* culture is necessary for targeted genetic insertions, deletions or mutations in stem cells prior to selection of the desired genotype, expansion of the cells and transplantation. Targeted mutagenesis in mouse spermatogonial stem cells has recently been described (Kanatsu-Shinohara *et al.*, 2006). Therefore, it appears reasonable to expect that similar progress will be made with germ cells from livestock species in the not too distant future. This would make introduction of genetic modifications into the male germ line of livestock animals a commercially interesting alternative to nuclear transfer/cloning.

Conclusions

Germ cell transplantation was initially developed in rodents. Application to a large domestic animal species was first reported in the pig and subsequently in goats and cattle. Important aspects of the approach, including isolation of donor cells, delivery to recipient testes and recipient animal preparation, have been established in livestock species while others such as long-term culture and expansion of germ cells *in vitro* and efficient introduction of stable genetic changes are still under investigation. Germ cell transplantation allows for preservation of male genetic material and introduction of genetic changes through the male germ line. It represents an approach to germ line manipulation through use of transgenic sperm for natural breeding that is potentially more efficient than currently existing strategies.

Acknowledgements

Work from the authors' laboratories presented here was supported by 5R01RR017359-04 (NCRR/NIH) and 2R42-HD044780-02 (NIH/NICHD) to ID and by CSIRO Flagship project 'Engineering new breeding systems' to JRH.

References

Avarbock, M.R., Brinster, C.J. and Brinster, R.L. (1996) Reconstitution of spermatogenesis from frozen spermatogonial stem cells. *Nature Medicine* 2, 693–696.

Brinster, R.L. and Avarbock, M.R. (1994) Germline transmission of donor haplotype following spermatogonial transplantation. *Proceedings of the National Academy of Sciences of the United States of America* 91, 11303–11307.

Brinster, R.L. and Zimmermann, J.W. (1994) Spermatogenesis following male germ-cell transplantation. *Proceedings of the National Academy of Sciences of the United States of America* 91, 11298–11302.

Brinster, C.J., Ryu, B.-Y., Avarbock, M.R., Karagenc, L., Brinster, R.L. and Orwig, K.E. (2003) Restoration of fertility by germ cell transplantation requires effective recipient preparation. *Biology of Reproduction* 69, 412–420.

Chuma, S., Kanatsu-Shinohara, M., Inoue, K., Ogonuki, N., Miki, H., Toyokuni, S., Hosokawa, M., Nakatsuji, N., Ogura, A. and Shinohara, T. (2005) Spermatogenesis from epiblast and primordial germ cells following transplantation into postnatal mouse testis. *Development* 132, 117–122.

Clermont, Y. (1972) Kinetics of spermatogenesis in mammals: seminiferous epithelium cycle and spermatogonial renewal. *Physiological Reviews* 52, 198–236.

Clouthier, D.E., Avarbock, M.R., Maika, S.D., Hammer, R.E. and Brinster, R.L. (1996) Rat spermatogenesis in mouse testis. *Nature* 381, 418–421.

Creemers, L.B., Meng, X., den Ouden, K., van Pelt, A.M.M., Izadyar, F., Santoro, M., de Rooij, D.G. and Sariola, H. (2002) Transplantation of germ cells from glial cell line-derived neurotrophic factor-overexpressing mice to host testes depleted of endogenous spermatogenesis by fractionated irradiation. *Biology of Reproduction* 66, 1579–1584.

Dirami, G., Ravindranath, N., Pursel, V. and Dym, M. (1999) Effects of stem cell factor and granulocyte macrophage-colony stimulating factor on survival of porcine type A spermatogonia cultured in KSOM. *Biology of Reproduction* 61, 225–230.

Dobrinski, I., Ogawa,T., Avarbock, M.R. and Brinster, R.L. (1999a) Computer assisted image analysis to assess colonization of recipient seminiferous tubules by spermatogonial stem cells from transgenic donor mice. *Molecular Reproduction and Development* 53, 142–148.

Dobrinski, I., Avarbock, M.R. and Brinster, R.L. (1999b) Transplantation of germ cells from rabbits and dogs into mouse testes. *Biology of Reproduction* 61, 1331–1339.

Dobrinski, I., Avarbock, M.R. and Brinster, R.L. (2000) Germ cell transplantation from large domestic animals into mouse testes. *Molecular Reproduction and Development* 57, 270–279.

Dobrinski, I., Ogawa, T., Avarbock, M.R. and Brinster, R.L. (2001) Effect of the GnRH-agonist leuprolide on colonization of recipient testes by donor spermatogonial stem cells after transplantation in mice. *Tissue Cell* 33, 200–207.

Enders, G.C. and May, I.I.J.J. (1994) Developmentally regulated expression of a mouse germ cell nuclear antigen examined from embryonic day 11 to adult in male and female mice. *Developmental Biology* 163, 331–340.

Feng, L-X., Chen, Y., Dettin, L., Reijo Pera, R.A., Herr, J.C., Goldberg, E. and Dym, M. (2002) Generation and *in vitro* differentiation of a spermatogonial cell line. *Science* 297, 392–395.

Franca, L.R., Ogawa, T., Avarbock, M.R., Brinster, R.L. and Russell, L.D. (1998) Germ cell genotype controls cell cycle during spermatogenesis in the rat. *Biology of Reproduction* 59, 1371–1377.

Goossens, E., Fredericks, V., De Block, G., Van Steirteghem, A. and Tournaye, H. (2003) Reproductive capacity of sperm obtained after germ cell transplantation in a mouse model. *Human Reproduction* 18, 1874–1880.

Guan, K., Nayernia, K., Maier, L.S., Wagner, S., Dressel, R., Lee, J.H., Nolte, J., Wolf, F., Li, M., Engel, W. and Hasenfuss, G. (2006) Pluripotency of spermatogonial stem cells from adult mouse testis. *Nature* 440, 1190–1203.

Hammer, R.E., Pursel, V.G., Rexroad, C.E., Wall, R.J., Bolt, D.J., Ebert, K.M., Palmiter, R.D. and Brinster, R.L. (1985) Production of transgenic rabbits, sheep and pigs by microinjection *Nature* 315, 680–683.

Hamra, F.K., Gatlin, J., Chapman, K.M., Grellhesl, D.M., Garcia, J.V., Hammer, R.E. and Garbers, D.L. (2002) Production of transgenic rats by lentiviral transduction of male germ-line stem cells. *Proceedings of the National Academy of Sciences of the United States of America* 99, 14931–14936.

Hamra, F.K., Schultz, N., Chapman, K.M., Grellhesl, D.M., Cronkhite, J.T., Hammer, R.E. and Garbers, D.L. (2004) Defining the spermatogonial stem cell. *Developmental Biology* 269, 393–410.

Hamra, F.K., Chapman, K.M., Nguyen, D.M., Williams-Stephens, A.A., Hammer, R.E. and Garbers, D.L. (2005) Self renewal, expansion, and transfection of rat spermatogonial stem cells in culture. *Proceedings of the National Academy of Sciences of the United States of America* 102, 17430–17435.

Hill, J.R. and Dobrinski, I. (2006) Male germ cell transplantation in livestock. *Reproduction, Fertility and Development* 18, 13–18.

Hofmann, M-C., Braydich-Stolle, L., Dettin, L., Johnson, E. and Dym, M. (2005) Immortalization of mouse germ line stem cells. *Stem Cells* 23, 200–210.

Honaramooz, A., Snedaker, A., Boiani, M., Schöler, H.R., Dobrinski, I. and Schlatt, S. (2002a) Sperm from neonatal mammalian testes grafted in mice. *Nature* 418, 778–781.

Honaramooz, A., Megee, S.O. and Dobrinski, I. (2002b) Germ cell transplantation in pigs. *Biology of Reproduction* 60, 21–28.

Honaramooz, A., Behboodi, E., Megee, S.O., Overton, S.A., Galantino-Homer, H.L., Echelard, Y. and Dobrinski, I. (2003a) Fertility and germline transmission of donor Haplotype following germ cell transplantation in immunocompetent goats. *Biology of Reproduction* 69, 1260–1264.

Honaramooz, A., Behboodi, E., Blash, S., Megee, S.O. and Dobrinski I. (2003b) Germ cell transplantation in goats. *Molecular Reproduction and Development* 64, 422–428.

Honaramooz, A., Megee, S.O., Foley, B. and Dobrinski, I. (2003c) Use of adeno-associated virus for transfection of male germ cells for transplantation in pigs *Theriogenology* 59, 536.

Honaramooz, A., Behboodi, E., Hausler, C.L., Blash, S., Ayres, S.L., Azuma, C., Echelard, Y. and Dobrinski, I. (2005) Depletion of endogenous germ cells in male pigs and goats in preparation for germ cell transplantation. *Journal of Andrology* 26, 698–705.

Huckins, C. (1971) The spermatogonial stem cell population in adult rats. I. Their morphology, proliferation, and maturation. *Anatomical Record* 160, 533–558.

Hyun, S., Lee, G., Kim, D., Kim, H., Lee, S., Nam, D., Jeong, Y., Kim, S., Yeom, S., Kang, S., Han, J., Lee, B. and Hwang, W. (2003) Production of nuclear transfer-derived piglets using porcine fetal fibroblasts transfected with the enhanced green fluorescent protein. *Biology of Reproduction* 69, 1060–1068.

Izadyar, F., Spierenberg, G.T., Creemers, L.B., den Ouden, K. and de Rooij, D.G. (2002) Isolation and purification of type A spermatogonia from the bovine testis. *Reproduction* 124, 85–94.

Izadyar, F., den Ouden, K., Stout, T.A.E., Stout, J., Coret, J., Lankveld, D.P.K., Spoormakers, T.J.P., Colenbrander, B., Oldenbroek, J.K., Van der Ploeg, K.D., Woelders, H., Kal, H.B. and de Rooij, D.G. (2003) Autologous and homologous transplantation of bovine spermatogonial stem cells. *Reproduction* 126, 774.

Jeong, D., McLean, D.J. and Griswold, M.D. (2003) Long-term culture and transplantation of murine testicular germ cells. *Journal of Andrology* 24, 661–669.

Jiang, F.X. and Short, R.V. (1995) Male germ cell transplantation in rats: apparent synchronization of spermatogenesis between host and donor seminiferous epithelia. *International Journal of Andrology* 18, 326–330.

Kanatsu-Shinohara, M., Ogonuki, N., Inoue, K., Ogura, A., Toyokuni, S., Honjo, T. and Shinohara, T. (2003a) Allogeneic offspring produced by male germ line stem cell transplantation into infertile mouse testis. *Biology of Reproduction* 68, 167–173.

Kanatsu-Shinohara, M., Ogonuki, N., Inoue, K., Miki, H., Ogura, A., Toyokuni, S. and Shinohara, T. (2003b) Long-term proliferation in culture and germline transmission of mouse male germline stem cells. *Biology of Reproduction* 69, 612–616.

Kanatsu-Shinohara, M., Ogonuki, N., Inoue, K., Ogura, A., Toyokuni, S. and Shinohara, T. (2003c) Restoration of fertility in infertile mice by transplantation of cryopreserved male germline stem cells. *Human Reproduction* 18, 2660–2667.

Kanatsu-Shinohara, M., Toyokuni, S. and Shinohara, T. (2004a) CD9 is a surface marker on mouse and rat male germline stem cells. *Biology of Reproduction* 70, 70–75.

Kanatsu-Shinohara, M., Inoue, K., Lee, J., Yoshimoto, M., Ogonuki, N., Miki, H., Baba, S., Kato, T., Kazuki, Y., Toyokuni, S., Toyoshima, M., Niwa, O., Oshimura, M., Heike, T., Nakahata, T., Ishino, F., Ogura, A. and Shinohara, T. (2004b) Generation of pluripotent stem cells from neonatal mouse testis. *Cell* 119, 1001–1012.

Kanatsu-Shinohara, M., Ogonuki, N., Iwano, T., Lee, J., Kazuki, Y., Inoue, K., Miki, H., Takehashi, M., Toyokuni, S., Shinkai, Y., Oshimura, M., Ishino, F., Ogura, A. and Shinohara, T. (2005a) Genetic and epigenetic properties of mouse male germline stem cells during long-term culture. *Development* 132, 4155–4163.

Kanatsu-Shinohara, M., Miki, H., Inoue, K., Ogonuki, N., Toyokuni S., Ogura A. and Shinohara, T. (2005b) Long-term culture of mouse male germline stem cells under serum or feeder-free conditions. *Biology of Reproduction* 72, 985–991.

Kanatsu-Shinohara, M., Ikawa, M., Takehashi, M., Ogonuki, N., Miki, H., Inoue, K., Kazuki, Y., Lee, J., Toyokuni, S., Oshimura, M., Ogura, A. and Shinohara, T. (2006) Production of knockout mice by random or targeted mutagenesis in spermatogonial stem cells. *Proceedings of the National Academy of Sciences of the United States of America* 103, 8018–8023.

Kim, Y., Selvaraj, V., Dobrinski, I., Lee, H., McEntee, M. and Travis, A.J. (2006) Recipient preparation and mixed germ cell isolation for spermatogonial stem cell transplantation in domestic cats. *Journal of Andrololgy* 27, 248–256.

Kolber-Simonds, D., Lai, L., Watt, S.R., Denaro, M., Arn, S., Augenstein, M.L., Betthauser, J., Carter, D.B., Greenstein, J.L., Hao, Y., Im, G.-S., Liu, Z., Mell, G.D., Murphy, C.N., Park, K.-W., Rieke, A., Ryan, D.J.J., Sachs, D.H., Forsberg, E.J., Prather, R.S. and Hawley, R.J. (2004) Production of *-1,3-galactosyltransferase null pigs by means of nuclear transfer with fibroblasts bearing loss of heterozygosity mutations. *Proceedings of the National Academy of Sciences of the United States of America* 101, 7335–7340.

Kubota, H., Avarbock, M.R. and Brinster, R.L. (2003) Spermatogonial stem cells share some, but not all, phenotypic and functional characteristics with other stem cells. *Proceedings of the National Academy of Sciences of the United States of America* 100, 6487–6492.

Kubota, H., Avarbock, M.R. and Brinster, R.L. (2004) Culture conditions and single growth factors affect fate determination of mouse spermatogonial stem cells. *Biology of Reproduction* 71, 722–731.

Lai, L., Kolber-Simonds, D., Park, K.-W., Cheong, H-T., Greenstein, J.L., Im, G.-S., Samuel, M., Bonk, A., Rieke, A., Day, B.N., Murphy, C.N., Carter, D.B., Hawley, R.J. and Prather, R.S. (2002) Production of *-1,3-galactosyltransferase knockout pigs by nuclear transfer cloning. *Science* 295, 1089–1092.

Lavitrano, M., Bacci, M.L., Forni, M., Lazzereschi, D., Di Stefano, C., Fioretti, D., Giancotti, P., Marfe, G., Pucci, L., Renzi, L., Wang, H., Stoppacciaro, A., Stassi, G., Sargiacomo, M., Sinibaldi, P., Turchi, V., Giovannoni, R., Della Casa, G., Seren, E. and Rossi, G. (2002) Efficient production by sperm-mediated gene transfer of human decay accelerating factor (hDAF) transgenic pigs for xenotransplantation. *Proceedings of the National Academy of Sciences of the United States of America* 99, 14230–14235.

Lavitrano, M., Forni, M., Bacci, M.L., Di Stefano, C., Varzi, V., Wang, H. and Seren, E. (2003) Sperm mediated gene transfer in pig: selection of donor boars and optimization of DNA uptake. *Molecular Reproduction and Development* 64, 284–291.

Lazzereschi, D., Forni, M., Cappello, F., Bacci, M.L., Di Stefano, C., Marfe, G., Giancotti, P., Renzi, L., Wang, H.J., Rossi, M., Della Casa, G., Pretagostini, R., Frati, G., Bruzzone, P., Stassi, G., Stoppacciaro, A., Turchi, V., Cortesini, R., Sinibaldi, P., Frati, L. and Lavitrano, M. (2000) Efficiency of transgenesis using sperm-mediated gene transfer: generation of hDAF transgenic pigs. *Transplantation Proceedings* 32, 892–894.

Luo, J., Megee, S., Rathi, R. and Dobrinski, I. (2006) Protein gene product 9.5 is a spermatogonia-specific marker in pig testis: application to enrichment and culture of porcine spermatogonia. *Molecular Reproduction and Development* 73, issue 12.

Meistrich, M.L., van Beek, M.E.A.B. (1993) Spermatogonial stem cells. In: Desjardins, C. and Ewing, L.L. (eds) *Cell and Molecular Biology of Testis*. University Press, New York, pp. 266–295.

Nagano, M., Avarbock, M.R., Leonida, E.B., Brinster, C.J. and Brinster, R.L. (1998) Culture of mouse spermatogonial stem cells. *Tissue Cell* 30, 389–397.

Nagano, M., Avarbock, M.R. and Brinster, R.L. (1999) Pattern and kinetics of mouse donor spermatogonial stem cell colonization in recipient testes. *Biology of Reproduction* 60, 1429–1436.

Nagano, M., McCarrey, J.R. and Brinster, R.L. (2001a) Primate spermatogonial stem cells colonize mouse testes. *Biology of Reproduction* 64, 1409–1416.

Nagano, M., Brinster, C.J., Orwig, K.E., Ryu, B.-Y., Avarbock, M.R. and Brinster, R.L. (2001b) Transgenic mice produced by retroviral transduction of male germ-line stem cells. *Proceedings of the National Academy of Sciences of the United States of America* 98, 13090–13095.

Nagano, M., Patrizio, P. and Brinster, R.L. (2002a) Long-term survival of human spermatogonial stem cells in mouse testes. *Fertility and Sterility* 78, 1225–1233.

Nagano, M., Watson, D.J., Ryu, B.-Y., Wolfe, J.H. and Brinster, R.L. (2002b) Lentiviral vector transduction of male germ line stem cells in mice. *FEBS Letters* 524, 111–115.

Oatley, J.M., Reeves, J.J. and McLean, D.J. (2004) Biological activity of cryopreserved bovine spermatogonial stem cells during in vitro culture. *Biology of Reproduction* 71, 942–947.

Ogawa, T., Arechaga, J.M., Avarbock, M.R. and Brinster, R.L. (1997) Transplantation of testis germinal cells into mouse seminiferous tubules. *International Journal of Developmental Biology* 41, 111–122.

Ogawa, T., Dobrinski, I., Avarbock, M.R. and Brinster, R.L. (1998) Leuprolide, a gonadotropin releasing hormone agonist, enhances colonization after spermatogonial transplantation into mouse testes. *Tissue and Cell* 30(5), 583–588.

Ogawa, T., Dobrinski, I. and Brinster, R.L. (1999a) Recipient preparation is critical for spermatogonial transplantation in the rat. *Tissue and Cell* 31, 461–472.

Ogawa, T., Dobrinski, I., Avarbock, M.R. and Brinster, R.L. (1999b) Xenogeneic spermatogenesis following transplantation of hamster germ cells to mouse testes. *Biology of Reproduction* 60, 515–521.

Okutsu, T., Suzuki, K., Takeuchi, Y., Takeuchi, T. and Yoshizaki, G. (2006) Testicular germ cells can colonize sexually undifferentiated embryonic gonad and produce functional eggs in fish. *Proceedings of the National Academy of Sciences of the United States of America* 103, 2725–2729.

Orwig, K.E., Avarbock, M.R. and Brinster, R.L. (2002) Retrovirus-mediated modification of male germline stem cells in rats. *Biology of Reproduction* 67, 874–879.

Park, K.-W., Lai, L., Cheong, H.-T., Cabot, R., Sun, Q.-Y., Wu, G., Rucker, E.B., Durtschi, D., Bonk, A., Samuel, M., Rieke, A., Day, B.N., Murphy, C.N., Carter, D.B. and Prather, R.S. (2002) Mosaic gene expression in nuclear transfer-derived embryos and the production of cloned transgenic pigs from ear-derived fibroblasts. *Biology of Reproduction* 66, 1001–1005.

Parreira, G.C., Ogawa, T., Avarbock, M.R., Franca, L.R., Brinster, R.L. and Russell, L.D. (1998) Development of germ cell transplants in mice. *Biology of Reproduction* 59, 1360–1370.

Phelps, C.J., Koike, C., Vaught, T.D., Boone, J., Wells, K.D., Chen, S.-H., Ball, S., Specht, S.M., Polejaeva, I.A., Monahan, J.A., Jobst, P.M., Sharma, S.B., Lamborn, A.E., Garst, A.S., Moore, M., Demetris, A.J., Rudert, W.A., Bottino, R., Bertera, S., Trucco, M., Starzl,T.E., Dai, Y. and Ayares, D.L. (2003) Production of *1,3-galactosyltransferase-deficient pigs. *Science* 299, 411–414.

Russell, L.D., Ettlin, R.A., SinhaHikim, A.P. and Clegg, E.D. (1990) Mammalian spermatogenesis. In: Russell, L.D., Ettlin, R.A., SinhaHikim, A.P. and Clegg, E.D. (eds) *Histological and Histopathological Evaluation of the Testis*. Cache River Press, Clearwater, FL, pp. 1–40.

Ryu, B.-Y., Kubota, H., Avarbock, M.R. and Brinster, R.L. (2005) Conservation of spermatogonial stem cell self-renewal signaling between mouse and rat. *Proceedings of the National Academy of Sciences of the United States of America* 102, 14302–14307.

Schlatt, S., Foppiani, L., Rolf, C., Weinbauer, G.F. and Nieschlag, E. (2002) Germ cell transplantation into X-irradiated monkey testes. *Human Reproduction* 17, 55–62.

Shinohara, T., Avarbock, M.R. and Brinster, R.L. (1999) β1- and α6-integrin are surface markers on mouse spermatogonial stem cells. *Proceedings of the National Academy of Sciences of the United States of America* 96, 5504–5509.

Shinohara, T., Avarbock, M.R. and Brinster, R.L. (2000a) Functional analysis of spermatogonial stem cells in Steel and cryptorchid infertile mouse models. *Developmental Biology* 220, 401–411.

Shinohara, T., Orwig, K.E., Avarbock, M.R. and Brinster, R.L. (2000b) Spermatogonial stem cell enrichment by multiparameter selection of mouse testis cells. *Proceedings of the National Academy of Sciences of the United States of America* 97, 8346–8351.

Shinohara, T., Orwig, K.E., Avarbock, M.R. and Brinster, R.L. (2001) Remodeling of the postnatal mouse testis is accompanied by dramatic changes in stem cell number and niche accessibility. *Proceedings of the National Academy of Sciences of the United States of America* 98, 6186–6191.

Shinohara, T., Orwig, K.E., Avarbock, M.R. and Brinster, R.L. (2003) Restoration of spermatogenesis in infertile mice by Sertoli cell transplantation. *Biology of Reproduction* 68, 1064–1071.

Takeuchi, Y., Yoshizaki, G. and Takeuchi, T. (2003) Generation of live fry from intraperitoneally transplanted primordial germ cells in rainbow trout. *Biology of Reproduction* 69, 1142–1149.

Tegelenbosch, R.A.J. and de Rooij, D.G. (1993) A quantitative study of spermatogonial multiplication and stem cell renewal in the C3H/101 F_1 hybrid mouse. *Mutation Research* 290, 193–200.

Toyooka, Y., Tsunekawa, N., Takahashi, Y., Matsui, Y., Satoh, M. and Noce, T. (2000) Expression and intracellular localization of mouse Vasa-homologue protein during germ cell development. *Mechanisms of Development* 93, 139–149.

van Pelt, A.M.M., Roepers-Gajadien, H.L., Gademan, I.S., Creemers, L.B., de Rooij, D.G. and van Dissel-Emiliani, F.M.F. (2002) Establishment of cell lines with rat spermatogonial stem cell characteristics. *Endocrinology* 143, 1845–1850.

Ventela, S., Ohta, H., Parvinen, M. and Nishimune, Y. (2002) Development of stages of the cycle in mouse seminiferous epithelium after transplantation of green fluorescent protein labeled spermatogonial stem cells. *Biology of Reproduction* 66, 1422–1429.

von Schonfeldt, V., Krishnamurthy, H., Foppiani, L. and Schlatt, S. (1999) Magnetic cell sorting is a fast and effective method of enriching viable spermatogonia from Djungarian hamster, mouse, and marmoset monkey testes. *Biology of Reproduction* 61, 582–589.

Yoshizaki, G., Tago, Y., Takeuchi, Y., Sawatari, E., Kobayashi, T. and Takeuchi, T. (2005) Green fluorescent protein labeling of primordial germ cells using a nontransgenic method and its application for germ cell transplantation in salmonidae. *Biology of Reproduction* 73, 88–93.

Zhang, Z., Renfree, M.B. and Short, R.V. (2003) Successful intra- and interspecific male germ cell transplantation in the rat. *Biology of Reproduction* 68, 961–967.

7 Cloning and Transgenesis to Redesign Livestock

D.N. Wells[1] and G. Laible[1]

[1]*Reproductive Technologies, AgResearch Ltd, New Zealand*

Abstract

An efficient animal cloning technology, using the procedure of nuclear transfer (NT), would provide many opportunities for livestock agriculture and biomedicine. Although the birth of 'Dolly', the first animal to be cloned from an adult cell, occurred back in 1996, it is still remarkable that NT using differentiated donor cells can produce physiologically normal cloned animals. But the process is very inefficient, poorly understood and appears highly prone to epigenetic errors resulting in abnormal phenotypes. Importantly, it appears that these deficiencies are not transmitted to offspring following sexual reproduction. This provides initial confidence in the first application of NT in agriculture, namely, the production of small numbers of cloned sires from genetically elite males, for natural mating, to effectively disseminate genetic gain. The continual advances in animal genomics towards the identification of genes that influence livestock production traits and impact on human health serve to increase the ability to genetically modify cultured cells, prior to NT, to generate livestock with increased productivity or that produce superior quality food and biomedical products for niche markets. The potential opportunities for animal agriculture are more challenging because of the greater demands on cost, efficiency, consumer acceptance and relative value of the product for commercial viability in contrast to biomedicine, which is the main driver for this technology platform. None the less, cloning and transgenesis are being used to redesign the genetic make-up of livestock; however, the integration of this technology into practical farming systems remains some time in the future.

Introduction

In this review, the current somatic cell cloning technology, its efficiency, present limitations and potential applications, especially in combination with transgenesis, to redesign livestock animals for agriculture and biomedicine is discussed, along with some of the challenges for adoption of these controversial technologies. The main emphasis is placed on cattle and also the other commonly farmed livestock species such as sheep, goats and pigs, but excludes poultry and fish. Relevant data from mouse models, which aid our understanding of the biology and consequences of nuclear cloning and transgenesis, are also presented.

The embryo manipulation procedure termed nuclear transfer (NT) is the most dramatic method of producing cloned animals because it effectively generates a viable animal from an individually selected cell (Wilmut *et al.*, 2002). Using a microsurgical approach, the nucleus of a donor cell is typically transferred into the cytoplasm of a mature oocyte following removal of its own maternal chromosomes. This reconstructed one-cell embryo is then stimulated to undergo embryonic development *in vitro*. Suitably developed embryos are then transferred to surrogate females for *in vivo* development to term, where some result in offspring with the same genomic make-up as the original donor cell (Fig. 7.1). The donor cell can come from a variety of sources, including early embryos (Willadsen, 1986; Campbell *et al.*, 1996) and various more differentiated cell types obtained from fetuses or, most significantly, adult animals as first demonstrated in the case of 'Dolly' the sheep – cloned from a mammary gland cell (Wilmut *et al.*, 1997). This breakthrough overturned a dogma in biology concerning nuclear totipotency from adult cells and consequently opened new directions in research and opportunities for agriculture and biomedicine. The use of differentiated cells for cloning has been termed somatic cell NT to distinguish it from NT with embryonic cells. Moreover, these somatic donor cells can be cultured *in vitro*, and maintained as primary cell lines, or cryopreserved, with access to millions of cells. Since Dolly, successful somatic cell NT has been demonstrated in the major livestock species including cattle (Cibelli *et al.*, 1998), goats (Baguisi *et al.*, 1999) and pigs (Polejaeva *et al.*, 2000), in addition to various laboratory rodents, equids, companion animals and some more exotic mammalian species. The donor cells may also be genetically modified *in vitro* using established

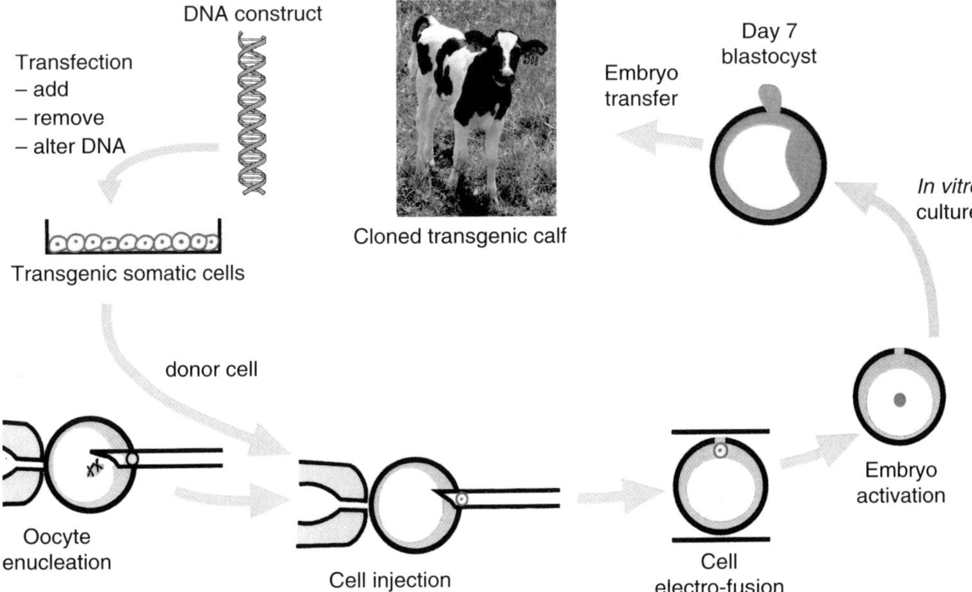

Fig. 7.1. Combining cloning and genetic modification. Donor cells are genetically modified *in vitro* and selected for use in nuclear transfer to generate cloned transgenic animals. The conventional nuclear transfer approach used for cloning cattle comprises a sequence of six main steps. First, the metaphase II stage chromosomes are physically removed from a mature, unfertilized oocyte in a process termed enucleation, producing a 'cytoplast'. An individual transgenic donor cell is then placed adjacent to the cytoplast and the two are electrically fused together. This provides the opportunity for the donor chromatin to be reprogrammed upon exposure to factors within the oocyte cytoplasm. Reconstructed one-cell embryos are then artificially activated to initiate development and cultured for 7 days *in vitro* until the blastocyst stage. Suitable quality embryos are transferred singularly to the reproductive tracts of appropriately synchronized recipient cows. Some may implant and develop to term for the eventual birth of cloned transgenic offspring, possessing essentially the same genotype as the original donor cell.

transfection methods, and when first used in combination with NT provided a novel and efficient route to produce transgenic animals (Schnieke *et al.*, 1997). Thus, the core technique of NT can produce either clones of existing genotypes or novel transgenic animals, depending upon the choice of donor cell.

Nuclear Cloning Efficiency

Overall, the current efficiency of NT with somatic cells is poor regardless of species. In our experience with cattle at AgResearch, while *in vitro* development of cloned embryos is comparable to *in vitro* fertilization (IVF; both around 40%), the more critical aspect of *in vivo* development is currently at best only one-third that following IVF. On average, 13% of transferred somatic cell NT embryos produced with our cloning procedures have resulted in calves delivered at full term (Wells *et al.*, 2004). Our data also show that the stage of the donor cell cycle at the time of NT affects subsequent calf viability, with an interaction between cell types. For cultured somatic cells obtained from conventional animals, the use of quiescent cells that are serum starved for a few days before NT results in significantly greater development to term and weaning compared to cloned embryos derived from cells in the G_1 stage of the cell cycle (Wells *et al.*, 2003a; Tucker *et al.*, 2004). However, for genetically modified cells it appears that the reverse is true and the selection of actively proliferating G_1 donor cells for NT significantly increases *in vivo* development (Wells *et al.*, 2003a). This information is useful for the two main applications of the technology (see below).

Although initial day 50 pregnancy rates in cattle following the transfer of single NT embryos can be as high as 65% and similar to both IVF embryos and artificial insemination, there is continual loss thereafter with the clones (Lee *et al.*, 2004). Moreover, peri- and post-natal mortality rates with cloned offspring are greater than normally expected, with, on average, only 64% of cloned calves born surviving to weaning (Wells *et al.*, 2004). In addition, annual mortality rates in adulthood can be much greater than those expected for conventional livestock; about 8% in cloned cattle, which is around threefold greater than normal (Wells *et al.*, 2004).

From an animal welfare perspective, in addition to farmer and consumer acceptance of the technology, these losses must be solved before any large-scale cloning opportunities are practicable or tolerated. It is important to remember, however, that somatic cell NT can be effective in producing some livestock that appear to be physiologically normal (Pace *et al.*, 2002; Wells *et al.*, 2004; Yonai *et al.*, 2005). This provides encouragement for eventually resolving technical issues and elucidating the molecular mechanisms responsible for effecting complete epigenetic reprogramming.

Epigenetic Reprogramming

During normal embryonic development and differentiation, there is a dynamic increase in the epigenetic modifications imposed on the DNA. This accommodates changes in the requirements of cells to access and use an appropriate subset of the total genomic information, enabling individual cell types to perform their specific functions. As a result, large parts of the genome become silenced to prevent the execution of inappropriate genetic programmes. For normal embryogenesis to occur following NT, the previous pattern of gene expression in the donor nucleus must first be reprogrammed to a state comparable to that in a fertilized zygote (Dean *et al.*, 2003). This involves changes to the pre-existing epigenetic modifications in the donor cell's chromatin, in order to inactivate those genes specific for the differentiated cellular phenotype and to enable embryonic genes to be reactivated in the correct tissues, in the correct abundance and at the correct times. Clearly, this is a highly orchestrated process and is understood poorly at present but there is increasing evidence of epigenetic errors in reprogramming following NT, leading to abnormal patterns of:

- DNA methylation (Bourchis *et al.*, 2001; Dean *et al.*, 2001; Kang *et al.*, 2001)
- chromatin modification (Santos *et al.*, 2003)
- X-chromosome inactivation (Xue *et al.*, 2002)
- expression of imprinted and non-imprinted genes (Rideout *et al.*, 2001; Humpherys *et al.*, 2002; Niemann *et al.*, 2002; Smith *et al.*, 2005).

The consequences of faulty reprogramming are high embryonic and fetal losses throughout pregnancy (Hill *et al.*, 2000; Lee *et al.*, 2004), excessive fluid accumulation in the placental membranes, prolonged gestation (Carter *et al.*, 2002; Peura *et al.*, 2003), heavier birthweight and difficult parturition, greater post-natal mortality (Wells *et al.*, 2004) and even some specific clone-associated phenotypes persisting in adulthood (Wilmut *et al.*, 2002). The pattern of mortality and clone phenotypes observed presumably reflects the inappropriate expression of various genes whose harmful effects are exerted at various different stages of development. Aberrations that occur early in embryonic or fetal development may impair health in adulthood (Ogura *et al.*, 2002) or increase the variation in patterns of gene expression, sometimes without apparent developmental compromise (Humpherys *et al.*, 2002; Archer *et al.*, 2003).

Various novel methods aimed at facilitating the nuclear reprogramming process have resulted in relatively modest, yet significant, increases in cloning success. These have included the use of cell extracts to prepare the donor nucleus for NT (Sullivan *et al.*, 2004), the use of pharmacological agents to modify chromatin structure (Enright *et al.*, 2003; Kishigami *et al.*, 2006) and the aggregation of cloned embryos (Boiani *et al.*, 2003). Further increases may be expected from a greater understanding and control over the mechanisms of epigenetic reprogramming.

Whilst there are phenotypic abnormalities in the cloned generation, it appears that the offspring of clones following sexual reproduction are essentially normal. This implies that the clone-associated phenotypes are epigenetic in nature and are corrected during gametogenesis. This conclusion is based on observations of progeny derived from matings between cloned females and cloned males in sheep (Wells, 2003), cattle (Wells, 2006, unpublished results), pigs (Martin *et al.*, 2004) and, most convincingly, mice (Shimozawa *et al.*, 2002; Tamashiro *et al.*, 2002). Whilst this provides some confidence for a major application of NT in agriculture, namely the generation of cloned sires for breeding, more detailed molecular studies are required to determine whether recessive genetic or epigenetic errors could still persist that would be heritable.

Cloning to Multiply Valuable Genetics

There are several important issues to be addressed before commercial opportunities for cloning in livestock agriculture can be realized. There are significant animal welfare concerns limiting the acceptability and applicability of the technology in its current form. There needs to be confidence in the long-term health status of cloned livestock and progeny in subsequent generations. In addition to ongoing animal assessment, continued modelling of the technology will identify where it is best suited in different farming systems (Smeaton *et al.*, 2003, 2004). Commercialization is awaiting regulatory approval on the safety of food products derived from clones and their offspring (FDA, 2003, 2005; Rudenko *et al.*, 2004). The available data comparing the composition of milk and meat in clones and conventional animals indicate that clone-derived food products are essentially identical with the conventional products (Walsh *et al.*, 2003; Takahashi and Ito, 2004; Tian *et al.*, 2005). But even with regulatory approval, there needs to be industry, farmer and consumer acceptance of the technology.

An efficient NT technology could enable the rapid dissemination of superior genotypes from nucleus breeding herds directly to commercial farmers. Genotypes could be provided that are ideally suited for specific product characteristics, disease resistance, or environmental conditions. Indeed, cloning can be used to help preserve indigenous breeds of livestock that have production traits and adaptability to local environments at risk of being lost from the global gene pool (Wells *et al.*, 1998). Unique opportunities exist in the meat industry for the resurrection of animals using cells recovered from carcasses with genetically superior meat characteristics to generate clones for either breeding or commercial meat production, thus rescuing these valuable genetics. Cloning could be extremely useful in multiplying outstanding F_1 crossbred animals, or composite breeds with otherwise complicated and expensive breeding strategies, to maximize the benefits of heterosis and to maintain favourable combinations of alleles in heterozygotes, which would otherwise segregate in the F_2 generation. The dissemination of genetic gain could be achieved through the controlled release of selected lines of elite cloned embryos or, most appropriately, by the production of cloned animals with superior genetics for breeding. These could be clones of progeny-tested animals, especially sires, so long as they remain reasonably faithful genomic copies of the original donor animal and that trans-generational epigenetic inheritance does not lead to deleterious long-term breeding consequences. This application is more relevant to the sheep and beef industries, which, to a large extent, rely on natural breeding rather than assisted reproductive technologies. Here, cloned sires could be used in widespread natural mating to provide an effective means of disseminating their superior genetics. This could provide a substitute for artificial insemination, which is expensive and inconvenient in more extensive farming systems, quite unlike the situation in dairy industries, where it has been a very successful technology to disseminate genetic gain. But even in the dairy industry, cloned bulls could generate extra semen to supply unmet international demands from top sires. To even more rapidly disseminate genetic gain, and reduce genetic lag by at least two generations, it will be advantageous to clone from embryonic blastomeres (or preferably, cultures of embryonic stem cells)(Wang *et al.*, 2005) following selection of genetically superior embryos, rather than using somatic cells from adult animals (Wells *et al.*, 2003b). This, however, will depend on the development of suitable genetic tests to capture its full potential.

In breeding from cloned sires, it is not the clone himself that is important but rather what is transmitted through his germ line; haploid copies of the donor animal's genome, in the form of sperm. In this regard, either germ line-transmitting male chimaeras (Nagashima *et al.*, 2004) or males in which their testes have been re-populated with spermatogonial stem cells (SSCs) obtained from an elite sire (Hill and Dobrinski, 2006; see Dobrinski and Hill, Chapter 6 this volume) are both attractive, alternative strategies to achieve the same goal of effectively disseminating genetic gain.

Combining Cloning and Genetic Modification

A major application of NT is to clone animals from cells that have been genetically modified *in vitro*, in order to produce transgenic livestock. This approach of cell-mediated transgenesis coupled with NT is, however, only one of several available to produce genetically modified animals (Clark and Whitelaw, 2003)(see below). Each transgenic method has its own set of unique features affording applicability in different situations. Despite the current problems with reprogramming, the use of NT to produce clones from genetically modified cells has a number of distinct advantages which collectively provide a comprehensive technological platform to generate transgenic animals. These include the ability to:

- Perform a wide repertoire of genetic modification in the cultured cell population. This includes not only the introduction of new DNA sequences from the same (Brophy *et al.*, 2003) or a different species (Schnieke *et al.*, 1997) and even artificial chromosomes (Kuroiwa *et al.*, 2002), but also to functionally delete an endogenous gene (Kuroiwa *et al.*, 2004), knock-down its expression (Golding *et al.*, 2006) or to subtly modify the DNA sequence in the regulatory or coding regions of an endogenous gene of interest (Laible *et al.*, 2006).

- Implement screening and selection strategies to identify those rare cells in the population that have undergone the desired and specific genetic modification, before producing the transgenic animal.
- Introduce a specific transgene into a desired genetic background of the chosen sex by using donor cells from a selected animal. This is particularly important for agricultural traits.
- Produce embryos or offspring that are all transgenic and where none should be mosaic (with a mixture of transgenic and non-transgenic cells in the same organism, which particularly complicates subsequent breeding).
- Produce small herds from each cell line in the first generation, rather than individual founder animals that need to be subsequently bred. However, with the present NT technology, it remains desirable to use assisted sexual reproduction to further multiply animals (Forsyth *et al.*, 2005) and circumvent potential epigenetic aberrations in the cloned generation.

Although it is still commonplace for introduced genes to be integrated at random locations within the genome, methods of 'gene targeting' through homologous and site-specific recombination are possible with a cell-mediated transgenic approach (Piedrahita, 2000; Sorrell and Kolb, 2005). This enables the introduction of a gene at a precise location in the genome (McCreath *et al.*, 2000; Kuroiwa *et al.*, 2004), or to knock-out or functionally delete an undesirable gene on an otherwise favourable genetic background (McCreath *et al.*, 2000; Kuroiwa *et al.*, 2004), or to alter the DNA sequence in a specific gene to improve or abolish a particular function in the resulting protein (Laible *et al.*, 2006). The combination of NT and gene targeting has the potential to be far more precise, extensive and rapid in terms of the genetic progress that can be achieved compared to traditional breeding and other available transgenic methods, especially conventional pronuclear injection of DNA into zygotes. Obstacles that remain, however, include the very low frequency of successful gene targeting events in primary cultures of somatic cells, the need to avoid the use of antibiotic resistance marker genes, commonly used to aid selection of transgenic cells, to alleviate some societal concerns with the technology, and the low efficiency of NT.

Despite some promising cell types having been isolated (Wang *et al.*, 2005), the absence of immortal, pluripotent embryonic stem cells in livestock, capable of germ line transmission in chimaeras (Prelle *et al.*, 2002), has so far limited cell-mediated transgenesis in farm animals to the modification of somatic cell types coupled with NT. Fetal fibroblasts are a popular choice to genetically modify; however, compared to embryonic stem cells these primary cells are limited in two key aspects. First, they only have a finite lifespan of around 35 population doublings in culture – perhaps less for clonal isolates. Second, dependent upon the loci, homologous recombination in somatic cells is often two orders of magnitude less than for embryonic stem cells (Sedivy and Dutriaux, 1999). Promoter–trap gene targeting strategies have therefore been favoured to increase efficiency of homologous recombination in somatic cells. However, this approach is limited to those genes that are expressed in the donor cells and on occasion might necessitate the culture of specific cell types that express the target gene. With primary cell cultures, a significant number of transfected colonies senesce before or during selection, expansion and characterization of the clonal sub-lines (Denning *et al.*, 2001b). In reality, this greatly reduces the ability of this approach to introduce multiple, complex changes into the genome in continuously cultured cells. In practice, this can be overcome by re-deriving fetal cell lines following NT with transfected cells and thus restoring proliferative lifespan. This enables sequential gene targeting and has been used to generate homozygous knockouts at two independent loci (Kuroiwa *et al.*, 2004). The ability to introduce a desired genetic modification into a specific site in the genome and the characterization of this integration event before generating the animal is regarded as a critical advantage, especially for the agricultural industry to have confidence in the new genotype and its pattern of inheritance. Nevertheless, there remains a considerable problem with extensive introgression of the genetic modification into the wider population without suffering inbreeding depression. This is especially true for situations requiring both alleles of an undesirable gene to be inactivated in large herds, but is less difficult for gain-of-function genotypes with effect in the hemizygous state.

Alternative Methodologies to Generate Transgenic Animals

The method used to produce the first transgenic livestock involved the direct injection of a DNA construct into one of the two pronuclei in recently fertilized zygotes (Hammer *et al.*, 1985). Although a relatively straightforward procedure, especially compared to NT, it is limited, however, by generally only allowing random introduction of new DNA into the genome at low efficiency; particularly in livestock, where 0.1–1% of injected zygotes typically result in a transgenic offspring (Wall *et al.*, 1997). Furthermore, only around half the transgenic animals born may express the transgene at an acceptable level, with gene silencing or low expression influenced by the neighbouring regulatory sequences at the site of integration. Additionally, many transgenic animals may simply be of the undesired gender or may be mosaic. The latter is a consequence of DNA integration occurring after the 1-cell stage and so not all cells in the animal carry the transgene, which, if it extends to the germ line, complicates subsequent breeding. Admittedly, genotyping embryos or early pregnancies to select those of the desired sex that carry the transgene can overcome some of these deficiencies. Moreover, the use of recombinase proteins might increase the efficiency of integration and possibly catalyse homologous recombination at targeted locations in the genome (Maga, 2001; Kaneko *et al.*, 2005). There is also the emergence of various other transgenic technologies to introduce and effectively disseminate new genetic change, which will prove highly applicable in many situations and circumvent the need for NT using genetically modified cells. The choice of method used will ultimately depend upon the specific application and nature of the intended genetic modification.

The injection or infection of oocytes or embryos with lentiviral vectors carrying relatively small (~10kb) DNA constructs (Hofmann *et al.*, 2004) represents a simple, efficient method of introducing exogenous DNA where up to 80% of the animals born may be transgenic. With lentiviral vector transgenesis being so efficient, it is feasible to consider infecting oocytes that are recovered from genetically elite females, which, at least for agricultural applications, is very important as it builds upon the existing genetic merit. This is simply not economic or practical with pronuclear microinjection. However, with lentivectors there tends to be multiple, different integration sites in each transgenic animal. For farm animals, with long generation intervals, this is more of a hindrance, in the time taken to characterize subsequent breeding lines from founders with differential segregation of multiple transgenes, rather than an advantage. The preference in many agricultural applications for the transgene to be introduced at a known, well-characterized site in the genome may therefore limit the utility of lentivectors. Although this aspect may be less important for biopharmaceutical applications, the size of the construct that can be introduced becomes limiting. Despite the continued development and safety evaluation of lentivectors used for human gene therapy (Thomas *et al.*, 2003), serious concern will no doubt be raised over the use of modified viral vectors in food-producing animals. None the less, the combination of lentivectors and RNA interference (RNAi) technology has been put forward as a powerful way to combat viral pathogens in livestock (Clark and Whitelaw, 2003). Progress towards exemplifying this goal has been reported recently where lentivectors were coupled with RNAi to stably introduce, into cultured goat fibroblasts, a transgene producing a short hairpin RNA to specifically knock-down the expression of the prion protein gene in the resulting cloned fetus (Golding *et al.*, 2006). This example emphasizes the convergence of a number of transgenic methods involving lentiviruses, RNAi and cell-mediated NT used in combination, rather than in exclusion.

Another attractive approach involves using sperm as vectors to introduce foreign DNA into oocytes, following either fertilization (Lavitrano *et al.*, 1989) or intracytoplasmic sperm injection (Perry *et al.*, 2001). However, these sperm-mediated transgenic methods are often inconsistent, limited in both the size of the construct and complexity of the genetic modification that can be introduced and commonly result in mosaic animals. A promising opportunity exists in the use of spermatogonial stem cells (SSCs), rather than sperm *per se*, to genetically modify the germ line of livestock. Access to cultured SSCs would potentially allow all the benefits of a cell-mediated transgenic approach as discussed above for somatic and embryonic stem cells. In essence, this has already been demonstrated in rodents (Brinster and Avarbock, 1994). SSCs can be isolated, cultured *in vitro* for long periods (up to 2 years, although they divide slowly) and maintain genetic and epigenetic stability (although telomeres do shorten)(Kanatsu-Shinohara *et al.*, 2005). They can be transfected *in vitro* and transgenic offspring generated following transplantation into the host testes of a recipient male, whose endogenous sperm has been depleted, to enable colonization and production of functional sperm of the new genotype (Hamra *et al.*, 2005). The prospect of extending this technology into livestock species is well advanced, with successful transplantation of SSCs already achieved in the buck, boar and bull (Honaramooz *et al.*, 2003; Dobrinski, 2005). Although limited to males, a distinct advantage of this strategy compared to NT is that genetic modifications could be transmitted directly in the form of sperm, after testes transplantation of transgenic SSCs. This circumvents the need to generate an entire cloned sire that has to attain puberty before conventional breeding and so the generation interval is greatly reduced (Hill and Dobrinski, 2006; see Dobrinski and Hill, Chapter 6 this volume).

Transgenic Applications

There are a wide range of opportunities available for redesigning livestock for either biomedical or agricultural applications, just depending upon the particular genes that are manipulated. Described below is a selection of the main applications of interest. These have utilized various transgenic methods; however, many projects were initiated with the pronuclear injection methodology, which until relatively recently was the only one available to produce transgenic livestock. To date, all examples except one are still at the research or testing phases to further develop and evaluate the technology, including the potential economic benefits, and to determine the safety and functionality of the various biomedical and food products as the more advanced projects seek to gain regulatory approval.

Livestock transgenics for biomedicine

There has been greater research effort, progress, economic incentive, ethical justification and public acceptance towards the generation of transgenic livestock for various high-value biomedical applications compared to those for agriculture and food production.

'Pharming'

The conventional production of rare human therapeutic proteins extracted from blood or other tissues is inefficient, expensive and has the risk of contamination with human pathogens. Recombinant forms of these therapeutic proteins can be manufactured in a variety of biological systems to alleviate these disadvantages. The choice of production system most appropriate to a specific pharmaceutical protein, ranging from fermentation culture with various microorganisms or mammalian cells through to 'pharming' plants and animals, is influenced by factors including regulatory issues, speed of production, scale-up costs and ability of the organism to perform the necessary post-translational modifications for biological activity of the protein (Dyck *et al.*, 2003; Ma *et al.*, 2003). Livestock are preferred when functional proteins are difficult to make in sufficient quantities, cost-effectively and safely by other methods.

Transgenic livestock have been engineered as bioreactors to produce valuable recombinant proteins in various bodily fluids (Houdebine, 2000). The mammary gland of dairy species has been favoured because of its ability to produce high amounts of protein that can be easily harvested in the milk. Numerous heterologous recombinant proteins have been produced in the mammary gland with secretion into the milk by regulating the expression of specific human genes, encoding medically important proteins, under the control of mammary-specific promoters (Clark, 1998; Brink *et al.*, 2000). Following extraction and purification, a number of therapeutic proteins have been evaluated in clinical trials to determine their safety and effectiveness in treating particular human diseases and disorders in the quest to gain regulatory approval. In harnessing the potential of the mammary gland to synthesize heterologous proteins, the choice of species (rabbits, pigs, sheep, goats or cows) depends upon the quantities required within specified timeframes (Rudolph, 1999). The mammary gland is, however, not a completely isolated system, with potential leakage of proteins into systemic circulation. Thus, for some highly bioactive proteins, it may be preferable that they are produced in the egg whites of laying hens (Ivarie, 2003), urine (Kerr *et al.*, 1998), or seminal fluid (Dyck *et al.*, 1999) to reduce the potential for pleiotropic side-effects on the animal.

It has been nearly two decades since both the exemplification of human proteins produced in the milk of livestock (Clark *et al.*, 1989) and the founding of the first animal biopharming companies centred on this technology. Yet it was only in June 2006 that the European Medicines Agency announced approval of the first drug produced in a transgenic animal: a recombinant form of human anti-thrombin III (ATIII; ATryn) developed by GTC Biotherapeutics and produced in the milk of transgenic dairy goats as an anticoagulant principally for treating patients suffering from hereditary anti-thrombin deficiency (GTC, 2006; Schmidt, 2006). However, there was a considerable delay in gaining market authorization as the regulators were initially concerned about the possible immunogenicity of the recombinant protein (Anon., 2006). The glycosylation patterns of recombinant proteins produced in livestock can differ slightly compared to those on the equivalent native human protein. This may affect their allergenicity, functionality or bioequivalence (Thomassen *et al.*, 2005; Anon, 2006) and hence, necessitates the strict regulatory requirements to evaluate the safety and efficacy of each individual product through clinical testing.

In other examples of livestock bioreactors, there has been some success in the use of recombinant human alpha-glucosidase, purified from rabbit milk, as an enzyme replacement therapy aiding the treatment of infantile Pompe disease (Van den Hout *et al.*, 2004). Recombinant human C1 inhibitor is currently in Phase III clinical trials for the treatment of hereditary angio-oedema (Pharming, 2006). Human blood coagulation factors VIII and IX have been produced in sheep milk for potential treatment of haemophilia A and B, respectively (Clark *et al.*, 1989; Schnieke *et al.*, 1997; Niemann *et al.*, 1999). The human protease inhibitor α1-anti-trypsin, produced in the milk of transgenic sheep, was aimed at treating emphysema and to potentially alleviate some symptoms of cystic fibrosis (Wright *et al.*, 1991), whereas human alpha-fetoprotein is under development as a biopharmaceutical for the treatment of autoimmune diseases (Parker *et al.*, 2004). Transgenic livestock are also being evaluated for the production of recombinant vaccines. Potential vaccines against malaria and rotavirus have been produced in milk of transgenic goats and rabbits, respectively (Behboodi *et al.*, 2005). Recombinant human therapeutic antibodies are also being produced in the milk (Pollock *et al.*, 1999) and blood (Grosse-Hovest *et al.*, 2004) of transgenic animals. Moreover, human polyclonal antibodies have been produced in the blood of cattle following the introduction of an artificial chromosome containing the entire sequence for human immunoglobulin heavy and light chain loci into bovine fetal fibroblasts and then creating calves via NT (Kuroiwa *et al.*, 2002). There has been an interesting development in the production of recombinant human butyrylcholinesterase, purified from transgenic goat's milk, as a potential bioscavenger of organophosphorus nerve agents for affected military personnel or civilian casualties of bioterrorism (Cerasoli *et al.*, 2005).

The technology for generating transgenic livestock as bioreactors is well established, and there are numerous examples. But the path to commercialization is fraught with financial difficulty especially in terms of the purification costs, the timeframes involved in rigorous clinical testing and regulatory evaluation and the risk that the therapeutic being developed may ultimately fail. The enormity of these challenges were exemplified with the recent failure of PPL Therapeutics, one of the pioneers of biopharming, caught up in economic constraints and regulatory uncertainties (Powell, 2003). It is unclear what the future holds for the general utility of farm animal bioreactors in converting forage into valuable biopharmaceuticals. The use of plants to directly produce certain therapeutics appears be a strong competing technology. Plant-based molecular pharming systems have successfully produced functional human pharmaceutical proteins, recombinant antibodies and vaccines and offer several advantages over livestock bioreactors, such as lower production costs, greater scale-up ability and enhanced safety due to absence of animal or human pathogens (Ma *et al.*, 2003). Moreover, progress is being made in mammalian-like, post-translational processing of proteins, where this is important for functionality, in biologically unrelated production systems such as plants (Joshi and Lopez, 2005) and microorganisms (Gerngross, 2004). Functional proteins can even be chemically synthesized *de novo* (Kochendoerfer *et al.*, 2003). In terms of ethical acceptance, the public are more likely to support the use of transgenic plants as bioreactors rather than animals (Einsiedel, 2005). A more significant environmental issue with pharming plants is, however, control over their physical containment and spread.

Xenotransplantation

The shortage of human donor organs to treat chronic organ failure and various degenerative tissue diseases could be overcome by targeting specific genetic modifications to generate herds of pigs whose organs would be immunologically compatible with humans following xenotransplantation (Lanza *et al.*, 1997). Pigs are favoured because of their similarities in organ size and physiology to humans, short generation interval and large litter size. Recently, pigs have been produced that completely lack the enzyme α-1,3-galactosyl-transferase (Phelps *et al.*, 2003) in attempts to eliminate the specific carbohydrate epitope found on the cell surface in pigs, but not humans, to counter the complement-based hyperacute immune rejection that occurs within minutes of transplantation. However, low but detectable levels of the gal antigen may still persist on the pig cells (Sharma *et al.*, 2003). None the less, initial evidence provides some encouragement with pig-to-baboon xenografts surviving for extended periods, up to 83 days for kidneys (Yamada *et al.*, 2005) and 2–6 months for hearts (Kuwaki *et al.*, 2005). Subsequent immunosuppressive drug therapy or additional genetic modifications could be used to manage the human body's other rejection processes to prolong the functionality of the organ. In using pig organs, the potential for cross-species viral infection, not so much for the individual but for the community at large, is a risk that needs to be managed (Fishman and Patience, 2004).

Disease models

In some cases it may be ethically acceptable to genetically modify farm animals to serve as models for various inherited human genetic diseases to aid research and preliminary evaluation of novel therapies (Petters, 1994). An ovine model of cystic fibrosis, for example, is considered superior to available mouse models because of the greater similarity in lung anatomy and physiology with humans (Harris, 1997); however, gene targeting at this non-expressed locus in fibroblasts has proved difficult (Williams *et al.*, 2003). Transgenic pigs have been developed as useful models for the rare human eye disease retinitis pigmentosa (Mahmoud *et al.*, 2003). Conventional livestock are not appropriate in all instances because of their long generation intervals, physical size and costs involved. Hence, more suitable laboratory models have been the motivation for developing NT from cultured somatic cells in species such as the rat (Zhou *et al.*, 2003), rabbit (Chesne *et al.*, 2002), and ferret (Li *et al.*, 2006) and now the opportunity exists to couple this with cell-mediated genetic modification.

Livestock transgenics for agriculture

As understanding of the genes that influence livestock production traits improves from the sequencing and comparison of the genomes of livestock species with mouse and human (Fadiel *et al.*, 2005; Womack, 2005), so does the knowledge to accurately modify the appropriate genes and regulatory sequences to generate new and desired animal products in the future. However, as most of the relevant livestock traits are complex and controlled by multiple genes, useful transgenic intervention is considerably more difficult than mere over-expression for a biopharming application. Agricultural applications of livestock transgenesis are primarily aimed at increasing: (i) animal productivity; (ii) the quality of valuable meat, milk and fibre components; (iii) disease resistance; and (iv) improving environmental sustainability. Genuine improvement in these attributes will have economic benefits for farmers and processors, and additional health benefits for consumers or the livestock themselves.

Meat

Many initial livestock transgenic experiments focused on modifying body composition by introducing growth hormone or insulin-like growth factor I (IGF-I). However, this pioneering work only resulted in slightly increased growth rates and was severely hampered by poor transcriptional regulation of the transgenes, resulting in high levels of these hormones in systemic circulation, with animals consequently suffering a number of deleterious side-effects including lameness, susceptibility to stress and reduced fertility (Pursel *et al.*, 1989; Pursel and Rexroad, 1993). More desirable effects on growth rate and body composition have been achieved without apparent abnormalities by restricted expression of the human IGF-I transgene to skeletal muscle (Pursel *et al.*, 1999). This manipulation resulted in female transgenic pigs having ~10% more carcass lean tissue and ~20% less total carcass fat, but there was no change in the body composition of transgenic males compared to conventional boars (Pursel *et al.*, 1999). An even tighter control of transgene expression can be achieved with inducible control elements which essentially function as on/off switches, such as the metallothionein promoter which can be regulated by manipulating the level of zinc in the diet (Nottle *et al.*, 1999). Using a similar promoter regulating ovine growth hormone, transgenic sheep grew significantly faster and were leaner but were burdened with a greater parasite fecal egg count, possibly indicative of a compromised immune system, and also displayed some foot problems similar to those observed in animals treated with exogenous growth hormone (Adams *et al.*, 2002).

Rather than attempting to manipulate primary endocrine signals, specific loss-of-function mutations in the myostatin gene or its regulatory sequences might confer an acceptable degree of double-muscling in livestock species similar to that observed from natural mutations in some cattle breeds (Kambadur *et al.*, 1997). Whereas the natural mutation is associated with major calving difficulties and the resulting welfare concerns, an engineered modification could avoid these problems by targeting the effects of the mutation to stages of major muscle growth in post-natal animals (Grobet *et al.*, 2003).

Efforts have been made to improve meat quality, specifically by improving the essential fatty acid composition (Saeki *et al.*, 2004). Meat enriched in healthier omega-3 unsaturated fatty acids has been accomplished recently in cloned transgenic pigs that express humanized copies of the *Caenorhabditis elegans* gene, *fat-1* (Lai *et al.*, 2006). This transgene encodes an *n*-3 fatty acid desaturase for the conversion of *n*-6 fatty acids, consumed in a grain-based diet, to the more beneficial *n*-3 fatty acids associated with a lower risk of morbidity and mortality from atherosclerosis and coronary heart disease (Lai *et al.*, 2006). This may prove to be an economical, safe and sustainable method to fortify meat without the need to provide a dietary source of *n*-3 fatty acids in the form of fishmeal.

Milk

The genetic modification of milk composition in dairy cattle especially has received considerable attention in efforts to improve productivity, aid human nutrition and alter various processing properties designed to suit the manufacture of specific dairy products (Wall *et al.*, 1997; Karatzas, 2003). Manipulations could include the over-expression of naturally occurring but minor endogenous milk proteins, known to have health-promoting properties, to generate functional foods or to enable the economic extraction of these valuable components as food ingredients. There may also be advantages in over-expressing major milk proteins, even for traditional commodity markets (Zuelke, 1998). Foreign genes could be introduced to produce novel proteins in milk for the generation of a nutraceutical, as with the addition of human lactoferrin with enhanced iron absorption properties and provision of passive immunity (van Berkel *et al.*, 2002). Additional phosphorylation sites could be introduced into the casein proteins, to increase calcium content and emulsification properties of milk, and extra proteolytic sites in the caseins might increase the rate of cheese ripening (Wall *et al.*, 1997). The targeted disruption of both alleles of β-lactoglobulin in cows could explore first the fundamental role of this ruminant specific whey protein and second, whether such milk reduces allergenicity, as human milk does not contain this protein.

There has been recent success in the prevention of mastitis caused by *Staphylococcus aureus* infection in dairy cattle, following the introduction of a biologically active form of lysostaphin (Wall *et al.*, 2005). Similarly, the demonstration of increased antimicrobial properties in the milk through the introduction of human lysozyme in goats further illustrates the potential of preventing mastitis in dairy species (Maga *et al.*, 2006). However, the concentration and specificity of these antimicrobials might be critically important to avoid compromising the biological function of milk as a food to rear young, or for other processing applications. For instance in mice, dams expressing lysostaphin in their milk at levels two- to 40-fold greater than that achieved in cattle (Wall *et al.*, 2005) had delayed mammary development post-partum and growth rates in suckled pups were also reduced, at least until weaning (Mitra *et al.*, 2003).

Different strategies have been tested to reduce the milk sugar lactose, as alternatives to the otherwise expensive post-harvest processing of milk. Lactose causes intestinal disorders in more than 70% of the adult human population which lack sufficient lactase enzyme activity for adequate lactose digestion after milk consumption (Sahi, 1994). Lactose is the major osmotic regulator of milk secretion and is synthesized by the lactose synthetase complex. This complex comprises two enzymes, one of which is the milk protein α-lactalbumin. The disruption of both copies of the α-lactalbumin gene in mice resulted in lactose-free milk, but strongly affected the osmotic regulation of milk, resulting in the production of a highly viscous secretion incapable of supporting adequate nutrition in suckling young (Stinnakre *et al.*, 1994). In the heterozygous state, with one inactive α-lactalbumin allele, milk had a modest 10–20% decrease in lactose content and a similar percentage increase in the concentration of total milk solids. An alternative approach of expressing a lactose hydrolysing enzyme in the mammary gland under the control of the α-lactalbumin promoter was more successful (Jost *et al.*, 1999). In heterozygous females, lactose was still produced and subsequently hydrolysed into the osmotically active monosaccharides glucose and galactose. Overall, milk lactose was reduced by 50–85% and there was only a slight increase in osmolarity. If extended to bovine milk, such a reduction in lactose could possibly ameliorate lactose intolerance (Jost *et al.*, 1999). Moreover, there was no apparent change in either fat or protein concentration and the milk successfully nourished offspring. Whilst the results from these transgenic mouse studies are promising, given the intrinsic differences in milk composition between mice and cows, the success of such a transgenic approach can only be assessed when applied to dairy cattle. Furthermore, the functional properties of any altered milk need to be evaluated to determine its suitability for processing into a range of dairy products. Aside from the issue of lactose intolerance, an acceptable reduction in lactose would be beneficial in concentrating milk solids, provided it did not compromise milk let-down, and lower the transportation costs of liquid milk from the farm to the dairy factory. This might be achieved by controlled regulation of lactase expression, α-lactalbumin gene dosage or RNAi knockdown.

Milk fat is very rich in saturated fatty acids, which have been associated with cardiovascular and coronary heart disease. The fatty acid composition of milk could be improved by modulating the enzymes involved in *de novo* lipid metabolism to decrease the unhealthy fats in favour of unsaturated fats (Wall *et al.*, 1997). However, manipulation of unsaturated fatty acid composition in milk may be more effectively achieved by varying the diet of the dairy cows, through the use of particular forage species or feed additives (Palmquist *et al.*, 1993) and has been a successful approach to produce spreadable butter (Fearon *et al.*, 2004).

Consideration has also been given to reducing the total fat content of milk (Wall *et al.*, 1997). Milk typically contains around 3.8% fat, 50% of which is synthesized in the mammary gland and is thus energetically expensive for the cow to make. If a strategy was used to knockdown the expression of acetyl-coenzyme A carboxylase in the mammary gland, for instance, *de novo* fatty acid synthesis might be reduced, resulting in a natural low-fat liquid milk desired by consumers (Wall *et al.*, 1997). This approach could also reduce the feed energy requirements for the cow, lowering the cost of milk production. It could also reduce the mobilization of adipose tissue during early lactation and likely improve conception rates to maintain a 365-day calving interval, important in many seasonal pastoral dairy farming systems.

Considering the importance of milk protein for the quality and yield of dairy products, there is strong interest by both farmers and milk processors to improve milk protein content. Cheese, for example, is essentially composed of casein, which accounts for about 80% of total milk protein, and milk fat. Any increase in the casein content is therefore an improvement for the manufacture of cheese. Moreover, the casein fractions, comprising four different casein proteins and numerous variants, are aggregated into large colloidal micelles which are a major determinant of the physicochemical properties of milk. One casein protein in particular, κ-casein, occupies the surface of the micelle and influences the size of this protein particle. This has been demonstrated in transgenic mice over-expressing bovine κ-casein, which reduced the size of the micelles. In addition, the milk from these transgenic mice formed a significantly stronger rennet-induced curd, clearly demonstrating that an increase in κ-casein affects the physicochemical properties of milk (Gutierrez-Adan *et al.*, 1996). Encouraged by these transgenic mouse studies we have applied these findings to cattle. Using a transgenic approach we have attempted to increase not only the casein content but also to reduce the size of the casein micelles in order to improve cheese yield. The introduction of additional copies of bovine β- and κ-casein genes into cloned dairy heifers resulted in a significant twofold increase in total κ-casein protein concentration in the milk from hormonally induced lactations (Brophy *et al.*, 2003). Moreover, this changed the ratio of β- to κ- casein and studies evaluating the functional properties of this altered milk are presently underway. Interestingly, this increase in κ-casein was at the expense of the production of some other milk proteins, emphasizing a biological ceiling in terms of protein output in the modern dairy cow. None the less, this work exemplifies the potential of transgenic technology to increase the production of specific endogenous milk proteins that may be required for particular niche (ingredient) markets and that this can be dramatically achieved in a single generation.

Wool

A limited number of research projects initiated by Australasian groups have explored the potential of genetic engineering to increase wool growth rates and generate fibre with improved processing and wearing qualities. In first-generation transgenic sheep, expressing an ovine IGF-1 cDNA driven by a mouse keratin promoter, clean fleece weight at 1 year of age was on average 6% greater (Damak *et al.*, 1996). However, at second- and third-year shearing, fleece weights were not significantly different (Su *et al.*, 1998). Moreover, there was a trend for transgenic animals to have coarser wool of lower staple strength, especially in the males. Additionally, the slight production advantage observed in animals of the first generation did not persist in the second generation (Su *et al.*, 1998). However, fleece weights in fifth-generation transgenic merino sheep expressing ovine growth hormone were greater (12%) but there was a significant interaction with breed type, with Poll-Dorset-cross sheep having 11% lower fleece weights (Adams *et al.*, 2002). Furthermore, fibre diameter was less desirably greater in these transgenic sheep. Cortical-specific expression of an intermediate filament keratin gene led to marked alterations in both the microstructure and macrostructure of wool fibres, which had higher lustre and reduced crimp (Bawden *et al.*, 1998). Attempts to introduce a functional cysteine biosynthetic pathway into sheep so that this rate-limiting essential amino acid may be synthesized *de novo* to enhance wool growth have only had limited success (Ward, 2000).

Sustainable Agriculture

As ironic as it may seem, the prospect of using transgenesis in livestock to aid sustainable agriculture is particularly appealing. The intention to improve animal health, reduce pollution and more effectively utilize both feed and animal resources might be better received by society. Although improved disease and pest resistance via genetic modification has long been an ambitious goal (Muller and Brem, 1998; Whitelaw and Sang, 2005), the result would be improved animal welfare and reduced reliance on animal remedies. The potential of RNAi technology to target the knockdown of specific pathogenic viruses, and/or their transcriptional products, and confer cellular resistance to infection has been highlighted (Clark and Whitelaw, 2003). By extrapolation from the mouse (Bueler *et al.*, 1993; Perrier *et al.*, 2002), introduction of mutated prion protein genes (Cyranoski, 2003), gene knockout (Denning *et al.*, 2001a; Kuroiwa *et al.*, 2004; Yu *et al.*, 2006) or suppression of prion protein by RNAi (Golding *et al.*, 2006) in livestock would all be expected to produce animals resistant to transmissible spongiform encephalopathies. Livestock resistant to such diseases would also be especially valuable in providing improved safeguards for biomedical applications. Another application, considered previously to improve animal welfare, aimed at the production and secretion of a chitinase protein in the sweat glands of transgenic sheep to kill parasitic larvae and provide resistance to cutaneous myiasis (flystrike)(Ward *et al.*, 1993). There is also the prospect of lactating females engineered to produce novel viral-neutralizing antibodies in their milk, so that the suckled young would gain passive immunity to potential infections of the gastro-intestinal tract, as demonstrated in mice (Castilla *et al.*, 1998). Improving the lactational performance of pigs may improve the survivability and health of piglets. Indeed, gilts over-expressing bovine α-lactalbumin in the mammary gland had higher milk lactose content, which increased milk yield by 20–50% in early lactation and consequently also increased the weight gain of suckling piglets (Wheeler *et al.*, 2001).

Novel solutions towards mitigating the adverse effects of environmental pollution from intensive farming systems are being developed. Transgenic pigs with a bacterial phytase gene expressed in their saliva more efficiently digest plant dietary phytate, the most abundant form of phosphorus in their formulated diet. This therefore lessens the need to provide inorganic phosphate supplements and also reduces, by up to 75%, the levels of phosphate in their excrement, which constitutes a major form of environmental pollution with the current pork industry in Canada and elsewhere (Golovan *et al.*, 2001).

Other bacterial genes that encode biochemical pathways that are non-functional in livestock could be introduced to increase the availability of specific nutrients that are rate limiting for production, so long as they did not cause pleiotropic side-effects. For instance, it has been proposed that increased feed utilization efficiency from dietary roughage might be achieved by introducing a glyoxylate cycle into ruminants, enabling the synthesis of glucose directly from acetate produced in the rumen (Ward, 2000). Equally, changes to the rumen microfloral community would be beneficial to increase feed utilization (Weimer, 1998). In parallel, advances in plant breeding continue to improve forage in terms of its digestibility, energy value, protein and mineral content, and enhance it's animal health attributes, by both conventional and transgenic means (Woodfield and Easton, 2004; Reddy *et al.*, 2005). Likewise, edible vaccines from transgenic plants could confer immune protection to animals against particular diseases (Dus Santos and Wigdorovitz, 2005). In some situations, it may be both easier and allow greater flexibility to simply modify the diet of the livestock, rather than genetically modifying the animals themselves, to increase productivity and alter the composition of animal food products.

The prospect of generating cloned bulls from progeny-tested sires that have been genetically modified to produce only male- or only female-bearing sperm as required, by rendering sperm of the opposite sex incapable of completing spermatogenesis (Forsberg, 2005), would be highly desirable. This would be especially true in more extensive farming situations that rely on natural mating to disseminate superior male genetics. Even the generation of bulls that produce sperm with a sex ratio deliberately skewed substantially one way or the other would be of practical use. In the artificial insemination industry, however, it would still be preferable to improve upon the various sperm sex-sorting or enrichment strategies available or possible. Such sperm-selection strategies have an animal welfare aspect, by minimizing the production of offspring of the unwanted gender. It would enable dairy farmers, for instance, to mate their top cows with X-bearing sperm from elite bulls to generate replacement heifers and, for the remainder of the herd, use Y-bearing sperm for dairy-beef production. Conversely, in the beef and sheep meat industries there would be a preference for terminal sires that only produce male offspring. Another advantage for market acceptability of this approach is that the offspring would not themselves be transgenic.

Regulatory Issues

The regulatory requirements for these emerging technologies in food-producing animals are still evolving. They mostly involve the appropriate regulatory bodies addressing environmental risks, animal welfare and food safety. Only once these policies have been implemented will transgenic animal technology in agriculture have the opportunity to move beyond the research realm.

Gene-flow to wild species has been a considerable issue in the transgenic plant debate. Generally, the ability to physically contain and control the breeding of farmed livestock is far greater than with plants and so the threat of inadvertent transmission of transgenes is considerably diminished. Animal welfare issues are discussed separately, below.

National food regulatory bodies are awaiting signals from international organizations such as the FAO/WHO, who are still in the process of broad consultation and defining guidelines for controlling the safety of genetically modified food products derived from livestock. The Codex Alimentarius Commission of the Joint FAO/WHO Food Standards Programme has established general principles for the risk analysis of food derived from modern biotechnology (FAO/WHO, 2003b) and, more specifically, is presently developing draft guidelines for safety assessment of animals produced via recombinant DNA technology, based largely on those established for plants (FAO/WHO, 2003a). General guidelines suggesting how to evaluate the safety of food products from cloned and transgenic animals are available from various regulatory agencies (Rudenko *et al.*, 2004; Kelly, 2005). The safety assessments are largely based on the premise that a healthy animal is likely to produce safe food and through the use of a comparative approach to evaluate the composition of food products. Moreover, guidelines are available to assess food for possible unintended health effects (Committee on Identifying and Assessing Unintended Effects of Genetically Engineered Foods on Human Health, 2004). To gain regulatory approval, the assessment of food safety in genetically modified animals will likely be a complex process. Whilst this is the norm for evaluating new drugs, it might be prohibitive for food products with commercial values that are orders of magnitude lower than those for human therapeutics. There is the expectation that every line of transgenic animals would have to be considered on a case-by-case basis, as the risk is ultimately associated with the particular genetic modification involved and the method used for its introduction. Whilst regulators might tackle the science-based assessment of food safety, the general public may not be satisfied, as a broader range of ethical and personal values are evoked by this technology (Einsiedel, 2005).

Consumer Acceptance

Transgenic technology is controversial in the public arena. There has been a great deal of discussion portrayed in the media over many years, expressing the views from both sides of the debate. Whilst proponents extol the potential benefits, opponents highlight their belief that the technology is 'unnatural' and of the uncertainty towards the long-term environmental and health impacts. With such polarized views, the general public are often left wondering who to believe. With a general erosion of public trust in scientists and companies developing these products, there remains a relatively low level of informed awareness in the wider community and scepticism in regulatory oversight (Einsiedel, 2005; Small, 2005). Moreover, distinctions between cloning and genetic engineering are not necessarily made by the public. This is not helped by the fact that NT is a method used to produce an animal from a transgenic cell – so the two technologies are intimately intertwined, further compounding the difficulty for the layperson to clearly distinguish them.

Public surveys conducted in New Zealand have shown that support for genetic modification has increased slightly over the last 5 years, especially in the biomedical area, with 33% of respondents having total support for this use of the technology (Small, 2005). Total support of New Zealanders in 2005 for food applications of genetic modification, however, was very low at only 9%. In fact, 26% were totally opposed. Most people (50–60%) hold the middle-ground and have conditional support for both food and medical applications of genetic modification (Small, 2005). That is, their decision depends upon the specific application of the technology. Individual case-by-case judgements are evident elsewhere in Europe and North America (Einsiedel, 2005). This reflects the complex nature of the technology and its wide array of potential uses – perceived by some as good or bad. Individuals need to evaluate each to assess whether the perceived benefits outweigh the possible risks. It is easier, for instance, to justify the generation of bioreactor organisms to produce therapeutic proteins to alleviate human disease and suffering than using genetic modification to increase productivity on farms. Unless there is some direct advantage to consumers, perhaps in terms of additional health benefits from the new food, they may be unlikely to change their views and purchase a potential product derived from transgenic livestock.

There are apparent hierarchies of acceptance with regard to the type of organism being modified for a particular purpose. For example, in molecular pharming, the genetic modification of plants is considered by the public at large to be preferable compared to the use of animals, where more objections are raised (Einsiedel, 2005). But if the recombinant proteins were to have greater functionality when produced by livestock bioreactors, because of more faithful post-translational modifications, then the justification for the use of farm animals must be clearly stated by researchers. Animal welfare concerns become more prevalent with the use of higher animals and the risk of unintended side-effects. Indeed the health and fitness of many transgenic livestock have been compromised, often due to inappropriate gene regulation in addition to perturbations from some of the reproductive technologies used, especially for NT.

Animal Welfare

Nuclear transfer is associated with animal welfare concerns due to decreased survival of both pregnancies and cloned offspring. These issues are far greater than those associated with *in vitro* fertilization and embryo culture used in some other transgenic methodologies. The major cause of the abnormalities associated with NT can be attributed to an incorrect epigenetic reprogramming of the donor cell genome. It is anticipated that ultimately an increased understanding of epigenetics and new strategies to improve reprogramming will lead to some increase in cloning success and lessen the current animal welfare burden.

The potential consequences on the welfare of transgenic animals depend upon the specific genetic modification and its effect in either a hemi- or homozygous state. These potential effects form a continuum from poor animal welfare (e.g. over-expression of human erythropoietin in rabbits; Massoud *et al.*, 1996) to neutral (e.g. increased expression of bovine milk proteins in cattle; Brophy *et al.*, 2003) to improved welfare (e.g. pest and disease resistance; Whitelaw and Sang, 2005). The consequences on the animal are also affected by the site of integration of the transgene, the degree of control over its expression (correct time, tissue, quantity) and the effects of exposure of the host animal to the biologically active, transgene-derived proteins. This was clearly observed with the inappropriate over-expression of growth hormone in pigs and other species (Pursel *et al.*, 1989). A greater understanding of gene function and regulation of expression will to some extent improve the predictability of the physiological consequences on the animal and reduce the incidence of compromised animal welfare. Even so, foreign proteins produced in the mammary gland and secreted into milk may still leak into blood and enter the general circulation (Carver *et al.*, 1992). The consequences of this on the physiology and welfare of the transgenic animal depends on the specific protein, its biological activities and the concentration in plasma. This further contributes to unintended effects and necessitates the requirement to monitor the health, fitness and behaviour of these transgenic animals throughout their lifetime as part of their higher standard of management and care (Van Reenen *et al.*, 2001).

Industry Adoption

Even, if at some time in the future regulatory approval and substantial public acceptance is gained, the prospect of incorporating transgenic animals into herds and flocks will pose challenges for animal industries, particularly if products still need to be segregated. This could be most difficult for dairy industries where milk is pooled and handled in bulk; it might be somewhat easier with meat and fibre products with individual animal identification and easier product segregation and traceability.

Milk is a commodity product which is processed by high-capacity facilities into a range of different products. Particular transgenic milk streams, tailored for specific purposes, might be unsuitable for general commodity dairy products. Modifying milk composition to benefit cheese manufacture for instance (Wall *et al.*, 1997; Brophy *et al.*, 2003) would be to the detriment of some other processed foods. Whilst the specific staphylolytic activity of lysostaphin used to engineer cows with resistance to mastitis (Wall *et al.*, 2005) is unlikely to affect the microorganisms used to manufacture dairy products, more general anti-microbial agents such as lysozyme or lactoferrin might do so. Conversely, however, they might slow the growth of bacterial contaminants in the milk, increasing the shelf-life of certain dairy products (Maga *et al.*, 2006). The over-expression of specific proteins might be at the expense of other endogenous milk proteins (Brophy *et al.*, 2003) and may affect the composition of the milk and hence its physicochemical and manufacturing properties (Maga *et al.*, 2006) in sometimes unanticipated ways. The point is that at present, through our limited understanding, the consequences of even simple genetic modifications on other characteristics of a complex biological fluid like milk, and the products derived from it, cannot be fully predicted and require rigorous evaluation in each instance. This prolongs the time before there would be any prospect of a particular genetic modification being introduced into the marketplace and the improvement afforded by the genetic modification would have to be substantial to compensate for this. This would be less important in situations where specific high-value endogenous milk components are over-expressed and extracted from the milk for the food ingredient or nutraceutical markets.

One possible scenario for the future is the generation of herds possessing different specific genetic modifications to tailor agricultural products for niche markets. In the dairy industry, transgenic milk from specific herds would need to be kept separate for manufacturing purposes, let alone for food-labelling compliance. Such a prospect would pose challenges for the structure of traditional commodity-based dairy industries processing bulk milk. The integration of transgenesis might necessitate regional herds producing milk of a similar type with specific processing capability available locally. Perhaps more importantly for adoption of the technology at an industry level, farmers need to be paid according to the specific products they are producing behind the farm gate.

The most efficient means of introducing a desired genetic modification into the wider livestock population is through low-cost artificial insemination or natural mating using transgenic males. It is a particular advantage of the cell-mediated transgenic approach that the genetic modification can be made on an already outstanding genetic background by using cells from an elite male. Ideally, the sire should be homozygous for the desired trait so all progeny receive a copy of the transgene. For widespread agricultural applications of gain-of-function transgenes, it is considered important that the integration site be well characterized and tested in hemizygous and homozygous states on a range of genetic backgrounds, as this can affect the phenotypic outcome in different breeds or strains (Siewerdt *et al.*, 1999;

Adams *et al.*, 2002). For loss-of-function transgenes, both alleles in progeny need to be inactivated. This will require an extra generation of breeding, with preferably another non-related transgenic sire harbouring the same genetic modification. The resulting homozygous knock-out individuals are then available for further selection and progress towards stably fixing the new genetic change into the herd or wider population. Animal industries may choose to regularly introduce the same transgene, ideally at the same locus, on a new genetic background using cell lines derived from the most recently selected progeny-tested sires, so as to capture the annual incremental genetic gains from conventional animal breeding and maintain genetic diversity. The economic benefits of a genetic modification affecting a production trait must be sufficiently large to compensate for the lag in genetic gain during the time taken to introduce the transgene and test its performance before wider dissemination. It has been estimated in some introgression schemes that the value of the transgene must be greater than 10% of the 'economic value' of the trait in question, to at least match what could have been achieved through conventional selection and breeding (Gama *et al.*, 1992).

Perspective

Since the domestication of livestock species, humans have been inadvertently redesigning the genetic make-up of their animals by selective breeding based on desirable phenotypic characteristics. This practice alters the frequency of many genes in an often unregulated manner. The new technologies of cloning from cultured cells and transgenesis with site-specific integration have the potential to allow a more targeted approach towards animal breeding. A tremendous opportunity exists with transgenesis to take production and health traits beyond the normal biological boundaries. Animals can be engineered for specific purposes in ways that are not possible with either conventional breeding or genomic selection, which exploits the natural variation amongst the existing population. However, major advances are still required to realize this vision. This includes improved reprogramming of the donor genome following NT, an increased frequency of gene targeting in somatic cells and an intimate understanding of the biological systems involved. The identification of genes and regulatory elements that influence livestock production traits will enable the effective utilization of cloning to duplicate entire genotypes and for transgenesis to introduce precise genetic enhancements to progress animal breeding in the 21st century.

Despite the advances being made in science, there is continued debate as to the value of transgenic animals in agriculture. It is important to note, however, that the majority of the public do appear to be conditionally supportive of genetic modification and they want to see good practical examples demonstrating the benefits of the technology with minimal risk. Benefits improving the health of farmed livestock and/or of the consumer, along with solutions to environmental impacts of agriculture will undoubtedly be met with the greatest acceptance by the public. But for any of these outcomes to be realized the producer needs to benefit also, at least financially. A difficulty is that the technology is complex and each application often has a unique set of pros and cons. Research data are required to evaluate each potential transgenic application before industry, regulatory, and consumer acceptance are sought. These results need to be communicated effectively to the public, producers and regulatory agencies by members of the scientific community. It has been more than 20 years since the first transgenic farm animals were created and 10 years since the birth of Dolly. And it will be some time yet before specific genetic modifications in livestock are possibly adopted on farms and accepted in segments of the marketplace. This might not happen in isolation and alternative genetic technologies in plants may emerge with the potential to complement modified livestock and significantly enhance some common agri-biotechnology goals. More specifically, it remains to be determined just how significant NT will become in agriculture and what the future holds for the use of cloned and transgenic livestock in (ph)farming systems.

Acknowledgement

This work was supported by the New Zealand Foundation for Research, Science and Technology and AgResearch.

References

Adams, N.R., Briegel, J.R. and Ward, K.A. (2002) The impact of a transgene for ovine growth hormone on the performance of two breeds of sheep. *Journal of Animal Science* 80, 2325–2333.

Anon. (2006) Transgenic milk prospects turn sour. *Nature Biotechnology* 24, 368.

Archer, G.S., Dindot, S., Friend, T.H., Walker, S., Zaunbrecher, G., Lawhorn, B. and Piedrahita, J.A. (2003) Hierarchical phenotypic and epigenetic variation in cloned swine. *Biology of Reproduction* 69, 430–436.

Baguisi, A., Behboodi, E., Melican, D.T., Pollock, J.S., Destrempes, M.M., Cammuso, C., Williams, J.L., Nims, S.D., Porter, C.A., Midura, P., Palacios, M.J., Ayres, S.L., Denniston, R.S., Hayes, M.L., Ziomek, C.A., Meade, H.M., Godke, R.A., Gavin, W.G., Overstrom, E.W. and Echelard, Y. (1999) Production of goats by somatic cell nuclear transfer. *Nature Biotechnology* 17, 456–461.

Bawden, C.S., Powell, B.C., Walker, S.K. and Rogers, G.E. (1998) Expression of a wool intermediate filament keratin transgene in sheep fibre alters structure. *Transgenic Research* 7, 273–287.

Behboodi, E., Ayres, S.L., Memili, E., O'Coin, M., Chen, L.H., Reggio, B.C., Landry, A.M., Gavin, W.G., Meade, H.M., Godke, R.A. and Echelard, Y. (2005) Health and reproductive profiles of malaria antigen-producing transgenic goats derived by somatic cell nuclear transfer. *Cloning Stem Cells* 7, 107–118.

Boiani, M., Eckardt, S., Leu, N.A., Scholer, H.R. and McLaughlin, K.J. (2003) Pluripotency deficit in clones overcome by clone-clone aggregation: epigenetic complementation? *Embo Journal* 22, 5304–5312.

Bourchis, D., Le Bourhis, D., Patin, D., Niveleau, A., Comizzoli, P., Renard, J.P. and Viegas-Pequignot, E. (2001) Delayed and incomplete reprogramming of chromosome methylation patterns in bovine cloned embryos. *Current Biology* 11, 1542–1546.

Brink, M.F., Bishop, M.D. and Pieper, F.R. (2000) Developing efficient strategies for the generation of transgenic cattle which produce biopharmaceuticals in milk. *Theriogenology* 53, 139–148.

Brinster, R.L. and Avarbock, M.R. (1994) Germline transmission of donor haplotype following spermatogonial transplantation. *Proceedings of the National Academy of Sciences of the United States of America* 91, 11303–11307.

Brophy, B., Smolenski, G., Wheeler, T., Wells, D., L'Huillier, P. and Laible, G. (2003) Cloned transgenic cattle produce milk with higher levels of beta-casein and kappa-casein. *Nature Biotechnology* 21, 157–162.

Bueler, H., Aguzzi, A., Sailer, A., Greiner, R.A., Autenried, P., Aguet, M. and Weissmann, C. (1993) Mice devoid of PrP are resistant to scrapie. *Cell* 73, 1339–1347.

Campbell, K.H., McWhir, J., Ritchie, W.A. and Wilmut, I. (1996) Sheep cloned by nuclear transfer from a cultured cell line. *Nature* 380, 64–66.

Carter, D.B., Lai, L., Park, K.W., Samuel, M., Lattimer, J.C., Jordan, K.R., Estes, D.M., Besch-Williford, C. and Prather, R.S. (2002) Phenotyping of transgenic cloned piglets. *Cloning Stem Cells* 4, 131–145.

Carver, A.G., Wright, D., Cottom, J., Cooper, M., Dalrymple, S., Temperley, M., Udell, D., Reeves, J., Percy, A., Scott, D., Barrass, Y., Gibson, Y., Jeffrey, C., Coleman, S.A. and Garner, I. (1992) Expression of human alpha 1 antitrypsin in transgenic sheep. *Cytotechnology* 9, 77–84.

Castilla, J., Pintado, B., Sola, I., Sanchez-Morgado, J.M. and Enjuanes, L. (1998) Engineering passive immunity in transgenic mice secreting virus-neutralizing antibodies in milk. *Nature Biotechnology* 16, 349–354.

Cerasoli, D.M., Griffiths, E.M., Doctor, B.P., Saxena, A., Fedorko, J.M., Greig, N.H., Yu, Q.S., Huang, Y., Wilgus, H., Karatzas, C.N., Koplovitz, I. and Lenz, D.E. (2005) In vitro and in vivo characterization of recombinant human butyrylcholinesterase (Protexia) as a potential nerve agent bioscavenger. *Chemico-Biological Interactions* 157–158, 363–365.

Chesne, P., Adenot, P.G., Viglietta, C., Baratte, M., Boulanger, L. and Renard, J.P. (2002) Cloned rabbits produced by nuclear transfer from adult somatic cells. *Nature Biotechnology* 20, 366–369.

Cibelli, J.B., Stice, S.L., Golueke, P.J., Kane, J.J., Jerry, J., Blackwell, C., Ponce de Leon, F.A. and Robl, J.M. (1998) Cloned transgenic calves produced from nonquiescent fetal fibroblasts. *Science* 280, 1256–1258.

Clark, A., Bessos, H., Bishop, J., Brown, P., Harris, S., Lathe, R., McClenaghan, M., Prowse, C., Simons, J., Whitehaw, C. and Wilmut, I. (1989) Expression of human anti-hemophilic factor IX in the milk of transgenic sheep. *Bio/Technology* 7, 487–492.

Clark, A.J. (1998) The mammary gland as a bioreactor: expression, processing, and production of recombinant proteins. *Journal of Mammary Gland Biology and Neoplasia* 3, 337–350.

Clark, J. and Whitelaw, B. (2003) A future for transgenic livestock. *Nature Reviews Genetics* 4, 825–833.

Committee on Identifying and Assessing Unintended Effects of Genetically Engineered Foods on Human Health (2004) *Safety of Genetically Engineered Foods: Approaches to Assessing Unintended Health Effects*. National Academies Press, Washington, DC.

Cyranoski, D. (2003) Koreans rustle up madness-resistant cows. *Nature* 426, 743.

Damak, S., Su, H., Jay, N.P. and Bullock, D.W. (1996) Improved wool production in transgenic sheep expressing insulin-like growth factor 1. *Biotechnology (NY)* 14, 185–188.

Dean, W., Santos, F., Stojkovic, M., Zakhartchenko, V., Walter, J., Wolf, E. and Reik, W. (2001) Conservation of methylation reprogramming in mammalian development: aberrant reprogramming in cloned embryos. *Proceedings of the National Academy of Sciences of the United States of America* 98, 13734–13738.

Dean, W., Santos, F. and Reik, W. (2003) Epigenetic reprogramming in early mammalian development and following somatic nuclear transfer. *Seminars in Cell and Developmental Biology* 14, 93–100.

Denning, C., Burl, S., Ainslie, A., Bracken, J., Dinnyes, A., Fletcher, J., King, T., Ritchie, M., Ritchie, W.A., Rollo, M., de Sousa, P., Travers, A., Wilmut, I. and Clark, A.J. (2001a) Deletion of the alpha(1,3)galactosyl transferase (GGTA1) gene and the prion protein (PrP) gene in sheep. *Nature Biotechnology* 19, 559–562.

Denning, C., Dickinson, P., Burl, S., Wylie, D., Fletcher, J. and Clark, A.J. (2001b) Gene targeting in primary fetal fibroblasts from sheep and pig. *Cloning Stem Cells* 3, 221–231.

Dobrinski, I. (2005) Germ cell transplantation and testis tissue xenografting in domestic animals. *Animal Reproduction Science* 89, 137–145.

Dus Santos, M.J. and Wigdorovitz, A. (2005) Transgenic plants for the production of veterinary vaccines. *Immunology and Cell Biology* 83, 229–238.

Dyck, M.K., Gagne, D., Ouellet, M., Senechal, J.F., Belanger, E., Lacroix, D., Sirard, M.A. and Pothier, F. (1999) Seminal vesicle production and secretion of growth hormone into seminal fluid. *Nature Biotechnology* 17, 1087–1090.

Dyck, M.K., Lacroix, D., Pothier, F. and Sirard, M.A. (2003) Making recombinant proteins in animals – different systems, different applications. *Trends in Biotechnology* 21, 394–399.

Einsiedel, E.F. (2005) Public perceptions of transgenic animals. *Revue Scientifique et Technique* 24, 149–157.

Enright, B.P., Kubota, C., Yang, X. and Tian, X.C. (2003) Epigenetic characteristics and development of embryos cloned from donor cells treated by trichostatin A or 5-aza-2′-deoxycytidine. *Biology of Reproduction* 69, 896–901.

Fadiel, A., Anidi, I. and Eichenbaum, K.D. (2005) Farm animal genomics and informatics: an update. *Nucleic Acids Research* 33, 6308–6318.

FAO/WHO (2003a) Guideline for the conduct of food safety assessment of foods derived from recombinant-DNA plants, www.codexalimentarius.net/download/standards/10021/CXG_045e.pdf (accessed 21 May 2006)

FAO/WHO (2003b) Principles for the risk analysis of foods derived from modern biotechnology, www.codexalimentarius.net/download/standards/10007/CXG_044e.pdf (accessed 21 May 2006)

FDA 2006. A Risk-Based Approach to Evaluate Animal Clones and Their Progeny - DRAFT. http://www.fda.gov/cvm/CloneRiskAssessment.htm [Date accessed 17 April 2007].

Fearon, A.M. Mayne, C.S., Beattie, J.A.M. and Bruce, D.W. (2004) Effect of level of oil inclusion in the diet of dairy cows at pasture on Animal performance and milk composition and properties. *Journal of the Science of Food and Agriculture* 84, 497–504.

Fishman, J.A. and Patience, C. (2004) Xenotransplantation: infectious risk revisited. *American Journal of Transplantation* 4, 1383–1390.

Forsberg, E.J. (2005) Commercial applications of nuclear transfer cloning: three examples. *Reproduction, Fertility and Development* 17, 59–68.

Forsyth, J.T., Troskie, H.E., Brophy, B., Wells, D.N. and Laible, G. (2005) Utilising preimplantation genetic diagnosis and OPU-IVP-ET to generate multiple progeny of predetermined genotype from cloned transgenic heifers. *Reproduction, Fertility and Development* 17, Abstract 330.

Gama, L.T., Smith, C. and Gibson, J.P. (1992) Transgene effects, introgression strategies and testing schemes in pigs. *Animal Production* 54, 427–440.

Gerngross, T.U. (2004) Advances in the production of human therapeutic proteins in yeasts and filamentous fungi. *Nature Biotechnology* 22, 1409–1414.

Golding, M.C., Long, C.R., Carmell, M.A., Hannon, G.J. and Westhusin, M.E. (2006) Suppression of prion protein in livestock by RNA interference. *Proceedings of the National Academy of Sciences of the United States of America* 103, 5285–5290.

Golovan, S.P., Meidinger, R.G., Ajakaiye, A., Cottrill, M., Wiederkehr, M.Z., Barney, D.J., Plante, C., Pollard, J.W., Fan, M.Z., Hayes, M.A., Laursen, J., Hjorth, J.P., Hacker, R.R., Phillips, J.P. and Forsberg, C.W. (2001) Pigs expressing salivary phytase produce low-phosphorus manure. *Nature Biotechnology* 19, 741–745.

Grobet, L., Pirottin, D., Farnir, F., Poncelet, D., Royo, L.J., Brouwers, B., Christians, E., Desmecht, D., Coignoul, F., Kahn, R. and Georges, M. (2003) Modulating skeletal muscle mass by postnatal, muscle-specific inactivation of the myostatin gene. *Genesis* 35, 227–238.

Grosse-Hovest, L., Muller, S., Minoia, R., Wolf, E., Zakhartchenko, V., Wenigerkind, H., Lassnig, C., Besenfelder, U., Muller, M., Lytton, S.D., Jung, G. and Brem, G. (2004) Cloned transgenic farm animals produce a bispecific antibody for T cell-mediated tumor cell killing. *Proceedings of the National Academy of Sciences of the United States of America* 101, 6858–6863.

GTC (2006) ATryn® – recombinant human anti-thrombin, www.transgenics.com/products/atryn.html (accessed 21 May 2006)

Gutierrez-Adan, A., Maga, E.A., Meade, H., Shoemaker, C.F., Medrano, J.F., Anderson, G.B. and Murray, J.D. (1996) Alterations of the physical characteristics of milk from transgenic mice producing bovine kappa-casein. *Journal of Dairy Science* 79, 791–799.

Hammer, R.E., Pursel, V.G., Rexroad, C.E. Jr, Wall, R.J., Bolt, D.J., Ebert, K.M., Palmiter, R.D. and Brinster, R.L. (1985) Production of transgenic rabbits, sheep and pigs by microinjection. *Nature* 315, 680–683.

Hamra, F.K., Chapman, K.M., Nguyen, D.M., Williams-Stephens, A.A., Hammer, R.E. and Garbers, D.L. (2005) Self renewal, expansion, and transfection of rat spermatogonial stem cells in culture. *Proceedings of the National Academy of Sciences of the United States of America* 102, 17430–17435.

Harris, A. (1997) Towards an ovine model of cystic fibrosis. *Human Molecular Genetics* 6, 2191–2194.

Hill, J.R. and Dobrinski, I. (2006) Male germ cell transplantation in livestock. *Reproduction, Fertility and Development* 18, 13–18.

Hill, J.R., Burghardt, R.C., Jones, K., Long, C.R., Looney, C.R., Shin, T., Spencer, T.E., Thompson, J.A., Winger, Q.A. and Westhusin, M.E. (2000) Evidence for placental abnormality as the major cause of mortality in first-trimester somatic cell cloned bovine fetuses. *Biology of Reproduction* 63, 1787–1794.

Hofmann, A., Zakhartchenko, V., Weppert, M., Sebald, H., Wenigerkind, H., Brem, G., Wolf, E. and Pfeifer, A. (2004) Generation of transgenic cattle by lentiviral gene transfer into oocytes. *Biology of Reproduction* 71, 405–409.

Honaramooz, A., Behboodi, E., Megee, S.O., Overton, S.A., Galantino-Homer, H., Echelard, Y. and Dobrinski, I. (2003) Fertility and germline transmission of donor haplotype following germ cell transplantation in immunocompetent goats. *Biology of Reproduction* 69, 1260–1264.

Houdebine, L.M. (2000) Transgenic animal bioreactors. *Transgenic Research* 9, 305–320.

Humpherys, D., Eggan, K., Akutsu, H., Friedman, A., Hochedlinger, K., Yanagimachi, R., Lander, E.S., Golub, T.R. and Jaenisch, R. (2002) Abnormal gene expression in cloned mice derived from embryonic stem cell and cumulus cell nuclei. *Proceedings of the National Academy of Sciences of the United States of America* 99, 12889–12894.

Ivarie, R. (2003) Avian transgenesis: progress towards the promise. *Trends in Biotechnology* 21, 14–19.

Joshi, L. and Lopez, L.C. (2005) Bioprospecting in plants for engineered proteins. *Current Opinion in Plant Biology* 8, 223–226.

Jost, B., Vilotte, J.L., Duluc, I., Rodeau, J.L. and Freund, J.N. (1999) Production of low-lactose milk by ectopic expression of intestinal lactase in the mouse mammary gland. *Nature Biotechnology* 17, 160–164.

Kambadur, R., Sharma, M., Smith, T.P. and Bass, J.J. (1997) Mutations in myostatin (GDF8) in double-muscled Belgian Blue and Piedmontese cattle. *Genome Research* 7, 910–916.

Kanatsu-Shinohara, M., Ogonuki, N., Iwano, T., Lee, J., Kazuki, Y., Inoue, K., Miki, H., Takehashi, M., Toyokuni, S., Shinkai, Y., Oshimura, M., Ishino, F., Ogura, A. and Shinohara, T. (2005) Genetic and epigenetic properties of mouse male germline stem cells during long-term culture. *Development* 132, 4155–4163.

Kaneko, T., Moisyadi, S., Suganuma, R., Hohn, B., Yanagimachi, R. and Pelczar, P. (2005) Recombinase-mediated mouse transgenesis by intracytoplasmic sperm injection. *Theriogenology* 64, 1704–1715.

Kang, Y.K., Koo, D.B., Park, J.S., Choi, Y.H., Chung, A.S., Lee, K.K. and Han, Y.M. (2001) Aberrant methylation of donor genome in cloned bovine embryos. *Nature Genetics* 28, 173–177.

Karatzas, C.N. (2003) Designer milk from transgenic clones. *Nature Biotechnology* 21, 138–139.

Kelly, L. (2005) The safety assessment of foods from transgenic and cloned animals using the comparative approach. *Revue Scientifique et Technique* 24, 61–74.

Kerr, D.E., Liang, F., Bondioli, K.R., Zhao, H., Kreibich, G., Wall, R.J. and Sun, T.T. (1998) The bladder as a bioreactor: urothelium production and secretion of growth hormone into urine. *Nature Biotechnology* 16, 75–79.

Kishigami, S., Mizutani, E., Ohta, H., Hikichi, T., Thuan, N.V., Wakayama, S., Bui, H.T. and Wakayama, T. (2006) Significant improvement of mouse cloning technique by treatment with trichostatin A after somatic nuclear transfer. *Biochemical and Biophysical Research Communications* 340, 183–189.

Kochendoerfer, G.G., Chen, S.Y., Mao, F., Cressman, S., Traviglia, S., Shao, H., Hunter, C.L., Low, D.W., Cagle, E.N., Carnevali, M., Gueriguian, V., Keogh, P.J., Porter, H., Stratton, S.M., Wiedeke, M.C., Wilken, J., Tang, J., Levy, J.J., Miranda, L.P., Crnogorac, M.M., Kalbag, S., Botti, P., Schindler-Horvat, J., Savatski, L., Adamson, J.W., Kung, A., Kent, S.B. and Bradburne, J.A. (2003) Design and chemical synthesis of a homogeneous polymer-modified erythropoiesis protein. *Science* 299, 884–887.

Kuroiwa, Y., Kasinathan, P., Choi, Y.J., Naeem, R., Tomizuka, K., Sullivan, E.J., Knott, J.G., Duteau, A., Goldsby, R.A., Osborne, B.A., Ishida, I. and Robl, J.M. (2002) Cloned transchromosomic calves producing human immunoglobulin. *Nature Biotechnology* 20, 889–894.

Kuroiwa, Y., Kasinathan, P., Matsushita, H., Sathiyaselan, J., Sullivan, E.J., Kakitani, M., Tomizuka, K., Ishida, I. and Robl, J.M. (2004) Sequential targeting of the genes encoding immunoglobulin-mu and prion protein in cattle. *Nature Genetics* 36, 775–780.

Kuwaki, K., Tseng, Y.L., Dor, F.J., Shimizu, A., Houser, S.L., Sanderson, T.M., Lancos, C.J., Prabharasuth, D.D., Cheng, J., Moran, K., Hisashi, Y., Mueller, N., Yamada, K., Greenstein, J.L., Hawley, R.J., Patience, C., Awwad, M., Fishman, J.A., Robson, S.C., Schuurman, H.J., Sachs, D.H. and Cooper, D.K. (2005) Heart transplantation in baboons using alpha1,3-galactosyltransferase gene-knockout pigs as donors: initial experience. *Nature Medicine* 11, 29–31.

Lai, L., Kang, J.X., Li, R., Wang, J., Witt, W.T., Yong, H.Y., Hao, Y., Wax, D.M., Murphy, C.N., Rieke, A., Samuel, M., Linville, M.L., Korte, S.W., Evans, R.W., Starzl, T.E., Prather, R.S. and Dai, Y. (2006) Generation of cloned transgenic pigs rich in omega-3 fatty acids. *Nature Biotechnology* 24, 435–436.

Laible, G., Wagner, S. and Alderson, J. (2006) Oligonucleotide-mediated gene modification and its promise for animal agriculture. *Gene* 366, 17–26.

Lanza, R.P., Cooper, D.K. and Chick, W.L. (1997) Xenotransplantation. *Scientific American* 277, 54–59.

Lavitrano, M., Camaioni, A., Fazio, V.M., Dolci, S., Farace, M.G. and Spadafora, C. (1989) Sperm cells as vectors for introducing foreign DNA into eggs: genetic transformation of mice. *Cell* 57, 717–723.

Lee, R.S., Peterson, A.J., Donnison, M.J., Ravelich, S., Ledgard, A.M., Li, N., Oliver, J.E., Miller, A.L., Tucker, F.C., Breier, B. and Wells, D.N. (2004) Cloned cattle fetuses with the same nuclear genetics are more variable than contemporary half-siblings resulting from artificial insemination and exhibit fetal and placental growth deregulation even in the first trimester. *Biology of Reproduction* 70, 1–11.

Li, Z., Sun, X., Chen, J., Liu, X., Wisely, S.M., Zhou, Q., Renard, J.P., Leno, G.H. and Engelhardt, J.F. (2006) Cloned ferrets produced by somatic cell nuclear transfer. *Developmental Biology* 293, 439–448.

Ma, J.K., Drake, P.M. and Christou, P. (2003) The production of recombinant pharmaceutical proteins in plants. *Nature Reviews Genetics* 4, 794–805.

Maga, E.A. (2001) The use of recombinase proteins to generate transgenic large animals. *Cloning Stem Cells* 3, 233–241.

Maga, E.A., Shoemaker, C.F., Rowe, J.D., Bondurant, R.H., Anderson, G.B. and Murray, J.D. (2006) Production and processing of milk from transgenic goats expressing human lysozyme in the mammary gland. *Journal of Dairy Science* 89, 518–524.

Mahmoud, T.H., McCuen, B.W. II, Hao, Y., Moon, S.J., Tatebayashi, M., Stinnett, S., Petters, R.M. and Wong, F. (2003) Lensectomy and vitrectomy decrease the rate of photoreceptor loss in rhodopsin P347L transgenic pigs. *Graefes Archive for Clinical and Experimental Ophthalmology* 241, 298–308.

Martin, M., Adams, C. and Wiseman, B. (2004) Pre-weaning performance and health of pigs born to cloned (fetal cell derived) swine versus non-cloned swine. *Theriogenology* 62, 113–122.

Massoud, M., Attal, J., Thepot, D., Pointu, H., Stinnakre, M.G., Theron, M.C., Lopez, C. and Houdebine, L.M. (1996) The deleterious effects of human erythropoietin gene driven by the rabbit whey acidic protein gene promoter in transgenic rabbits. *Reproduction, Nutrition, Development* 36, 555–563.

McCreath, K.J., Howcroft, J., Campbell, K.H., Colman, A., Schnieke, A.E. and Kind, A.J. (2000) Production of gene-targeted sheep by nuclear transfer from cultured somatic cells. *Nature* 405, 1066–1069.

Mitra, A., Hruska, K.S., Wellnitz, O., Kerr, D.E., Capuco, A.V. and Wall, R.J. (2003) Expression of lysostaphin in milk of transgenic mice affects the growth of neonates. *Transgenic Research* 12, 597–605.

Muller, M. and Brem, G. (1998) Transgenic approaches to the increase of disease resistance in farm animals. *Revue Scientifique et Technique* 17, 365–378.

Nagashima, H., Giannakis, C., Ashman, R.J. and Nottle, M.B. (2004) Sex differentiation and germ cell production in chimeric pigs produced by inner cell mass injection into blastocysts. *Biology of Reproduction* 70, 702–707.

Niemann, H., Halter, R., Carnwath, J.W., Herrmann, D., Lemme, E. and Paul, D. (1999) Expression of human blood clotting factor VIII in the mammary gland of transgenic sheep. *Transgenic Research* 8, 237–247.

Niemann, H., Wrenzycki, C., Lucas-Hahn, A., Brambrink, T., Kues, W.A. and Carnwath, J.W. (2002) Gene expression patterns in bovine in vitro-produced and nuclear transfer-derived embryos and their implications for early development. *Cloning Stem Cells* 4, 29–38.

Nottle, M.B., Nagashima, H., Verma, P.J., Du, Z.T., Grupen, C.G., McIlfatrick, S.M., Ashman, R.J., Harding, M.P., Giannakis, C., Wigley, P.L., Lyons, I.G., Harrison, D.T., Luxford, B.G., Campbell, R.G., Crawford, R.J. and Robins, A.J. (1999) Production and analysis of transgenic pigs containing a metallothionein porcine growth hormone gene construct. In: Murray, J.D., Anderson, G.B., Oberbauer, A.M. and McGloughlin, M.M. (eds) *Transgenic Animals in Agriculture*. CABI Publishing, Wallingford, Oxfordshire, pp. 145–156.

Ogura, A., Inoue, K., Ogonuki, N., Lee, J., Kohda, T. and Ishino, F. (2002) Phenotypic effects of somatic cell cloning in the mouse. *Cloning Stem Cells* 4, 397–405.

Pace, M.M., Augenstein, M.L., Betthauser, J.M., Childs, L.A., Eilertsen, K.J., Enos, J.M., Forsberg, E.J., Golueke, P.J., Graber, D.F., Kemper, J.C., Koppang, R.W., Lange, G., Lesmeister, T.L., Mallon, K.S., Mell, G.D., Misica, P.M., Pfister-Genskow, M., Strelchenko, N.S., Voelker, G.R., Watt, S.R. and Bishop, M.D. (2002) Ontogeny of cloned cattle to lactation. *Biology of Reproduction* 67, 334–339.

Palmquist, D.L., Beaulieu, A.D. and Barbano, D.M. (1993) Feed and animal factors influencing milk fat composition. *Journal of Dairy Science* 76, 1753–1771.

Parker, M.H., Birck-Wilson, E., Allard, G., Masiello, N., Day, M., Murphy, K.P., Paragas, V., Silver, S. and Moody, M.D. (2004) Purification and characterization of a recombinant version of human alpha-fetoprotein expressed in the milk of transgenic goats. *Protein Expression and Purification* 38, 177–183.

Perrier, V., Kaneko, K., Safar, J., Vergara, J., Tremblay, P., DeArmond, S.J., Cohen, F.E., Prusiner, S.B. and Wallace, A.C. (2002) Dominant-negative inhibition of prion replication in transgenic mice. *Proceedings of the National Academy of Sciences of the United States of America* 99, 13079–13084.

Perry, A.C., Rothman, A., de las Heras, J.I., Feinstein, P., Mombaerts, P., Cooke, H.J. and Wakayama, T. (2001) Efficient metaphase II transgenesis with different transgene archetypes. *Nature Biotechnology* 19, 1071–1073.

Petters, R.M. (1994) Transgenic livestock as genetic models of human disease. *Reproduction, Fertility and Development* 6, 643–645.

Peura, T.T., Hartwich, K.M., Hamilton, H.M. and Walker, S.K. (2003) No differences in sheep somatic cell nuclear transfer outcomes using serum-starved or actively growing donor granulosa cells. *Reproduction, Fertility and Development* 15, 157–165.

Pharming (2006) Pharming confirms positive outlook for rhC1INH, www.pharming.com/index.php?act=show&pg=265 (accessed 21 May 2006)

Phelps, C.J., Koike, C., Vaught, T.D., Boone, J., Wells, K.D., Chen, S.H., Ball, S., Specht, S.M., Polejaeva, I.A., Monahan, J.A., Jobst, P.M., Sharma, S.B., Lamborn, A.E., Garst, A.S., Moore, M., Demetris, A.J., Rudert, W.A., Bottino, R., Bertera, S., Trucco, M., Starzl, T.E., Dai, Y. and Ayares, D.L. (2003) Production of alpha 1,3-galactosyltransferase-deficient pigs. *Science* 299, 411–414.

Piedrahita, J.A. (2000) Targeted modification of the domestic animal genome. *Theriogenology* 53, 105–116.

Polejaeva, I.A., Chen, S.H., Vaught, T.D., Page, R.L., Mullins, J., Ball, S., Dai, Y., Boone, J., Walker, S., Ayares, D.L., Colman, A. and Campbell, K.H. (2000) Cloned pigs produced by nuclear transfer from adult somatic cells. *Nature* 407, 86–90.

Pollock, D.P., Kutzko, J.P., Birck-Wilson, E., Williams, J.L., Echelard, Y. and Meade, H.M. (1999) Transgenic milk as a method for the production of recombinant antibodies. *Journal of Immunological Methods* 231, 147–157.

Powell, K. (2003) Barnyard biotech – lame duck or golden goose? *Nature Biotechnology* 21, 965–967.

Prelle, K., Zink, N. and Wolf, E. (2002) Pluripotent stem cells – model of embryonic development, tool for gene targeting, and basis of cell therapy. *Anatomia Histologia Embryologia* 31, 169–186.

Pursel, V.G. and Rexroad, C.E. Jr (1993) Status of research with transgenic farm animals. *Journal of Animal Science* 71 Suppl 3, 10–19.

Pursel, V.G., Pinkert, C.A., Miller, K.F., Bolt, D.J., Campbell, R.G., Palmiter, R.D., Brinster, R.L. and Hammer, R.E. (1989) Genetic engineering of livestock. *Science* 244, 1281–1288.

Pursel, V.G., Wall, R.J., Mitchell, A.D., Elasser, T.H., Solomon, M.B., Coleman, M.E., DeMayo, F. and Schwartz, R.J. (1999) Expression of insulin-like growth factor-1 in skeletal muscle of transgenic swine. In: Murray, J.D., Anderson, G.B., Oberbauer, A.M. and McGloughlin, M.M. (eds) *Transgenic Animals in Agriculture*. CABI Publishing, Wallingford, Oxfordshire, pp. 131–144.

Reddy, M.S., Chen, F., Shadle, G., Jackson, L., Aljoe, H. and Dixon, R.A. (2005) Targeted down-regulation of cytochrome P450 enzymes for forage quality improvement in alfalfa (*Medicago sativa* L.). *Proceedings of the National Academy of Sciences of the United States of America* 102, 16573–16578.

Rideout, W.M. III, Eggan, K. and Jaenisch, R. (2001) Nuclear cloning and epigenetic reprogramming of the genome. *Science* 293, 1093–1098.

Rudenko, L., Matheson, J.C., Adams, A.L., Dubbin, E.S. and Greenlees, K.J. (2004) Food consumption risks associated with animal clones: what should be investigated? *Cloning Stem Cells* 6, 79–93.

Rudolph, N.S. (1999) Biopharmaceutical production in transgenic livestock. *Trends in Biotechnology* 17, 367–374.

Saeki, K., Matsumoto, K., Kinoshita, M., Suzuki, I., Tasaka, Y., Kano, K., Taguchi, Y., Mikami, K., Hirabayashi, M., Kashiwazaki, N., Hosoi, Y., Murata, N. and Iritani, A. (2004) Functional expression of a Delta12 fatty acid desaturase gene from spinach in transgenic pigs. *Proceedings of the National Academy of Sciences of the United States of America* 101, 6361–6366.

Sahi, T. (1994) Genetics and epidemiology of adult-type hypolactasia. *Scandinavian Journal of Gastroenterology* 202, (suppl.) 7–20.

Santos, F., Zakhartchenko, V., Stojkovic, M., Peters, A., Jenuwein, T., Wolf, E., Reik, W. and Dean, W. (2003) Epigenetic marking correlates with developmental potential in cloned bovine preimplantation embryos. *Current Biology* 13, 1116–1121.

Schmidt, C. (2006) Belated approval of first recombinant protein from animal. *Nature Biotechnology* 24, 877.

Schnieke, A.E., Kind, A.J., Ritchie, W.A., Mycock, K., Scott, A.R., Ritchie, M., Wilmut, I., Colman, A. and Campbell, K.H. (1997) Human factor IX transgenic sheep produced by transfer of nuclei from transfected fetal fibroblasts. *Science* 278, 2130–2133.

Sedivy, J.M. and Dutriaux, A. (1999) Gene targeting and somatic cell genetics – a rebirth or a coming of age? *Trends in Genetics* 15, 88–90.

Sharma, A., Naziruddin, B., Cui, C., Martin, M.J., Xu, H., Wan, H., Lei, Y., Harrison, C., Yin, J., Okabe, J., Mathews, C., Stark, A., Adams, C.S., Houtz, J., Wiseman, B.S., Byrne, G.W. and Logan, J.S. (2003) Pig cells that lack the gene for alpha1-3 galactosyltransferase express low levels of the gal antigen. *Transplantation* 75, 430–436.

Shimozawa, N., Ono, Y., Kimoto, S., Hioki, K., Araki, Y., Shinkai, Y., Kono, T. and Ito, M. (2002) Abnormalities in cloned mice are not transmitted to the progeny. *Genesis* 34, 203–207.

Siewerdt, F., Eisen, E.J. and Murray, J.D. (1999) Direct and correlated responses to short-term selection for 8-week body weight in lines of transgenic (oMt1a-oGH) mice. In: Murray, J.D., Anderson, G.B., Oberbauer, A.M. and McGloughlin, M.M. (eds) *Transgenic Animals in Agriculture*. CABI Publishing, Wallingford, Oxfordshire, pp. 231–250.

Small, B. (2005) Genetic engineering: New Zealand public attitudes 2001, 2003, and 2005. *Proceedings of the Talking Biotechnology – Reflecting on Science in Society Conference*. 29 November–2 December, University of Wellington, Wellington, New Zealand.

Smeaton, D.C., Harris, B.L., Xu, Z.Z. and Vivanco, W.H. (2003) Factors affecting commercial application of embryo technologies in New Zealand: a modelling approach. *Theriogenology* 59, 617–634.

Smeaton, D.C., Woodward, S.J.R., Wells, D.N. and Smith, J.F. (2004) Gross margin analyses of using cloned embryos in cattle. *Proceedings of the New Zealand Society of Animal Production* 64, 262–267.

Smith, S.L., Everts, R.E., Tian, X.C., Du, F., Sung, L.Y., Rodriguez-Zas, S.L., Jeong, B.S., Renard, J.P., Lewin, H.A. and Yang, X. (2005) Global gene expression profiles reveal significant nuclear reprogramming by the blastocyst stage after cloning. *Proceedings of the National Academy of Sciences of the United States of America* 102, 17582–17587.

Sorrell, D.A. and Kolb, A.F. (2005) Targeted modification of mammalian genomes. *Biotechnology Advances* 23, 431–469.

Stinnakre, M.G., Vilotte, J.L., Soulier, S. and Mercier, J.C. (1994) Creation and phenotypic analysis of alpha-lactalbumin-deficient mice. *Proceedings of the National Academy of Sciences of the United States of America* 91, 6544–6548.

Su, H.Y., Jay, N.P., Gourley, T.S., Kay, G.W. and Damak, S. (1998) Wool production in transgenic sheep: results from first-generation adults and second-generation lambs. *Animal Biotechnology* 9, 135–147.

Sullivan, E.J., Kasinathan, S., Kasinathan, P., Robl, J.M. and Collas, P. (2004) Cloned calves from chromatin remodeled in vitro. *Biology of Reproduction* 70, 146–153.

Takahashi, S. and Ito, Y. (2004) Evaluation of meat products from cloned cattle: biological and biochemical properties. *Cloning Stem Cells* 6, 165–171.

Tamashiro, K.L., Wakayama, T., Akutsu, H., Yamazaki, Y., Lachey, J.L., Wortman, M.D., Seeley, R.J., D'Alessio, D.A., Woods, S.C., Yanagimachi, R. and Sakai, R.R. (2002) Cloned mice have an obese phenotype not transmitted to their offspring. *Nature Medicine* 8, 262–267.

Thomas, C.E., Ehrhardt, A. and Kay, M.A. (2003) Progress and problems with the use of viral vectors for gene therapy. *Nature Reviews Genetics* 4, 346–358.

Thomassen, E.A., van Veen, H.A., van Berkel, P.H., Nuijens, J.H. and Abrahams, J.P. (2005) The protein structure of recombinant human lactoferrin produced in the milk of transgenic cows closely matches the structure of human milk-derived lactoferrin. *Transgenic Research* 14, 397–405.

Tian, X.C., Kubota, C., Sakashita, K., Izaike, Y., Okano, R., Tabara, N., Curchoe, C., Jacob, L., Zhang, Y., Smith, S., Bormann, C., Xu, J., Sato, M., Andrew, S. and Yang, X. (2005) Meat and milk compositions of bovine clones. *Proceedings of the National Academy of Sciences of the United States of America* 102, 6261–6266.

Tucker, F., Eichenlaub, M., Miller, A., Oliver, J., Forsyth, J., Berg, M., Cockrem, K., Oback, B. and Wells, D. (2004) Increased cattle cloning efficiency with serum-starved compared to G1 somatic donor cells. *Biology of Reproduction, Special Issue: The 37th Annual Meeting of the Society for the Study of Reproduction*. 1–4 August 2004, Vancouver, Canada, Abstract 577.

van Berkel, P.H., Welling, M.M., Geerts, M., van Veen, H.A., Ravensbergen, B., Salaheddine, M., Pauwels, E.K., Pieper, F., Nuijens, J.H. and Nibbering, P.H. (2002) Large scale production of recombinant human lactoferrin in the milk of transgenic cows. *Nature Biotechnology* 20, 484–487.

Van den Hout, J.M., Kamphoven, J.H., Winkel, L.P., Arts, W.F., De Klerk, J.B., Loonen, M.C., Vulto, A.G., Cromme-Dijkhuis, A., Weisglas-Kuperus, N., Hop, W., Van Hirtum, H., Van Diggelen, O.P., Boer, M., Kroos, M.A., Van Doorn, P.A., Van der Voort, E., Sibbles, B., Van Corven, E.J., Brakenhoff, J.P., Van Hove, J., Smeitink, J.A., de Jong, G., Reuser, A.J. and Van der Ploeg, A.T. (2004) Long-term intravenous treatment of Pompe disease with recombinant human alpha-glucosidase from milk. *Pediatrics* 113, e448–457.

Van Reenen, C.G., Meuwissen, T.H., Hopster, H., Oldenbroek, K., Kruip, T.H. and Blokhuis, H.J. (2001) Transgenesis may affect farm animal welfare: a case for systematic risk assessment. *Journal of Animal Science* 79, 1763–1779.

Wall, R.J., Kerr, D.E. and Bondioli, K.R. (1997) Transgenic dairy cattle: genetic engineering on a large scale. *Journal of Dairy Science* 80, 2213–2224.

Wall, R.J., Powell, A.M., Paape, M.J., Kerr, D.E., Bannerman, D.D., Pursel, V.G., Wells, K.D., Talbot, N. and Hawk, H.W. (2005) Genetically enhanced cows resist intramammary *Staphylococcus aureus* infection. *Nature Biotechnology* 23, 445–451.

Walsh, M.K., Lucey, J.A., Govindasamy-Lucey, S., Pace, M.M. and Bishop, M.D. (2003) Comparison of milk produced by cows cloned by nuclear transfer with milk from non-cloned cows. *Cloning and Stem Cells* 5, 213–219.

Wang, L., Duan, E., Sung, L.Y., Jeong, B.S., Yang, X. and Tian, X.C. (2005) Generation and characterization of pluripotent stem cells from cloned bovine embryos. *Biology of Reproduction* 73, 149–155.

Ward, K.A. (2000) Transgene-mediated modifications to animal biochemistry. *Trends in Biotechnology* 18, 99–102.

Ward, K.A., Brownlee, A.G., Leish, Z. and Bonsing, J. (1993) Genetic manipulation for disease control. *Proceedings of the VII World Conference on Animal Production*. 28 June–2 July, Edmonton, Canada 1, 267–280.

Weimer, P.J. (1998) Manipulating ruminal fermentation: a microbial ecological perspective. *Journal of Animal Science* 76, 3114–3122.

Wells, D.N. (2003) Cloning in livestock agriculture. *Reproduction* Supplement 61, 131–150.

Wells, D.N., Misica, P.M., Tervit, H.R. and Vivanco, W.H. (1998) Adult somatic cell nuclear transfer is used to preserve the last surviving cow of the Enderby Island cattle breed. *Reproduction, Fertility and Development* 10, 369–378.

Wells, D.N., Laible, G., Tucker, F.C., Miller, A.L., Oliver, J.E., Xiang, T., Forsyth, J.T., Berg, M.C., Cockrem, K., L'Huillier, P.J., Tervit, H.R. and Oback, B. (2003a) Coordination between donor cell type and cell cycle stage improves nuclear cloning efficiency in cattle. *Theriogenology* 59, 45–59.

Wells, D.N., Oback, B. and Laible, G. (2003b) Cloning livestock: time to return to embryonic cells. *Trends in Biotechnology* 21, 428–432.

Wells, D.N., Forsyth, J.T., McMillan, V. and Oback, B. (2004) Review: the health of somatic cell cloned cattle and their offspring. *Cloning Stem Cells* 6, 101–110.

Wheeler, M.B., Bleck, G.T. and Donovan, S.M. (2001) Transgenic alteration of sow milk to improve piglet growth and health. *Reproduction* 58, suppl. 313–324.

Whitelaw, C.B. and Sang, H.M. (2005) Disease-resistant genetically modified animals. *Revue Scientifique et Technique* 24, 275–283.

Willadsen, S.M. (1986) Nuclear transplantation in sheep embryos. *Nature* 320, 63–65.

Williams, S.H., Sahota, V., Palmai-Pallag, T., Tebbutt, S.J., Walker, J. and Harris, A. (2003) Evaluation of gene targeting by homologous recombination in ovine somatic cells. *Molecular Reproduction and Development* 66, 115–125.

Wilmut, I., Schnieke, A.E., McWhir, J., Kind, A.J. and Campbell, K.H. (1997) Viable offspring derived from fetal and adult mammalian cells. *Nature* 385, 810–813.

Wilmut, I., Beaujean, N., De Sousa, P.A., Dinnyes, A., King, T.J., Paterson, L.A., Wells, D.N. and Young, L.E. (2002) Somatic cell nuclear transfer. *Nature* 419, 583–587.

Womack, J.E. (2005) Advances in livestock genomics: opening the barn door. *Genome Research* 15, 1699–1705.

Woodfield, D.R. and Easton, H.S. (2004) Advances in pasture plant breeding for animal productivity and health. *New Zealand Veterinary Journal* 52, 300–310.

Wright, G., Carver, A., Cottom, D., Reeves, D., Scott, A., Simons, P., Wilmut, I., Garner, I. and Colman, A. (1991) High level expression of active human alpha-1-antitrypsin in the milk of transgenic sheep. *Biotechnology (NY)* 9, 830–834.

Xue, F., Tian, X.C., Du, F., Kubota, C., Taneja, M., Dinnyes, A., Dai, Y., Levine, H., Pereira, L.V. and Yang, X. (2002) Aberrant patterns of X chromosome inactivation in bovine clones. *Nature Genetics* 31, 216–220.

Yamada, K., Yazawa, K., Shimizu, A., Iwanaga, T., Hisashi, Y., Nuhn, M., O'Malley, P., Nobori, S., Vagefi, P.A., Patience, C., Fishman, J., Cooper, D.K., Hawley, R.J., Greenstein, J., Schuurman, H.J., Awwad, M., Sykes, M. and Sachs, D.H. (2005) Marked prolongation of porcine renal xenograft survival in baboons through the use of alpha1,3-galactosyltransferase gene-knockout donors and the cotransplantation of vascularized thymic tissue. *Nature Medicine* 11, 32–34.

Yonai, M., Kaneyama, K., Miyashita, N., Kobayashi, S., Goto, Y., Bettpu, T. and Nagai, T. (2005) Growth, reproduction, and lactation in somatic cell cloned cow's with short telomeres. *Journal of Animal Science* 88, 4097–4110.

Yu, G., Chen, J., Yu, H., Liu, S., Chen, J., Xu, X., Sha, H., Zhang, X., Wu, G., Xu, S. and Cheng, G. (2006) Functional disruption of the prion protein gene in cloned goats. *Journal of General Virology* 87, 1019–1027.

Zhou, Q., Renard, J.P., Le Friec, G., Brochard, V., Beaujean, N., Cherifi, Y., Fraichard, A. and Cozzi, J. (2003) Generation of fertile cloned rats by regulating oocyte activation. *Science* 302, 1179.

Zuelke, K.A. (1998) Transgenic modification of cow's milk for value-added processing. *Reproduction, Fertility and Development* 10, 671–676.

8 Transforming Livestock with Transgenics: is Designed Resistance to Viral Diseases the Way Forward?

Tim Doran[1] and Luke Lambeth[1]
[1]CSIRO Livestock Industries, Australia

Abstract
A great deal of high-profile transgenic technology development for livestock species has occurred since the production of the first transgene in a livestock species was reported in 1985 (Hammer *et al.*, 1985). Despite these research efforts, the practical use of these animals has been limited and restricted to medical applications, such as producing pharmaceutical proteins in milk or eggs, rather than the agricultural applications that were originally envisioned. In this chapter we review the technology that is predicted to underpin the resurgence of research using transgenic livestock, namely DNA-delivered RNAi (ddRNAi). We also summarize the latest developments for engineering RNAi transgenes into the genomes of livestock species. Finally, we focus on how these technologies may come together to generate the first commercial transgenic livestock species, poultry that are resistant to avian influenza (AI). Many major challenges to public health have occurred as a result of the transfer of diseases from domestic farm animals to humans. The possibility of genetically engineering poultry to make them resistant to AI is attracting attention and has now become a real possibility with improved methods for genetic modification and the emergence of RNAi as an antiviral strategy. We make a case that this application may well be the first commercial reality for the use of transgenics to modify the genomes of livestock species.

Introduction

Clark and Whitelaw (2003) proposed in their review 'A future for transgenic livestock' that the advent of new methods such as RNA interference (RNAi) for modifying genomes will underpin a resurgence of research using transgenic livestock. They suggest that this will not only lead to an increase in our understanding of basic biology in commercial species, but may also be an important alternative to traditional breeding to enhance productivity traits and might lead to the generation of farm animals that are more resistant to infectious disease such as foot-and-mouth disease or influenza. They did not have to wait long. In early 2006, researchers demonstrated that they could 'knock down' the production of infectious prion proteins in livestock by using an innovative approach based on a combination of transgenic technology and RNAi (Golding *et al.*, 2006). This much anticipated union of the two transformational technologies could now enable scientists to genetically engineer livestock that are naturally resistant to prion-associated diseases such as mad cow disease, bovine spongiform encephalopathy (BSE). Gregory J. Hannon, one of the senior authors of the study, agrees with Clark and Whitelaw, by saying; 'I think it will be possible to do more quickly what selective breeders have been doing for a long time – creating animals with disease resistance and more advantageous properties for agriculture.'

A great deal of high-profile transgenic technology development for livestock species has occurred since the production of the first transgene in a livestock species was reported in 1985 (Hammer *et al.*, 1985). Despite these research efforts, the practical use of these animals has been limited and restricted to medical applications, such as producing pharmaceutical proteins in milk or eggs, rather than the agricultural applications that were originally envisioned. Many proof-of-principle studies have been carried out, such as the prion protein knock-down, but a commercial application of this technology has yet to make it to market. In this chapter we review the technology that was predicted to underpin the resurgence of research using transgenic livestock, namely DNA-delivered RNAi (ddRNAi). We also summarize the latest developments for engineering RNAi transgenes into the genomes of livestock species. Finally, we focus on how these technologies may come together to generate the first production transgenes, poultry that are resistant to avian influenza (AI). Many major challenges to public health have occurred as a result of the transfer of diseases from domestic farm animals to humans. The possibility of genetically engineering poultry to make them resistant to AI is attracting attention and has now become a real possibility with improved methods for genetic modification and the emergence of RNAi as an antiviral strategy. We make a case that this application may well be the first commercial reality for the use of transgenics to modify the genomes of livestock species.

Transforming Technologies

Developments in the generation of transgenic livestock

Ever since animals were first domesticated several thousand years ago, mankind has been indirectly genetically modifying these species for our benefit through selective breeding. To a large extent, the differences between today's livestock and their progenitors are testament to how successful this programme of selective improvement has been. Although robust and successful, conventional selection and breeding is limited because animals produced by mating selected individuals are a genetic mixture of their parents. Unknown or undesirable traits can be inadvertently co-selected. In addition, only those genetic loci that are present in the parents can be selected, which limits the range and extent of genetic improvement. Some traits such as fertility and disease resistance, compared to traits such as body weight or milk yield, are very difficult to quantify and improve upon using conventional selection and breeding. Gene addition through the use of transgenic technology has been seen for some time as a transformational way forward to overcome these limitations.

Several transgenic technologies have been developed and advanced for the genetic modification of livestock species in recent years (Clark and Whitelaw, 2003; Hunter *et al.*, 2005).

Pronuclear injection of DNA

This technique was developed in the mouse (Gordon and Ruddle, 1981) and has been adapted for the production of transgenic sheep, cattle, pigs and goats. The technique involves the direct introduction of a DNA construct into one of the two pronuclei of the fertilized egg. The zygotes are then transferred to pseudopregnant recipient females and the offspring screened for integration and expression of the transgene DNA. The efficiency of this method is low and usually only 3–5% of the animals born as a result carry the transgene (Nottle *et al.*, 2001). In addition, this technique results in the random integration of genes into the host chromosomes and variable expression of the transgenes as a result of positional effects related to the chromosomal sites of integration (Chicas and Macino, 2001). More recently, nuclear transfer techniques have been adapted to allow more precise modifications of the genome, such as the disruption of specific endogenous genes (Wilmut *et al.*, 1997).

Nuclear transfer technology

Since Dolly the sheep, nuclear transfer technology has been developed as another method for the production of transgenic animals. Nuclear transfer followed by the successful production of transgenic animals has now been achieved in cattle and pigs (Wilmut *et al.*, 2002) and was also the method used to develop the prion protein knock-down goat (Golding *et al.*, 2006). The technique of nuclear transfer in mammals involves removing the nucleus from an egg and replacing it with the nucleus from a donor cell. The reconstructed eggs are activated to induce the events that follow fertilization in a normal egg, and after a short period of *in vitro* culture, the resulting embryo in transferred to recipient females, where a low proportion of transferred embryos develop to term (Wilmut *et al.*, 1997). Animals that develop from eggs generated by nuclear transfer have the genotype of the cell used as the nuclear donor. Tissue culture cells can be used as nuclear donors. If these cells are genetically modified in culture, before nuclear transfer, resulting animals will carry the genetic modification. Nuclear transfer, as with pronuclear injection, is an inefficient method for modifying livestock with only 2–3% of cloned embryos surviving to term. The method, however, has significant advantages over pronuclear injection because specific genetic changes can be selected in donor cells before nuclear transfer and every liveborn animal will be transgenic.

Retroviruses

Retroviruses integrate into the chromosomes of their host cells and so became obvious candidates from transgene vectors (Barquinero *et al.*, 2004). All retroviral vector systems are basically similar in design. The *gag*, *pol* and *env* genes within the virus that encode essential viral proteins are removed and replaced with the specific transgene. Viral particles are then generated by co-expression of the vector-RNA genome and the genes encoding the viral proteins in tissue culture cells, followed by recovery of viral particles from the culture medium. Initially these recombinant viral particles were used to produce transgenic mice via injection of concentrated viral preparations into the perivitelline space of zygotes that were then transferred to recipient mothers (Barquinero *et al.*, 2004). Recently, major advances in the use of retroviral vectors have been made using vectors derived from the lentivirus class of retroviruses (Buchschacher and Wong-Staal, 2000). These vectors, developed mainly for gene therapy, can transduce non-dividing cells. More importantly, from the perspective of producing transgenic livestock, marked increases in the efficiency of transgene delivery, plus reliable transgene expression, were demonstrated when these vectors were used to generate transgenic mice (Lois *et al.*, 2002). An additional improvement was the development of the process of ‘pseudotyping’, in which the lentiviral coat protein is replaced with that from another virus, which can alter the host range of the vector and increase the efficiency of transduction of the target cells (Kang *et al.*, 2002). Such improvements and the demonstration of high production efficiencies of transgenic mice led to testing of these vectors for the production of transgenic livestock and chickens. Recent successes include the generation of transgenic cattle by the infection of bovine oocytes with lentiviral vectors followed by *in vitro* fertilization (Hofmann *et al.*, 2003), transgenic pigs by the injection of virus into the perivitelline space of zygotes (Whitelaw *et al.*, 2004), transgenic chickens by the transduction of embryos in newly laid eggs (McGrew *et al.*, 2004) and transgenic goats by infection of donor tissue culture cells prior to nuclear transfer (Golding *et al.*, 2006). Overall, the efficiencies of production of transgenic animals achieved using these lentiviral methods are significantly higher than those achieved using the other described methods. In pigs and cattle, frequencies of production of 10–30% are possible, in contrast to 1–5% by pronuclear injection. In the chick, the production frequency of transgenic founders using lentiviral vectors is estimated to be ten- to 100-fold higher than previous methods (McGrew *et al.*, 2004).

The use of lentiviral vectors does have some limitations, such as the possible integration of retroviruses in close proximity to potential oncogenes, followed by the activation of these oncogenes to convert a normal cell into a tumour cell (although the statistical chances of such an insertion are estimated to be in the range of 10^{-5})(Baum *et al.*, 2004).

Sperm-mediated gene transfer

Sperm-mediated gene transfer (SMGT) is a very appealing transgenic technology as it is far simpler than the other methods. However, until recent developments, SMGT has not been reliable enough to attract the more general interest of the other methods. SMGT was first suggested as early as 1971 (Brackett *et al.*, 1971) and since then numerous reports have been made showing successful sperm-mediated transfer of foreign DNA into both non-mammalian and mammalian animals, but with inefficient germ-line transmission and unreliable transgene expression (Smith, 1999). This has now been greatly improved upon by ways of increasing DNA binding to sperm without interfering with fertilization. Chang *et al.* (2002) reported the development of a sperm-reactive monoclonal antibody that can be used as a cross linker to facilitate the binding of exogenous DNA to sperm (linker-based sperm-mediated gene transfer – LB-SMGT). This antibody is a basic protein that binds to DNA through ionic interaction, allowing exogenous DNA to be linked specifically to sperm. After fertilization of an egg with cross-linked sperm, the DNA is shown to be successfully integrated into the genome of viable pig offspring with germ-line transfer to the F1 generation at the highly efficient rate of 37.5%. This linker antibody is reactive to a surface antigen on sperm of not only pigs but other livestock species including chicken, cow, goat and sheep. Preliminary data using LB-SMGT through artificial insemination in chickens indicated the presence of a transgene in 49% (44/90) of chicken embryos by PCR analysis. Further to this, expression of the transgene was detected in 53% (18/34) of the chicks in the F0 generation.

Resistance mechanisms of livestock to viruses – genetic engineering strategies

A number of approaches have been developed with the potential that insertion of a transgene into the genome of an animal may confer resistance to specific viruses. The goal of developing resistance to viral infection is to engineer animals that express molecules that block productive infection and therefore reduce the risk of transfer from animal to animal or animal to human. The extent of the research effort to genetically engineer new resistance mechanisms in animals is much smaller than that in plants. The major focus in animals has been to block viral attachment and penetration into a host cell by developing transgenes that: (i) produce viral antireceptor proteins to block cellular receptors; or (ii) replace host receptor genes with a modified form that is able to perform the receptor's physiological function but does not allow the attachment of the virus (Gavora, 1996).

In a recent review by Hunter *et al.* (2005), alternative strategies were suggested specifically for engineering resistance to avian influenza in poultry and these are described in the following sections.

The Mx gene

The Mx genes of vertebrates were first discovered in mice through their ability to confer a potent antiviral state in response to influenza virus (Staeheli *et al.*, 1984). The mechanism of action is not fully understood, but some evidence indicates that Mx interacts with the viral polymerase during influenza virus infection. Mx genes have been characterized in pigs and cattle and the use of transgenesis to introduce functional Mx genes to increase resistance against viral diseases that are susceptible to Mx function has been investigated in pigs (Muller *et al.*, 1992). An Mx gene has been identified in the chicken but the allele present in most commercial lines is apparently not functional, owing to a single amino acid substitution (Ko *et al.*, 2002). The mouse *Mx1* gene expressed in chick-embryo fibroblasts inhibited influenza virus replication, suggesting that Mx genes from different species are functional in the chick (Garber *et al.*, 1991). The transgenic introduction of a functional Mx gene into chickens might increase the resistance of commercial poultry to influenza virus.

RNA decoys

RNA decoys are short RNA molecules that mimic the binding sites for RNAs on RNA binding proteins. If they are expressed at high levels they prevent the function of the protein by acting as competitive inhibitors to the natural RNA partner molecule. Influenza virus RNA polymerase consists of three viral proteins that form an RNA-binding-protein complex (Tiley *et al.*, 1994). Decoy RNAs that mimic binding sites that are essential for the formation of this complex are potent inhibitors of the viral RNA polymerase in tissue culture cells (Luo *et al.*, 1997). Decoy RNAs can be expressed from a single DNA molecule, therefore, transgenes designed to express decoy RNAs can be introduced into the chicken genome using transgenic methods described above.

DNA-delivered RNA interference (ddRNAi)

The recent discovery of the sequence-specific method of gene suppression known as RNA interference (RNAi) has been met with a revitalized interest in the development of nucleic acid-based gene inhibition approaches, which has been reflected in a wave of publications. First discovered in mammalian cells in 2001 (Elbashir *et al.*, 2001a), due to its highly specific and effective nature, RNAi has fast become the primary means by which researchers target specific genes for degradation both in mammalian cell culture and *in vivo*. The use of this technology has allowed the rapid identification and validation of gene function. But beyond this, the use of RNAi has great promise in the development of sequence-specific therapeutics against a wide range of diseases including viral disease by the specific targeting of viral transcripts.

In this section we give an overview of the various aspects of RNAi, including the origins, discovery, the cellular mechanism, and the means by which researchers have utilized this technology to initiate gene knockdown. Of major focus will be the development of DNA-delivered RNAi (ddRNAi) for expression of RNAi molecules from a transgene. The use of RNA polymerase III promoters for this purpose and the properties of these promoters that make their use particularly favourable for RNAi transgene development will also be reviewed.

RNA interference (RNAi)

The discovery of RNAi

RNAi is a naturally occurring cellular process that is highly conserved among multicellular organisms as diverse as plants, worms, yeast, and humans, and involves the processing of dsRNAs to initiate gene suppression. Also known as post-transcriptional gene silencing (PTGS), the term RNAi refers to the pathway of dsRNA-induced silencing in animals (Fire *et al.*, 1998), but is also referred to as quelling in the filamentous fungus *Neurospora crassa* (reviewed in Nakayashiki, 2005), and when it was first described in plants was termed co-suppression (reviewed in Tijsterman *et al.*, 2002). This initial discovery of PTGS in plants occurred in 1990 when Napoli and colleagues attempted to deepen the purple colour of petunias by over-expressing the gene that encodes the enzyme chalcone synthase that synthesizes this pigment. Unexpectedly, plants with the introduced gene produced either white and/or patterned flowers instead of the predicted purple colour (Napoli *et al.*, 1990).

The mechanism of action remained elusive for some time, until the key breakthrough came in 1998, when in two separate studies, gene silencing initiated by double-stranded RNA (dsRNA) molecules was discovered in the nematode *Caenorhabditis elegans* (Fire *et al.*, 1998), and then in tobacco plants (Waterhouse *et al.*, 1998). In the study by Fire and colleagues, it was found that by injecting a mixture of sense and antisense strands into *C. elegans* much more efficient silencing was achieved compared to either sense or antisense stands alone (Fire *et al.*, 1998). Waterhouse *et al.* (1998) showed that transformed plants that were capable of producing RNAs capable of duplex formation conferred specific viral immunity or gene silencing. Although prior to these experiments interference of *C. elegans* gene expression had been described using both sense and antisense RNAs, it was the discovery that the potency of dsRNA far exceeded that of ssRNAs, and the phenomenon was named RNA interference.

After the discovery of RNAi, studies of its mechanism immediately followed. Initial work reported that tomato plants undergoing PTGS contained RNAs that were 21–25 nucleotide (nt) in length, complementary to both strands of the silenced gene and had been processed from a long dsRNA precursor (Hamilton and Baulcombe, 1999). The processing of long dsRNA to 21–23 nt RNAs, now known as small interfering RNAs (siRNAs), in *C. elegans* was then shown *in vitro* in *Drosophila melanogaster* embryo extracts (Zamore *et al.*, 2000), and in a *Drosophila* cell line (Bernstein *et al.*, 2001). Others reported very similar findings both in *Drosophila* and in *C. elegans* (Parrish *et al.*, 2000; Yang *et al.*, 2000; Hammond *et al.*, 2000) and sequencing of these short RNAs revealed that they had a very specific structure: 21–23 nt with two nt 3' overhangs (Elbashir *et al.*, 2001b). Evidence for the specificity of these molecules for RNAi came from studies in *Drosophila*, when small RNAs that were isolated from cells undergoing RNAi were sufficient to induce specific gene knockdown in naïve *Drosophila* embryo lysates and a *Drosophila* cell line (Yang *et al.*, 2000; Zamore *et al.*, 2000; Elbashir *et al.*, 2001b).

With this work from *Drosophila* in mind, researchers then analysed the potential for RNAi in mammalian cells. It was thought, however, that the use of RNAi in mammalian cells was not possible due to the presence of an innate immune defence mechanism directed against the entry of dsRNA into cells, such as occurs during a viral infection. Although toll-like receptor 3 had been identified as a dsRNA response protein (Alexopoulou *et al.*, 2001), the two best-characterized dsRNA-activated pathways signal through the dsRNA recognition proteins PKR (dsRNA-dependent protein kinase) and the 2',5'-oligoadenylate synthetase. The activation of PKR by dsRNA results in its autophosphorylation and the phosphorylation of its substrates, ultimately leading to non-specific translation inhibition (Williams, 1997). In addition, PKR activation also induces a signalling cascade that increases the intracellular production of interferons, which leads to an up-regulation of the interferon-stimulated genes, which are mediators of antiviral and antiproliferative and pro-apoptotic activity (Kumar *et al.*, 1994). The dsRNA activation of 2',5'-oligoadenylate synthetase results in the activation of the enzyme RNAse L, which in turn cleaves both cellular and viral RNAs to inhibit translation (Khabar *et al.*, 2003).

The solution that allowed researchers to circumvent this intolerant cellular activity came as a direct result of the biochemical understanding of the RNAi pathway. It was discovered that the delivery of chemically synthesized small interfering RNA (siRNA) duplexes designed to mimic the naturally occurring products of dsRNA processing could evade the immune response and elicit potent and specific gene silencing (Elbashir *et al.*, 2001a). Elbashir and colleagues showed that by introducing siRNAs designed to target a reporter gene and an endogenous gene in a number of human cell lines efficient suppression of their expression could be achieved (Elbashir *et al.*, 2001a). These findings have led to the widespread use of RNAi to study gene function including important clinically relevant and viral genes, alluding to the potential therapeutic application of RNAi-based technologies.

The natural role of RNAi

It appears that the natural functions of RNAi and its related processes are for the protection of the genome against mobile genetic elements such as integrative viruses and transposons, and importantly in the orchestrated functioning of the developmental programmes of eukaryotic organisms (reviewed in Zamore and Haley, 2005; Caplen, 2004). At the discovery of RNAi, the role of this natural pathway was thought to largely exist to degrade aberrant RNA such as those of transposable elements. Since the integrity of a cell's genome can be rapidly compromised by the presence of such elements, RNAi may serve as an important factor to help combat their movement (reviewed in Rao and Sockanathan, 2005).

More recently, the role of small RNAs in controlling gene expression has emerged as an extremely diverse and important aspect of both plant and animal cellular development. In addition to siRNAs that are derived from exogenous dsRNAs, microRNAs (miRNAs) are a second class of small RNA that can induce silencing by targeting RNA. However, in contrast to siRNAs, miRNAs are derived from endogenously transcribed long hairpin precursor RNAs and achieve sequence-specific RNA silencing using some common cellular factors. Although the first miRNA, *lin-4*, was discovered in the early 1990s (Wightman *et al.*, 1993; Lee *et al.*, 2004), the importance of these molecules in controlling gene expression has only recently been recognized in important regulatory activities. The role of RNA silencing pathways in animals, particularly in the context of embryonic development, is only now becoming clear and it is estimated that acting via the RNAi pathway, miRNAs may regulate up to a third of all human genes (Lewis *et al.*, 2005). These include, but are not limited to, cellular differentiation, apoptosis, developmental timing and regulation of transposable elements (reviewed in Kim, 2005), although the exact function of most miRNAs is unknown.

The important role of RNAi in viral-directed immunity has been predicted since its initial discovery. This is most clear-cut for plants, especially since Ratcliff *et al.* (1997) showed that previous observations of transgene-induced viral immunity (Lindbo *et al.*, 1993; Baulcombe, 1996) were due to PTGS initiated by the expression of the viral transcript. Since these experiments, the importance of RNAi in viral immunity has been further demonstrated, and considering that upon infection all viruses have the potential to produce dsRNA, it is possible that the RNAi pathways play an important role in viral immunity in vertebrates.

Recent evidence supports this conclusion as it appears that a number of viruses encode and utilize factors that act via the RNAi pathway (reviewed in Schutz and Sarnow, 2006). In particular, miRNAs are expressed by a number of viruses including SV40 (Sullivan *et al.*, 2005), Herpesvirus family members (Pfeffer *et al.*, 2004; Cai *et al.*, 2005; Samols *et al.*, 2005) and HIV-1 (Bennasser *et al.*, 2005). In addition, host-encoded miRNAs are also known to affect retrovirus primate foamy virus type-1 (PFV-1) (Lecellier *et al.*, 2005) and hepatitis C virus (Jopling *et al.*, 2005), and it is believed that saturation of RNAi pathway components to prevent viral degradation is achieved by the adenoviral-expressed VA RNAs (Gwizdek *et al.*, 2003). Fundamental to both the further explanation of these complex interactions and also the successful harnessing of this technology for gene manipulation is an understanding of the underlying cellular pathways by which these small RNAs interact.

The RNAi mechanism

A simplified model for the RNAi mechanism is based on a two-step intracellular pathway, which involves a ribonuclease machine to achieve the nucleolytic destruction of the targeted mRNA. In the first step, the trigger RNA, usually exogenous dsRNA of greater than ~30 base pairs (bp), is processed into the 21–23 nt active siRNA by an RNAse III enzyme termed Dicer (Fig. 8.1). Like all RNAse III enzymes, Dicer shows specificity for dsRNA (Nicholson, 1999) and cleaves them to leave 3' overhangs of two or three nucleotides with 5'-phosphate and 3'-hyrdroxyl termini (Elbashir *et al.*, 2001b). This enzyme was originally isolated from *Drosophila* cell extracts as it was capable of producing fragments of 22 nucleotides

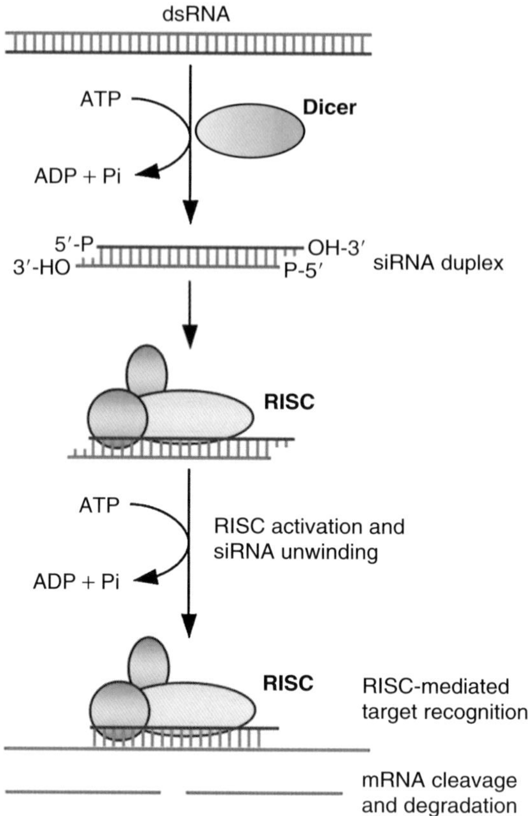

Fig. 8.1. *The cellular mechanism of RNAi.* Endogenous dsRNA from repetitive sequences such as transposons and exogenous dsRNA from viral origin trigger the RNAi pathway. These dsRNA substrates are processed by Dicer into siRNA duplexes with 3′ overhangs and 5′-phosphate and 3′-hydroxyl termini. The siRNA then binds to the multiprotein complex to form the RISC. Activation of RISC is accompanied by siRNA unwinding of the siRNA duplex, and the antisense strand is guided to the target mRNA. Binding of the homologous siRNA strand results in the RISC-mediated cleavage of the target mRNA.

(Bernstein *et al.*, 2001), and is evolutionarily conserved in worms, flies, fungi, plants and mammals. A proposed model for Dicer involves the ATP-dependent translocation of the enzyme along its target prior to dsRNA cleavage, the efficiency of which has been shown to be directly proportional to the length of the target (Bernstein *et al.*, 2001). Dicer has four distinct domains: an amino-terminal helicase domain, dual RNAse III motifs, a dsRNA binding domain, and a PAZ domain (named after its component proteins Piwi, Argo, and Zwile/Pinhead) (Tabara *et al.*, 1999; Catalanotto *et al.*, 2000).

Following the cleavage of dsRNA into siRNAs by Dicer the second important stage of mRNA degradation involves the formation of a ribonucleic protein complex termed RNA-induced silencing complex (RISC), which guides the siRNAs to their target RNAs (Fig. 8.1). The process is very similar to the processing of miRNAs, except that the miRNA precursors are first processed in the nucleus by an RNAse III enzyme termed Drosher prior to Dicer cleavage, and the pairing of processed miRNA to its target RNA by RISC often contains mismatches, which leads to translational repression of target RNAs (Doench *et al.*, 2003; Bartel and Chen, 2004).

RISC was originally identified based on its requirement for siRNA binding for gene silencing in cultured *Drosophila* cells

(Hammond *et al.*, 2000). Binding of the siRNA to RISC results in the ATP-dependent activation of this complex (Nykanen *et al.*, 2001) and unwinding of the siRNA. Interestingly, a bias in the formation of active RISC with a particular RNA strand has been observed and is thought to reflect how the siRNAs are processed, unwound, and loaded into RISC, which is critically determined by the nucleotide interactions within the siRNA (Khvorova *et al.*, 2003; Schwarz *et al.*, 2003). This bias is also evident in naturally occurring miRNAs whose strands accumulate in RISC in unequal proportions (Bartel, 2004). It is presumed that this strand bias occurs by the rate-limiting unwinding step that occurs during the transition of the siRNA duplex to the RISC, which allows the most weakly paired end to enter first (reviewed in Meister and Tuschl, 2004).

Upon binding of the unwound siRNA within RISC to the target homologous mRNA, cleavage occurs at a single site in the centre of the duplex region, 10 nt from the 5' end of the siRNA (Elbashir *et al.*, 2001a). Although this process does not require ATP (Nykanen *et al.,* 2001), multiple rounds of mRNA cleavage are more efficient in the presence of ATP (Hutvagner and Zamore, 2002). The loaded RISC complex catalyses hydrolysis of the target RNA phosphodiester linkage and resembles the reaction that occurs when Dicer generates siRNA from dsRNAs (Martinez and Tuschl, 2004; Schwarz *et al.*, 2004). This understanding of the RNAi pathway has allowed researchers to use and optimize a number of different approaches to harness this innate pathway for the silencing of genes of interest.

Methods used to induce RNAi in animal cells

Long dsRNA

RNAi mediated by the introduction of long dsRNA has been used extensively to investigate gene function in various organisms including planaria (Alvarado and Newmark, 1999), hydras (Lohmann *et al.*, 1999), trypanosomes (Ngo *et al.*, 1998), *Drosophila* (Kennerdell and Carthew, 1998), mosquitoes (Caplen *et al.*, 2002), *C. elegans* (Barstead, 2001) and mouse oocytes (Svoboda *et al.*, 2000). This technique has been very effective, especially since the cellular processing of long dsRNA results in the presentation of various siRNA sequences to the target mRNA (Fig. 8.2). In higher order eukaryotes including mammals and birds, instead of dsRNA, the use of synthetic siRNAs designed to emulate natural Dicer cleavage products has avoided induction of dsRNA-activated immune defence pathways and permitted researchers to exploit the RNAi pathway for specific gene knockdown (Elbashir *et al.*, 2001a) (Fig. 8.2).

Synthetic siRNAs

Important developments in the design, production and delivery of siRNAs have seen these processes become both well defined and highly effective, resulting in the extremely widespread use of this technology (Fig. 8.2). Since there is a bias in siRNA strand selection for RISC assembly and as siRNA duplexes are functionally asymmetric (Schwarz *et al.*, 2003), with only one of the strands having the ability to trigger RNAi, it is preferable to design an siRNA duplex with an antisense strand that can enter RISC. For efficient siRNA-mediated RNAi, consideration of the siRNA sequence is also crucial. Sequence-dependent variability of siRNA efficacy has been shown by siRNAs that when targeted at different regions of the same mRNA varied markedly in their effectiveness (Holen *et al.*, 2002; Miyagishi and Taira, 2002; Hemann *et al.*, 2003). Thus the design of the effector molecules is of fundamental importance to the desired outcome. Statistical analyses of large groups of siRNA sequences have resulted in the identification of design rules that substantially improve the frequency of effective sequences (Schwarz *et al.*, 2003; Khvorova *et al.*, 2003; Reynolds *et al.*, 2004; Amarzguioui and Prydz, 2004; Ui-Tei *et al.*, 2004). In general, siRNAs have been 19 or 21 nt in length with two nt 3' overhangs of uridine residues, although increased potency has been reported using siRNAs of up to 27 bp in length (Kim *et al.*, 2005).

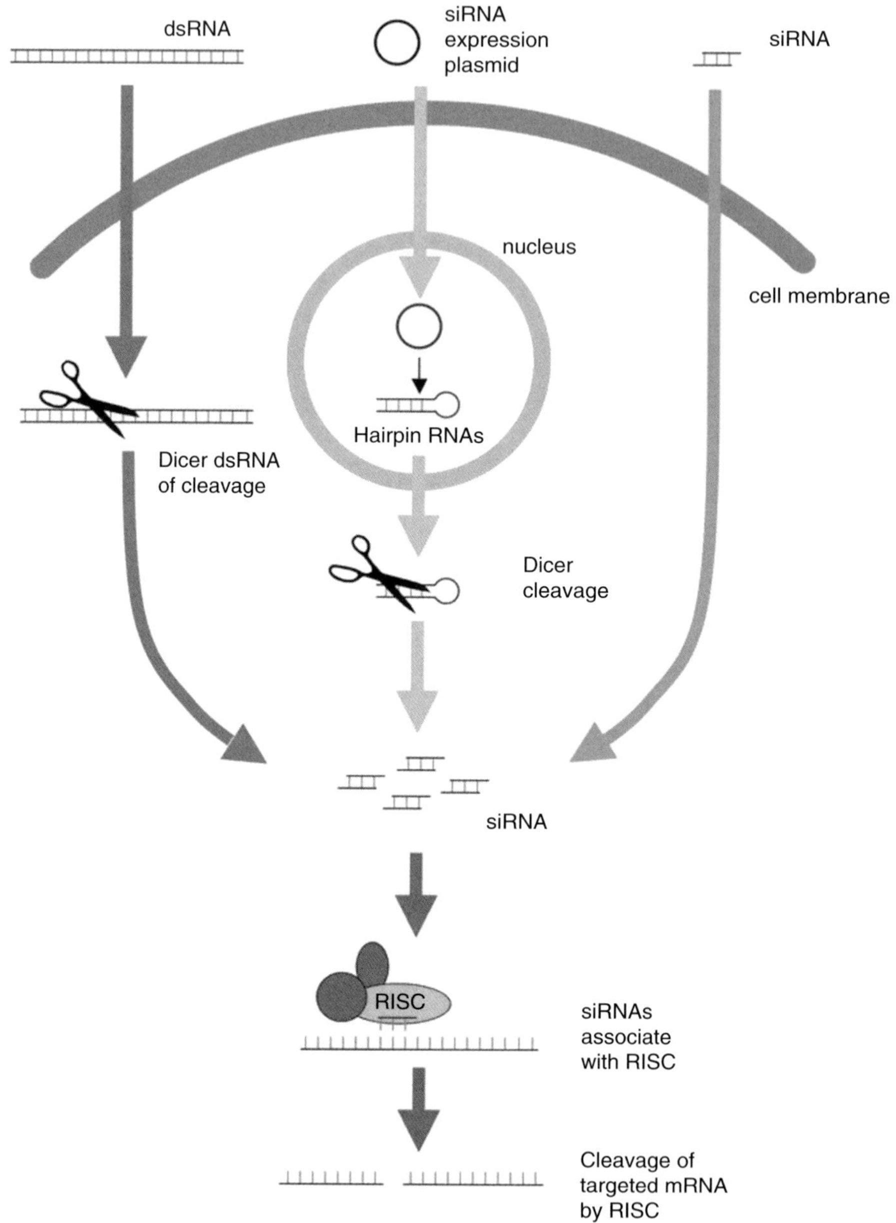

Fig. 8.2. *The methods used to initiate RNAi in animal cells.* The introduction of long dsRNA, siRNA expression plasmids or synthetic siRNA into cells all can be used for RNAi. Long dsRNA is processed into active siRNAs, but is not used in mammalian cells. Plasmids express RNAs that can form hairpin structures within the cell, which are then processed into active siRNAs, and synthetic siRNAs are directly introduced into cells. All methods involved result in the production of an active siRNA that can enter the RNAi pathway to initiate gene knockdown.

There are a number of ways to produce siRNA for transfection into cells. The most popular is chemically synthesized molecules, and these are widely available from a number of manufacturers and offer the considerable advantage of a high level of uniform composition, are of high concentration and can be synthesized with a range of modifications that may increase stability and improve efficiency (Chiu and Rana, 2002; Amarzguioui *et al.*, 2003; Braasch *et al.*, 2003).

DNA-delivered siRNA expression

Suppression of gene expression mediated by siRNAs is transient and usually lasts for 3–5 days in cell culture (Holen *et al.*, 2002). Although this may be sufficient for many applications, in some situations the use of transient RNAi is not satisfactory. The expression of siRNAs from DNA constructs introduced into cells offers a viable alternative (Fig. 8.2). This allows the generation of stable transformants, thereby alleviating the restriction of transient knockdown and can facilitate the study of proteins with long half-lives or for other long-term studies such as viral infection. Stable integration of DNA molecules expressing siRNA can also overcome the variability in transfection efficiency of synthetic siRNA. Other benefits include the ability of selection for transfected cells via antibiotic markers or reporter genes, vector construction can be less expensive than chemical synthesis of siRNA, and this approach also presents the option of inducible siRNA transcription.

The most commonly used approach for siRNA expression involves the transcription of short hairpin RNA (shRNA) molecules by RNA polymerase III promoters (see Promotors for DNA-based shRNA expression). It is predicted that when these shRNAs are transcribed within a cell they form structures that are reminiscent of naturally expressed miRNAs and as such can enter the RNAi pathway to initiate gene knockdown (Fig. 8.3A). The use of siRNA sequences embedded in naturally occurring miRNAs has been recently demonstrated with a high level of success. In particular, siRNA sequences targeting the human SOD1 gene (Zhou *et al.*, 2005), the human p53 gene (Dickins *et al.*, 2005) and reporter target genes (Zeng *et al.*, 2002, 2005) have been substituted into human miRNA *mir*-30 resulting in efficient RNAi.

The limitation of constitutive gene knockdown of an essential gene targets can be circumvented by the induction of shRNA expression. A number of inducible promoter systems have been developed, including the tetracycline-inducible system (tetO) (van de Wetering *et al.*, 2003), a Cre-LoxP recombination-based system (Ventura *et al.*, 2004; Coumoul *et al.*, 2004; Tiscornia *et al.*, 2004; Kasim *et al.*, 2004) and the Ecdysone (muristerone A)-inducible system (Gupta *et al.*, 2004). The tetO system requires the addition of tetracycline into stably transfected cells, which results in the dissociation of a repressor and thereby allows transcription to proceed. In the Cre-LoxP system, the insertion of LoxP-flanked stuffer sequence in the promoter region results in promoter inactivation, and re-activation is achieved by expression of a Cre protein, which results in the removal of the intervening sequence. For the Ecdysone (muristerone A)-inducible system, the addition of ecdysone analogue initiates an activation cascade, which results in activation of a GAL4-Oct-2^{P} transactivator expression and the subsequent activation of a U6 promoter.

To overcome limitations of plasmid-based shRNA such as transfection efficiency, a number of viral-based systems for shRNA expression have been devised. Systems have been described that feature RNA pol III promoters and include the use of retroviral (Devroe and Silver, 2004; Kronke *et al.*, 2004; Lu *et al.*, 2006), adenoviral (Shen *et al.*, 2003; Zhao *et al.*, 2003) and lentiviral systems (Abbas-Terki *et al.*, 2002; Rubinson *et al.*, 2003; Van den Haute *et al.*, 2003). Less commonly, vectors based on simian foamy virus type-1 (SFV-1) (Park *et al.*, 2005) and baculovirus vectors (Gonzalez-Alegre *et al.*, 2005) have been described. Viral vectors are particularly useful in cell lines that are refractory to transfection and are capable of infecting both dividing and non-dividing cells. The development of these vectors has allowed much more efficient and rapid

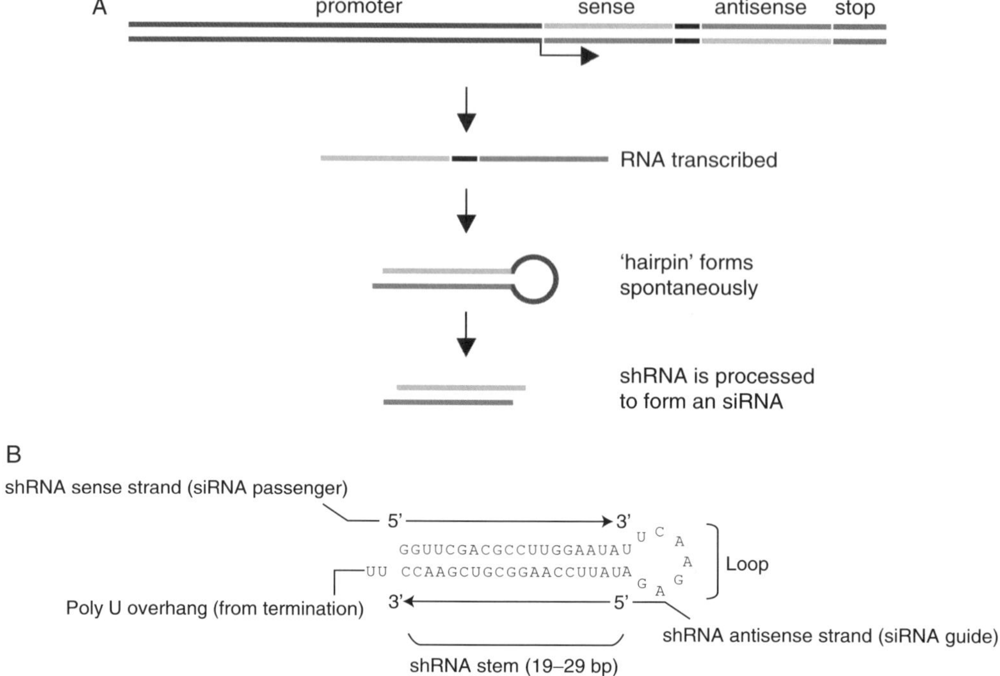

Fig. 8.3. *The DNA-based shRNA expression vector process and the structure of an shRNA.* (A) An shRNA template that features the sense, loop, antisense and termination sequence is engineered downstream of a promoter sequence. When introduced into the cell, transcription of the RNA results in the formation of a small hairpin structure that is processed by Dicer and enters the RNAi pathway. (B) Typical shRNA designs feature an siRNA core sequence that ranges from 19–29 bp in length that forms the shRNA stem. The loop sequence connects the 3′ end of the upper siRNA stand (shRNA sense strand) to the 5′ end of the lower siRNA (shRNA antisense strand). The shRNA sense strand is identical to the target mRNA and gives rise to the siRNA 'passenger' strand, and the shRNA antisense strand is complementary to the target mRNA and gives rise to the siRNA 'guide' strand. The poly uridine overhang of the shRNA antisense strand is derived as part of the termination of transcription.

analysis of gene function in both animal cells and tissues and shows promise in the development of novel gene therapies.

All of this preliminary work in construction of ddRNAi systems for plasmid and viral vectors has now led to the development of transgenes featuring pol III promoters for the expression of shRNAs in transgenic animals (Golding *et al.*, 2006).

shRNA-mediated ddRNAi

Methods for expression of shRNA have advanced rapidly since the initial discovery of RNAi in animal cells. There are a number of aspects of shRNA molecules and vector design that have been tested to generate more effective expression systems. Although many subtle variations exist, the anatomy of the expressed shRNA molecules has remained relatively constant and features a number of characteristic components that are essential for function (Fig. 8.3B). Other factors that are crucial for activity have been considered, including the loop sequence and the use of a number of promoter types for shRNA expression.

Effector molecule design

Although a large statistical analysis of factors affecting shRNA efficacy has not yet been published, it appears that the same rules that determine siRNA activity can also be applied to selection of shRNA sequences. Most design rules for siRNA, however, have been based on duplexes of 19 nt with two nt 3'-overhangs, whereas the lengths of effective expressed shRNAs range in size from 19 to 29 bp (Brummelkamp *et al.*, 2002; Paddison *et al.*, 2002; Harborth *et al.*, 2003). The effect of stem length on siRNA activity has also been the subject of a number of reports, resulting in a mixture of sometimes contrasting findings.

Recently, it was shown that increasing the duplex length to 27–29 bp can enhance siRNA and synthetic shRNA activity, presumably through enhanced dicer processing (Siolas *et al.*, 2005; Kim *et al.*, 2005; Rose *et al.*, 2005). Other reports have specifically analysed the effect of shRNA stem length on gene silencing activity. It was reported that increasing the stem length of a 19 bp shRNA to 28 bp could improve upon poor initial activity, but an already active 19 bp could not be further improved by a similar extension (Yu *et al.*, 2003). In another report that tested 22 to 29 bp shRNAs, it was found that the most active stem length was 25 to 29 bp (Paddison *et al.*, 2002), while in contrast, in a separate study it was found that when shRNAs from 19 to 25 bp were tested, 21 bp was the most active (Miyagishi *et al.*, 2004).

Although the use of a number of different loop sequences for shRNAs have been successfully used (Brummelkamp *et al.*, 2002; Lee *et al.*, 2002; Paddison *et al.*, 2002; Paul *et al.*, 2002), no particular benefit has been described using one or another. The most commonly used sequence has been 5'-UUCAAGAGA-3', which was originally tested by Brummelkamp *et al.* (2002). It has, however, recently been suggested that the use of these unrelated sequences can be improved upon. In particular, the naturally occurring miRNA loop of human miRNA *mir*-23 (Lagos-Quintana *et al.*, 2001) has been used in a number of instances (Kawasaki and Taira, 2003; Carmona *et al.*, 2006).

Promoters for DNA-based shRNA expression

Promoter choice for shRNA expression has been shown to be an important aspect for effective ddRNAi. Early experiments described the transcription of shRNAs by the human U6 small nuclear RNA (snRNA) promoter, a member of the RNA polymerase III (pol III) family, resulting in the efficient knockdown of firefly luciferase (Paddison *et al.*, 2002) and neuron-specific β-tubulin in mouse cells (Yu *et al.*, 2002). Since then, the use of human RNA pol III promoters for shRNA transcription has been the most frequently used approach for DNA vector-mediated RNAi. In particular, the U6 promoter (Lee *et al.*, 2002; Paul *et al.*, 2002), and the H1 promoter (Brummelkamp *et al.*, 2002) have been by far the most commonly used for shRNA transcription. In addition, although less commonly described, the use of the 7SK promoter (Czauderna *et al.*, 2003; Koper-Emde *et al.*, 2004; Grimm *et al.*, 2005), the $tRNA^{val}$ promoter (Kawasaki and Taira, 2003), and the BC1 promoter (Kobayashi *et al.*, 2005) for shRNA expression have also resulted in highly efficient gene knockdown. An enhanced $tRNA^{val}$ promoter, named the modified $tRNA^{val}$-derived promoter (MTD), was used to express shRNA that induced increased HIV-1 suppression (Boden *et al.*, 2003).

RNA pol III promoters are classified into three different categories (Type 1, 2 or 3) based upon the composition of the promoter elements sequences and, their position relative to the transcriptional start site (Gunnery *et al.*, 1999). The tRNA promoter is a typical type 2 promoter, and, as such, transcription from these promoters leads to the production of considerably longer products, which may stimulate the non-specific antiviral response. As a result, their use has been less frequent than other pol III promoters. The U6, H1 and 7SK are type 3 promoters, which are self-contained promoters and possess numerous advantageous features that have resulted in their increasingly prevalent and widespread use for shRNA expression. Type 3 promoters naturally direct the synthesis of small, highly abundant non-coding RNA transcripts known as small nuclear RNAs (snRNAs). They have a compact and relatively simple organization and are located entirely upstream of the transcribed region (reviewed in Paule and White, 2000; Schramm and Hernandez, 2002). Transcription is terminated within a stretch of four to five thymidine residues, yielding small RNA transcripts with a defined 3' overhang. Since this feature has been shown to be critical for shRNA activity (Elbashir *et al.*, 2001b), and considering that these promoters are ideal for synthesis of small and defined RNAs, they have been an understandable choice for a wide range of applications in the expression of shRNA.

Recently, the characterization and validation of pol III promoters from livestock species such as chickens and cows has been reported for shRNA expression from DNA vectors (Kudo and Sutou, 2005; Lambeth *et al.*, 2005). The use of these promoters in the development of species-specific transgenes for the expression of shRNA molecules may well be critical for the future acceptance of designed animals engineered for disease resistance.

Application of ddRNAi as an antiviral strategy

One of the earliest and most obvious uses of DNA-expressed shRNAs was to target viral transcripts for the suppression of infection. The specific nature of RNAi is attractive for therapeutic applications as it should not affect any host gene functions and holds promise for the development of effective antiviral therapies. As such, the targeting of many prevalent human viruses such as hepatitis B (HBV), hepatitis C (HCV), influenza A and human immunodeficiency virus type 1 (HIV-1) has been a major research focus; however, the successful suppression of many other viruses has also been demonstrated.

The successful targeting of HCV has been described using both plasmid-expressed HVC-specific shRNAs (Yokota *et al.*, 2003; Kronke *et al.*, 2004) and retrovirus- and lentivirus-expressed shRNAs (Takigawa *et al.*, 2004) in cell culture targeting HCV-replicon systems. Expressed shRNAs were capable of reducing HBV replication in tissue culture models (Chen *et al.*, 2003; Liu *et al.*, 2004; Moore *et al.*, 2005; Wu *et al.*, 2005a) and in transfected immunodeficient mice (McCaffrey *et al.*, 2003; Wu *et al.*, 2005b; Carmona *et al.*, 2006), the suppressive effects of which lasted approximately 1 month (Peng *et al.*, 2005). A marked reduction in tumour growth associated with hepatocellular carcinoma caused by HBV infection has also been observed as a result of the expression of shRNA in nude mice (Chan and Ng, 2006).

After testing siRNAs *in vitro*, inhibition of Japanese encephalitis virus (JEV) replication in mice by intraperitoneal injection of shRNA plasmids was achieved (Murakami *et al.*, 2005). A number of reports also describe the use of DNA-expressed shRNAs for the suppression of HIV-1 (Lee *et al.*, 2002; Boden *et al.*, 2004; Nishitsuji *et al.*, 2004; M.J. Li *et al.*, 2005). DNA-expressed shRNAs suppressed the expression of Epstein-Barr virus latent membrane protein-1 (LMP-1) (X.P. Li *et al.*, 2004). A potential approach for the treatment of severe acute respiratory syndrome (SARS) patients was demonstrated by targeting DNA-expressed shRNAs to coronavirus RNA-dependent RNA polymerase (RDRP), leading to a significant reduction in the plaque formation of SARS coronaviruses in cell culture (Lu *et al.*, 2004). The replication of a number of other viruses has been effectively suppressed using pol III-expressed shRNAs, including enterovirus 71 (Lu *et al.*, 2004) and simian immunodeficiency virus (Park *et al.*, 2005).

ddRNAi has also been used effectively to inhibit the replication of influenza virus. U6-transcribed shRNAs delivered by retroviral vectors prevented and treated influenza A virus infection in the lungs of mice (Ge *et al.*, 2003). Also, plasmids featuring a U6 promoter for transcription of shRNAs specific for influenza A were shown to inhibit replication of the avian H5N1 virus strain in cultured chicken cells and also embryonated chicken eggs (Y.C. Li *et al.*, 2005).

Avian Influenza-resistant Transgenic Chickens – a Case Study

Influenza A virus is a member of the *Orthomyxoviridae* family of viruses. Influenza A viruses cause infection in humans, a limited range of mammals, and birds. There are 15 known haemagglutinin (H) serotypes and 9 known neurominidase (N) serotypes. All 15 known influenza A virus H subtypes can be isolated from avian species. Only the H5 and H7 haemagglutinin subtypes are considered to be highly pathogenic in avian species. Birds are believed to be particularly important reservoirs of influenza A viruses, generating pools of genetically diverse viruses through gene segment re-assortment, which may then transfer back to the human population via close contact between humans and birds, in particular poultry.

What are the drivers for avian influenza-resistant chickens?

There are three main drivers for developing transgenic chickens resistant to avian influenza (AI). The first driver is the impact of AI on the world's poultry industry. Chicken meat consumption around the world is increasing. To meet this demand, 50 billion meat chickens are hatched annually, representing nearly 40% of all meat consumed worldwide and with a retail value of approximately US$250 billion. The retail market has been growing at 5–10% annually in some developed countries where there is a consumption rate of approximately 30 kg per person per year. In developing countries such as China, chicken meat consumption is currently approximately 1 kg/person/year. This is beginning to increase rapidly as both the wealth of the country and the demand for protein in the diet grows. One of the biggest economic threats currently to the poultry industry, let alone meeting this increased demand for chicken meat in the developing world, is the massive impact that the H5 and H7 subtypes of AI has on poultry production and trade.

Since the H5N1 AI outbreak in Hong Kong in 1997, there have been consistently further outbreaks of the H5N1 subtype in Asian countries including China, Japan, South Korea, Vietnam, Thailand, Indonesia, Cambodia, and Laos, and now more recently in Europe, Africa and the Middle East. There is currently no effective vaccine to help control the spread of such outbreaks in poultry and therefore the mass slaughter of infected birds is currently used in control programmes. Even so, despite the wholesale slaughter of poultry in Hong Kong in 1997, it was impossible to eradicate H5N1 as the precursor viruses continue to circulate in migratory birds. The economic impact of such mass slaughter of animals is obvious; however, the consequences of mass slaughter for both animal welfare issues and the environmental issues associated with the disposal of millions of carcasses, as seen recently during the foot-and-mouth outbreak in the United Kingdom, forms the second major driver for engineering disease-resistant birds.

The third driver is the need to reduce the threat of avian influenza subtypes as a source of new influenza A virus for pandemics of human populations. The direct transmission of avian H5N1 influenza viruses to humans in 1997 changed our understanding of the transmission of avian influenza viruses to humans and their virulence in humans. The natural reservoirs of influenza A viruses are aquatic birds such as ducks and it is thought that a bridging host is involved in the transfer of the virus from these birds to humans. The bridging reservoir is likely to be intensively reared livestock such as poultry, where re-assortment between different strains of avian influenza and human influenza virus may occur and result in a new strain that can readily infect humans via close contact (Nicholson *et al.*, 2003). Because of this threat to human health, the World Health Organization advocates that we reduce the virus in the bridging reservoir in order to reduce the risk of a human influenza pandemic. The emergence of multiple genotypes of avian H5N1 viruses in recent years indicates the continued co-circulation of these viruses between wild aquatic birds and poultry in Asia. After the initial H5N1 infections in humans in 1997, fatal human H5N1 disease reappeared in Hong Kong in 2003, showing that H5N1 infection of humans was not an isolated event. Since 2003, H5N1 has caused 220 reported cases of human infection from infected poultry with a greater than 50% mortality rate. There is no doubt that the H5N1 influenza A subtype poses a significant threat of a human influenza pandemic.

The ultimate transgene – sequence specific, species specific

The development of transgenic poultry that are resistant to AI would certainly be of benefit to the poultry industry and would also help to reduce the risk of a human influenza pandemic by reducing the potential for virus re-assortment within intensively reared birds. Both the engineering and validation of AI-resistant chickens would be a significant test of the efficacy and value of using RNAi transgenes as a transformational approach to controlling viral disease in livestock species.

What might the ultimate RNAi transgene components be? Based on what is known about ddRNAi for expression of shRNAs, the transgene will encode a chicken RNA polymerase III promoter, such as one of the U6 promoters or the 7SK promoter. It will be important that the promoter is of chicken origin, so as not to introduce DNA from a foreign species on the transgene. The shRNA sequences will be targeted to conserved regions of all influenza A viruses. Influenza A virus has a segmented RNA genome. Three of the eight RNA segments encode three components of the RNA transcriptase (PA, PB1 and PB2). Three additional RNA segments encode the major glycoproteins: haemagglutinin (HA), neurominidase (NA) and nucleocapsid protein (NP). Each of the remaining two RNA segments encodes two proteins, either matrix protein M1 and M2, or the non-structural proteins NS1 and NS2, which function either as viral structural proteins or in the viral life cycle respectively. Not surprisingly for an RNA virus, there are extensive differences in the nucleotide sequences of influenza A genes among isolates from different species. The major sequence differences are seen in the HA and NA genes, which manifest as changes in the amino acid sequence, causing antigenic drift. It will be important to design shRNAs to regions of the viral genome that are conserved among the different subtypes and strains of virus from different species. Ge *et al.* (2003) have recently done this, and were able to design and test 20 siRNAs of 21 bp in size that were specific for all sequenced NP, PA, PB1, PB2, M and NS genes. They were unable to design siRNAs for HA and NA because these genes contain no stretch of conserved 21 nucleotides as a result of extensive variations in these genes among different virus isolates from different species. Of the 20 siRNAs tested, a number were very effective at inhibiting replication of two influenza A strains both *in vitro* and *in vivo* (Ge *et al.*, 2003; Tompkins *et al.*, 2004). These sequences could be expressed as shRNAs from chicken pol III promoters encoded on the transgene.

It will be of significant benefit to express multiple shRNAs from the one transgene to even further reduce the risk of viral target sequence variability to an RNAi strategy. This 'Multi-Warhead' transgene would comprise multiple transcription units, each with a different chicken pol III promoter (there are four chicken U6 promoters and one 7SK) expressing individual shRNA molecules targeting conserved sequences of different influenza A genes. The promoter sequences are native to chickens and the small 21 bp shRNA sequences would already be present in AI-infected or vaccinated birds. The RNAi targets are absolutely specific to influenza A viruses and so there would be no off-target effects from such a specific transgene. The transgene is also not protein coding and therefore will not have the potential of causing an adverse allergic response to a foreign protein in someone who eats a chicken engineered to contain the RNAi transgene. These features will play an important role in the ultimate test of such a transgenic chicken, that is, consumer acceptance of such a chicken for consumption.

Consumer Acceptance – the Ultimate Test

While it will be important for any transgene to have sequence- and species-specific features to be accepted, the ultimate road to consumer acceptance will be based on a complex range of issues that will require open dialogue and education so that the public can make an informed decision. This will no doubt lead to a debate in which perceived risks of genetically modified food will need to be balanced against the very real risks of outbreaks of infectious diseases in livestock populations, such as mad cow disease, foot-and-mouth or avian influenza. If creating transgenic livestock that are innately resistant to such diseases helps to prevent the mass slaughter of animals and also helps to prevent the environmental consequences associated with the disposal of very large numbers of carcasses, then in the public's eye an alternative strategy based on designed resistance and genetic modification may well be the lesser of two evils. The reason why an influenza-resistant chicken may be the first transgenic animal to tip the balance in favour of public acceptance is the vital role it could play in helping to prevent the spread of disease from animals to humans. There is a pressing issue of global importance to reduce the reservoir of influenza virus in farmed poultry, so as to stem the spread of viruses such as the H5N1 subtype to humans, and to ultimately lessen the risk of a human influenza pandemic such as the one in 1918 that killed up to 50 million people. The potential benefit of this technology to human health will be a major positive factor in public perception towards transforming food biotechnology with transgenics.

References

Abbas-Terki, T., Blanco-Bose, W., Deglon, N., Pralong, W. and Aebischer, P. (2002) Lentiviral-mediated RNA interference. *Human Gene Therapy* 13, 2197–2201.

Alexopoulou, L., Holt, A.C., Medzhitov, R. and Flavell, R.A. (2001) Recognition of double-stranded RNA and activation of NF-kappaB by Toll-like receptor 3. *Nature* 413, 732–738.

Alvarado, A.S. and Newmark, P.A. (1999) Double-stranded RNA specifically disrupts gene expression during planarian regeneration. *Proceedings of the National Academy of Sciences of the United States of America* 96, 5049–5054.

Amarzguioui, M. and Prydz, H. (2004) Algorithm for selection of functional siRNA sequences. *Biochemical and Biophysical Research Communications* 316, 1050–1058.

Amarzguioui, M., Holen, T., Babaie, E. and Prydz, H. (2003) Tolerance for mutations and chemical modifications in a siRNA. *Nucleic Acids Research* 31, 589–595.

Barquinero, J., Eixarch, H. and Perez-Melgosa, M. (2004) Retroviral vectors: new applications for an old tool. *Gene Therapy* 11, S3–S9.

Barstead, R. (2001) Genome-wide RNAi. *Current Opinion in Chemical Biology* 5, 63–66.

Bartel, D.P. (2004) MicroRNAs: genomics, biogenesis, mechanism, and function. *Cell* 116, 281–297.

Bartel, D.P. and Chen, C.Z. (2004) Micromanagers of gene expression: the potentially widespread influence of metazoan microRNAs. *Nature Review Genetics* 5, 396–400.

Baulcombe, D.C. (1996) RNA as a target and an initiator of post-transcriptional gene silencing in transgenic plants. *Plant Molecular Biology* 32, 79–88.

Baum, C., Von Kalle, C., Staal, F.J.T., Li, Z.X., Fehse, B., Schmidt, M., Weerkamp, F., Karlsson, S., Wagemaker, G. and Williams, D.A. (2004) Chance or necessity? Insertional mutagenesis in gene therapy and its consequences. *Molecular Therapy* 9, 5–13.

Bennasser, Y., Le, S.Y., Benkirane, M. and Jeang, K.T. (2005) Evidence that HIV-1 encodes an siRNA and a suppressor of RNA silencing. *Immunity* 22, 773.

Bernstein, E., Caudy, A.A., Hammond, S.M. and Hannon, G.J. (2001) Role for a bidentate ribonuclease in the initiation step of RNA interference. *Nature* 409, 363–366.

Boden, D., Pusch, O., Lee, F., Tucker, L., Shank, P.R. and Ramratnam, B. (2003) Promoter choice affects the potency of HIV-1 specific RNA interference. *Nucleic Acids Research* 31, 5033–5038.

Boden, D., Pusch, O., Lee, F., Tucker, L. and Ramratnam, B. (2004) Efficient gene transfer of HIV-1-specific short hairpin RNA into human lymphocytic cells using recombinant adeno-associated virus vectors. *Molecular Therapy* 9, 396–402.

Braasch, D.A., Jensen, S., Liu, Y.H., Kaur, K., Arar, K., White, M.A. and Corey, D.R. (2003) RNA interference in mammalian cells by chemically-modified RNA. *Biochemistry* 42, 7967–7975.

Brackett, B.G., Baranska, W., Sawicki, W. and Koprowski, H. (1971) Uptake of heterologous genome by mammalian spermatozoa and its transfer to ova through fertilization. *Proceedings of the National Academy of Sciences of the United States of America* 68, 353–357.

Brummelkamp, T.R., Bernards, R. and Agami, R. (2002) A system for stable expression of short interfering RNAs in mammalian cells. *Science* 296, 550–553.
Buchschacher, G.L. and Wong-Staal, F. (2000) Development of lentiviral vectors for gene therapy for human diseases. *Blood* 95, 2499–2504.
Cai, X.Z., Lu, S.H., Zhang, Z.H., Gonzalez, C.M., Damania, B. and Cullen, B.R. (2005) Kaposi's Sarcoma-associated herpesvirus expresses an array of viral microRNAs in latently infected cells. *Proceedings of the National Academy of Sciences of the United States of America* 102, 5570–5575.
Caplen, N.J. (2004) Gene therapy progress and prospects. Downregulating gene expression: the impact of RNA interference. *Gene Therapy* 11, 1241–1248.
Caplen, N.J., Zheng, Z.L., Falgout, B. and Morgan, R.A. (2002) Inhibition of viral gene expression and replication in mosquito cells by dsRNA-triggered RNA interference. *Molecular Therapy* 6, 243–251.
Carmona, S., Ely, A., Crowther, C., Moolla, N., Salazar, F.H., Marion, P.L., Ferry, N., Weinberg, M.S. and Arbuthnot, P. (2006) Effective inhibition of HBV replication in vivo by anti-HBx short hairpin RNAs. *Molecular Therapy* 13, 411–421.
Catalanotto, C., Azzalin, G., Macino, G. and Cogoni, C. (2000) Gene silencing in worms and fungi. *Nature* 404, 245.
Chan, D.W. and Ng, O.I. (2006) Knock-down of hepatitis B virus X protein reduces the tumorigenicity of hepatocellular carcinoma cells. *Journal of Pathology* 208, 372–380.
Chang, K., Qian, J., Jiang, M., Liu, Y.H., Wu, M.C., Chen, C.D., Lai, C.K., Lo, H.L., Hsiao, C.T., Brown, L., Bolen, J., Huang, H.I., Ho, P.Y., Shih, P.Y., Yao, C.W., Lin, W.J., Chen, C.H., Wu, F.Y., Lin, Y.J., Xu, J. and Wang, K. (2002) Effective generation of transgenic pigs and mice by linker based sperm-mediated gene transfer. *BMC Biotechnology* 2, 5.
Chen, Y., Du, D., Wu, J., Chan, C.P., Tan, Y., Kung, H.F. and He, M.L. (2003) Inhibition of hepatitis B virus replication by stably expressed shRNA. *Biochemical and Biophysical Research Communications* 311, 398–404.
Chicas, A. and Macino, G. (2001) Characteristics of post-transcriptional gene silencing. *Embo Reports* 2, 992–996.
Chiu, Y.L. and Rana, T.M. (2002) RNAi in human cells: basic structural and functional features of small interfering RNA. *Molecular Cell* 10, 549–561.
Clark, J. and Whitelaw, B. (2003) A future for transgenic livestock. *Nature Reviews Genetics* 4, 825–833.
Coumoul, X., Li, W., Wang, R.H. and Deng, C. (2004) Inducible suppression of Fgfr2 and Survivin in ES cells using a combination of the RNA interference (RNAi) and the Cre-LoxP system. *Nucleic Acids Research* 32, e85.
Czauderna, F., Santel, A., Hinz, M., Fechtner, M., Durieux, B., Fisch, G., Leenders, F., Arnold, W., Giese, K., Klippel, A. and Kaufmann, J. (2003) Inducible shRNA expression for application in a prostate cancer mouse model. *Nucleic Acids Research* 31, e127.
Devroe, E. and Silver, P.A. (2004) Therapeutic potential of retroviral RNAi vectors. *Expert Opinion on Biological Therapy* 4, 319–327.
Dickins, R.A., Hemann, M.T., Zilfou, J.T., Simpson, D.R., Ibarra, I., Hannon, G.J. and Lowe, S.W. (2005) Probing tumor phenotypes using stable and regulated synthetic microRNA precursors. *Nature Genetics* 37, 1289–1295.
Doench, J.G., Petersen, C.P. and Sharp, P.A. (2003) siRNAs can function as miRNAs. *Genes in Development* 17, 438–442.
Elbashir, S.M., Harborth, J., Lendeckel, W., Yalcin, A., Weber, K. and Tuschl, T. (2001a) Duplexes of 21-nucleotide RNAs mediate RNA interference in cultured mammalian cells. *Nature* 411, 494–498.
Elbashir, S.M., Lendeckel, W. and Tuschl, T. (2001b) RNA interference is mediated by 21-and 22-nucleotide RNAs. *Genes in Development* 15, 188–200.
Fire, A., Xu, S., Montgomery, M.K., Kostas, S.A., Driver, S.E. and Mello, C.C. (1998) Potent and specific genetic interference by double-stranded RNA in *Caenorhabditis elegans*. *Nature* 391, 806–811.
Garber, E.A., Chute, H.T., Condra, J.H., Gotlib, L., Colonno, R.J. and Smith, R.G. (1991) Avian cells expressing the murine Mx1 protein are resistant to influenza-virus infection. *Virology* 180, 754–762.
Gavora, J.S. (1996) Resistance of livestock to viruses: mechanisms and strategies for genetic engineering. *Genetics Selection Evolution* 28, 385–414.
Ge, Q., McManus, M.T., Nguyen, T., Shen, C.H., Sharp, P.A., Eisen, H.N. and Chen, J. (2003) RNA interference of influenza virus production by directly targeting mRNA for degradation and indirectly inhibiting all viral RNA transcription. *Proceedings of the National Academy of Sciences of the United States of America* 100, 2718–2723.

Golding, M.C., Long, C.R., Carmell, M.A., Hannon, G.J. and Westhusin, M.E. (2006) Suppression of prion protein in livestock by RNA interference. *Proceedings of the National Academy of Sciences of the United States of America* 103, 5285–5290.

Gonzalez-Alegre, P., Bode, N., Davidson, B.L. and Paulson, H.L. (2005) Silencing primary dystonia: lentiviral-mediated RNA interference therapy for DYT1 dystonia. *Journal of Neuroscience* 25, 10502–10509.

Gordon, J.W. and Ruddle, F.H. (1981) Integration and stable germ line transmission of genes injected into mouse pronuclei. *Science* 214, 1244–1246.

Grimm, D., Pandey, K. and Kay, M.A. (2005) Adeno-associated virus vectors for short hairpin RNA expression. *Methods in Enzymology* 392, 381–405.

Gunnery, S., Ma, Y. and Mathews, M.B. (1999) Termination sequence requirements vary among genes transcribed by RNA polymerase III. *Journal of Molecular Biology* 286, 745–757.

Gupta, S., Schoer, R.A., Egan, J.E., Hannon, G.J. and Mittal, V. (2004) Inducible, reversible, and stable RNA interference in mammalian cells. *Proceedings of the National Academy of Sciences of the United States of America* 101, 1927–1932.

Gwizdek, C., Ossareh-Nazari, B., Brownawell, A.M., Doglio, A., Bertrand, E., Macara, I.G. and Dargemont, C. (2003) Exportin-5 mediates nuclear export of minihelix-containing RNAs. *Journal of Biological Chemistry* 278, 5505–5508.

Hamilton, A.J. and Baulcombe, D.C. (1999) A species of small antisense RNA in posttranscriptional gene silencing in plants. *Science* 286, 950–952.

Hammer, R.E., Pursel, V.G., Rexroad, C.E., Wall, R.J., Bolt, D.J., Ebert, K.M., Palmiter, R.D. and Brinster, R.L. (1985) Production of transgenic rabbits, sheep and pigs by microinjection. *Nature* 315, 680–683.

Hammond, S.M., Bernstein, E., Beach, D. and Hannon, G.J. (2000) An RNA-directed nuclease mediates post-transcriptional gene silencing in drosophila cells. *Nature* 404, 293–296.

Harborth, J., Elbashir, S.M., Vandenburgh, K., Manninga, H., Scaringe, S.A., Weber, K. and Tuschl, T. (2003) Sequence, chemical, and structural variation of small interfering RNAs and short hairpin RNAs and the effect on mammalian gene silencing. *Antisense Nucleic Acid Drug Development* 13, 83–105.

Hemann, M.T., Fridman, J.S., Zilfou, J.T., Hernando, E., Paddison, P.J., Cordon-Cardo, C., Hannon, G.J. and Lowe, S.W. (2003) An epi-allelic series of p53 hypomorphs created by stable RNAi produces distinct tumor phenotypes in vivo. *Nature Genetics* 33, 396–400.

Hofmann, A., Kessler, B., Ewerling, S., Weppert, M., Vogg, B., Ludwig, H., Stojkovic, M., Boelhauve, M., Brem, G., Wolf, E. and Pfeifer, A. (2003) Efficient transgenesis in farm animals by lentiviral vectors. *Embo Reports* 4, 1054–1060.

Holen, T., Amarzguioui, M., Wiiger, M.T., Babaie, E. and Prydz, H. (2002) Positional effects of short interfering RNAs targeting the human coagulation trigger tissue factor. *Nucleic Acids Research* 30, 1757–1766.

Hunter, C.V., Tiley, L.S. and Sang, H.M. (2005) Developments in transgenic technology: applications for medicine. *Trends in Molecular Medicine* 11, 293–298.

Hutvagner, G. and Zamore, P.D. (2002) A microRNA in a multiple-turnover RNAi enzyme complex. *Science* 297, 2056–2060.

Jopling, C.L., Yi, M.K., Lancaster, A.M., Lemon, S.M. and Sarnow, P. (2005) Modulation of hepatitis C virus RNA abundance by a liver-specific microRNA. *Science* 309, 1577–1581.

Kang, Y.B., Stein, C.S., Heth, J.A., Sinn, P.L., Penisten, A.K., Staber, P.D., Ratliff, K.L., Shen, H., Barker, C.K., Martins, I., Sharkey, C.M., Sanders, D.A., McCray, P.B. and Davidson, B.L. (2002) In vivo gene transfer using a nonprimate lentiviral vector pseudotyped with Ross River virus glycoproteins. *Journal of Virology* 76, 9378–9388.

Kasim, V., Miyagishi, M. and Taira, K. (2004) Control of siRNA expression using the Cre-loxP recombination system. *Nucleic Acids Research* 32, e66.

Kawasaki, H. and Taira, K. (2003) Short hairpin type of dsRNAs that are controlled by tRNA(Val) promoter significantly induce RNAi-mediated gene silencing in the cytoplasm of human cells. *Nucleic Acids Research* 31, 700–707.

Kennerdell, J.R. and Carthew, R.W. (1998) Use of dsRNA-mediated genetic interference to demonstrate that frizzled and frizzled 2 act in the wingless pathway. *Cell* 95, 1017–1026.

Khabar, K.S., Siddiqui, Y.M., al-Zoghaibi, F., al-Haj, L., Dhalla, M., Zhou, A., Dong, B., Whitmore, M., Paranjape, J., al-Ahdal, M.N., al-Mohanna, F., Williams, B.R.G. and Silverman, R.H. (2003) RNase L mediates transient control of the interferon response through modulation of the double-stranded RNA-dependent protein kinase PKR. *Journal of Biological Chemistry* 278, 20124–20132.

Khvorova, A., Reynolds, A. and Jayasena, S.D. (2003) Functional siRNAs and miRNAs exhibit strand bias. *Cell* 115, 209–216.

Kim, D.H., Behlke, M.A., Rose, S.D., Chang, M.S., Choi, S. and Rossi, J.J. (2005) Synthetic dsRNA dicer substrates enhance RNAi potency and efficacy. *Nature Biotechnology* 23, 222–226.

Kim, V.N. (2005) Small RNAs: classification, biogenesis, and function. *Molecular Cells* 19, 1–15.

Ko, J.H., Jin, H.K., Asano, A., Takada, A., Ninomiya, A., Kida, H., Hokiyama, H., Ohara, M., Tsuzuki, M., Nishibori, M., Mizutani, M. and Watanabe, T. (2002) Polymorphisms and the differential antiviral activity of the chicken Mx gene. *Genome Research* 12, 595–601.

Kobayashi, S., Higuchi, T. and Anzai, K. (2005) Application of the BC1 RNAgene promoter for short hairpin RNA expression in cultured neuronal cells. *Biochemical and Biophysical Research Communications* 334, 1305–1309.

Koper-Emde, D., Herrmann, L., Sandrock, B. and Benecke, B.J. (2004) RNA interference by small hairpin RNAs synthesized under control of the human 7SK RNA promoter. *Biological Chemistry* 385, 791–794.

Kronke, J., Kittler, R., Buchholz, F., Windisch, M.P., Pietschmann, T., Bartenschlager, R. and Frese, M. (2004) Alternative approaches for efficient inhibition of hepatitis C virus RNA replication by small interfering RNAs. *Journal of Virology* 78, 3436–3446.

Kudo, T. and Sutou, S. (2005) Usage of putative chicken U6 promoters for vector-based RNA interference. *Journal of Reproductive Development* 51, 411–417.

Kumar, A., Haque, J., Lacoste, J., Hiscott, J. and Williams, B.R. (1994) Double-stranded RNA-dependent protein kinase activates transcription factor NF-kappa B by phosphorylating I kappa B. *Proceedings of the National Academy of Sciences of the United States of America* 91, 6288–6292.

Lagos-Quintana, M., Rauhut, R., Lendeckel, W. and Tuschl, T. (2001) Identification of novel genes coding for small expressed RNAs. *Science* 294, 853–858.

Lambeth, L.S., Moore, R.J., Muralitharan, M., Dalrymple, B.P., McWilliam, S. and Doran, T.J. (2005) Characterization and application of a bovine U6 promoter for expression of short hairpin RNAs. *BMC Biotechnology* 5, 13.

Lecellier, C.H., Dunoyer, P., Arar, K., Lehmann-Che, J., Eyquem, S., Himber, C., Saib, A. and Voinnet, O. (2005) A cellular microRNA mediates antiviral defense in human cells. *Science* 308, 557–560.

Lee, N.S., Dohjima, T., Bauer, G., Li, H., Li, M.J., Ehsani, A., Salvaterra, P. and Rossi, J. (2002) Expression of small interfering RNAs targeted against HIV-1 rev transcripts in human cells. *Nature Biotechnology* 20, 500–505.

Lee, Y., Kim, M., Han, J., Yeom, K.H., Lee, S., Baek, S.H. and Kim, V.N. (2004) MicroRNA genes are transcribed by RNA polymerase II. *EMBO Journal* 23, 4051–4060.

Lewis, B.P., Burge, C.B. and Bartel, D.P. (2005) Conserved seed pairing, often flanked by adenosines, indicates that thousands of human genes are microRNA targets. *Cell* 120, 15–20.

Li, M.J., Kim, J., Li, S., Zaia, J., Yee, J.K., Anderson, J., Akkina, R. and Rossi, J.J. (2005) Long-term inhibition of HIV-1 infection in primary hematopoietic cells by lentiviral vector delivery of a triple combination of anti-HIV shRNA, anti-CCR5 ribozyme, and a nucleolar-localizing TAR decoy. *Molecular Therapy* 12, 900–909.

Li, X.P., Li, G., Peng, Y., Kung, H.F. and Lin, M.C. (2004) Suppression of Epstein-Barr virus-encoded latent membrane protein-1 by RNA interference inhibits the metastatic potential of nasopharyngeal carcinoma cells. *Biochemical and Biophysical Research Communications* 315, 212–218.

Li, Y.C., Kong, L.H., Cheng, B.Z. and Li, K.S. (2005) Construction of influenza virus siRNA expression vectors and their inhibitory effects on multiplication of influenza virus. *Avian Diseases* 49, 562–573.

Lindbo, J.A., Silvarosales, L., Proebsting, W.M. and Dougherty, W.G. (1993) Induction of a highly specific antiviral state in transgenic plants – implications for regulation of gene-expression and virus-resistance. *Plant Cell* 5, 1749–1759.

Liu, J., Guo, Y., Xue, C.F., Li, Y.H., Huang, Y.X., Ding, J., Gong, W.D. and Zhao, Y. (2004) Effect of vector-expressed siRNA on HBV replication in hepatoblastoma cells. *World Journal of Gastroenterology* 10, 1898–1901.

Lohmann, J.U., Endl, I. and Bosch, T.C.G. (1999) Silencing of developmental genes in hydra. *Developmental Biology* 214, 211–214.

Lois, C., Hong, E.J., Pease, S., Brown, E.J. and Baltimore, D. (2002) Germline transmission and tissue-specific expression of transgenes delivered by lentiviral vectors. *Science* 295, 868–872.

Lu, A., Zhang, H., Zhang, X., Wang, H., Hu, Q., Shen, L., Schaffhausen, B.S., Hou, W. and Li, L. (2004) Attenuation of SARS coronavirus by a short hairpin RNA expression plasmid targeting RNA-dependent RNA polymerase. *Virology* 324, 84–89.

Lu, L., Ho, Y. and Kwang, J. (2006) Suppression of porcine arterivirus replication by baculovirus-delivered shRNA targeting nucleoprotein. *Biochemical and Biophysical Research Communications* 340, 1178–1183.

Luo, G.X., Danetz, S. and Krystal, M. (1997) Inhibition of influenza viral polymerases by minimal viral RNA decoys. *Journal of General Virology* 78, 2329–2333.

Martinez, J. and Tuschl, T. (2004) RISC is a 5′ phosphomonoester-producing RNA endonuclease. *Genes and Development* 18, 975–980.

McCaffrey, A.P., Nakai, H., Pandey, K., Huang, Z., Salazar, F.H., Xu, H., Wieland, S.F., Marion, P.L. and Kay, M.A. (2003) Inhibition of hepatitis B virus in mice by RNA interference. *Nature Biotechnology* 21, 639–644.

McGrew, M.J., Sherman, A., Ellard, F.M., Lillico, S.G., Gilhooley, H.J., Kingsman, A.J., Mitrophanous, K.A. and Sang, H. (2004) Efficient production of germline transgenic chickens using lentiviral vectors. *Embo Reports* 5, 728–733.

Meister, G. and Tuschl, T. (2004) Mechanisms of gene silencing by double-stranded RNA. *Nature* 431, 343–349.

Miyagishi, M. and Taira, K. (2002) U6 promoter-driven siRNAs with four uridine 3′ overhangs efficiently suppress targeted gene expression in mammalian cells. *Nature Biotechnology* 20, 497–500.

Miyagishi, M., Sumimoto, H., Miyoshi, H., Kawakami, Y. and Taira, K. (2004) Optimization of an siRNA-expression system with an improved hairpin and its significant suppressive effects in mammalian cells. *Journal of Gene Medicine* 6, 715–723.

Moore, M.D., McGarvey, M.J., Russell, R.A., Cullen, B.R. and McClure, M.O. (2005) Stable inhibition of hepatitis B virus proteins by small interfering RNA expressed from viral vectors. *Journal of Gene Medicine* 7, 918–925.

Muller, M., Brenig, B., Winnacker, E.L. and Brem, G. (1992) Transgenic pigs carrying CDNA copies encoding the murine Mx1 protein which confers resistance to influenza-virus infection. *Gene* 121, 263–270.

Murakami, M., Ota, T., Nukuzuma, S. and Takegami, T. (2005) Inhibitory effect of RNAi on Japanese encephalitis virus replication in vitro and in vivo. *Microbiology and Immunology* 49, 1047–1056.

Nakayashiki, H. (2005) RNA silencing in fungi: mechanisms and applications. *FEBS Letters* 579, 5950–5957.

Napoli, C., Lemieux, C. and Jorgensen, R. (1990) Introduction of chimeric chalcone synthase gene into petunia results in reversible co-suppression of homologous gene *in trans*. *Plant Cell* 2, 279–289.

Ngo, H., Tschudi, C., Gull, K. and Ullu, E. (1998) Double-stranded RNA induces mRNA degradation in *Trypanosoma brucei*. *Proceedings of the National Academy of Sciences of the United States of America* 95, 14687–14692.

Nicholson, A.W. (1999) Function, mechanism and regulation of bacterial ribonucleases. *Fems Microbiology Reviews* 23, 371–390.

Nicholson, K.G., Wood, J.M. and Zambon, M. (2003) Influenza. *Lancet* 362, 1733–1745.

Nishitsuji, H., Ikeda, T., Miyoshi, H., Ohashi, T., Kannagi, M. and Masuda, T. (2004) Expression of small hairpin RNA by lentivirus-based vector confers efficient and stable gene-suppression of HIV-1 on human cells including primary non-dividing cells. *Microbes and Infection* 6, 76–85.

Nottle, M.B., Haskard, K.A., Verma, P.J., Du, Z.T., Grupen, C.G., Mcilfatrick, S.M., Ashman, R.J., Harrison, S.J., Barlow, H., Wigley, P.L., Lyons, I.G., Cowan, P.J., Crawford, R.J., Tolstoshev, P.L., Pearse, M.J., Robins, A.J. and d'Apice, A.J.F. (2001) Effect of DNA concentration on transgenesis rates in mice and pigs. *Transgenic Research* 10, 523–531.

Nykanen, A., Haley, B. and Zamore, P.D. (2001) ATP requirements and small interfering RNA structure in the RNA interference pathway. *Cell* 107, 309–321.

Paddison, P.J., Caudy, A.A., Bernstein, E., Hannon, G.J. and Conklin, D.S. (2002) Short hairpin RNAs (shRNAs) induce sequence-specific silencing in mammalian cells. *Genes and Development* 16, 948–958.

Park, J., Nadeau, P., Zucali, J.R., Johnson, C.M. and Mergia, A. (2005) Inhibition of simian immunodeficiency virus by foamy virus vectors expressing siRNAs. *Virology* 343, 275–282.

Parrish, S., Fleenor, J., Xu, S., Mello, C. and Fire, A. (2000) Functional anatomy of a dsRNA trigger: differential requirement for the two trigger strands in RNA interference. *Molecular Cell* 6, 1077–1087.

Paul, C.P., Good, P.D., Winer, I. and Engelke, D.R. (2002) Effective expression of small interfering RNA in human cells. *Nature Biotechnology* 20, 505–508.

Paule, M.R. and White, R.J. (2000) Transcription by RNA Polymerases I and III. *Nucleic Acids Research* 28, 1283–1298.

Peng, J., Zhao, Y., Mai, J., Pang, W.K., Wei, X., Zhang, P. and Xu, Y. (2005) Inhibition of hepatitis B virus replication by various RNAi constructs and their pharmacodynamic properties. *Journal of General Virology* 86, 3227–3234.

Pfeffer, S., Zavolan, M., Grasser, F.A., Chien, M.C., Russo, J.J., Ju, J.Y., John, B., Enright, A.J., Marks, D., Sander, C. and Tuschl, T. (2004) Identification of virus-encoded microRNAs. *Science* 304, 734–736.

Rao, M. and Sockanathan, S. (2005) Molecular mechanisms of RNAi: implications for development and disease. *Birth Defects Research Communications* 75, 28–42.

Ratcliff, F., Harrison, B.D. and Baulcombe, D.C. (1997) A similarity between viral defense and gene silencing in plants. *Science* 276, 1558–1560.

Reynolds, A., Leake, D., Boese, Q., Scaringe, S., Marshall, W.S. and Khvorova, A. (2004) Rational siRNA design for RNA interference. *Nature Biotechnology* 22, 326–330.

Rose, S.D., Kim, D.H., Amarzguioui, M., Heidel, J.D., Collingwood, M.A., Davis, M.E., Rossi, J.J. and Behlke, M.A. (2005) Functional polarity is introduced by Dicer processing of short substrate RNAs. *Nucleic Acids Research* 33, 4140–4156.

Rubinson, D.A., Dillon, C.P., Kwiatkowski, A.V., Sievers, C., Yang, L., Kopinja, J., Zhang, M., McManus, M.T., Gertler, F.B., Scott, M.L. and Parijs, L.V. (2003) A lentivirus-based system to functionally silence genes in primary mammalian cells, stem cells and transgenic mice by RNA interference. *Nature Genetics* 33, 401–406.

Samols, M.A., Hu, J.H., Skalsky, R.L. and Renne, R. (2005) Cloning and identification of a miRNA cluster within the latency-associated region of Kaposi's sarcoma-associated herpesvirus. *Journal of Virology* 79, 9301–9305.

Schramm, L. and Hernandez, N. (2002) Recruitment of RNA polymerase III to its target promoters. *Genes and Development* 16, 2593–2620.

Schutz, S. and Sarnow, P. (2006) Interaction of viruses with the mammalian RNA interference pathway. *Virology* 344, 151–157.

Schwarz, D.S., Hutvagner, G., Du, T., Xu, Z.S., Aronin, N. and Zamore, P.D. (2003) Asymmetry in the assembly of the RNAi enzyme complex. *Cell* 115, 199–208.

Schwarz, D.S., Tomari, Y. and Zamore, P.D. (2004) The RNA-induced silencing complex is a Mg2+-dependent endonuclease. *Current Biology* 14, 787–791.

Shen, C., Buck, A.K., Liu, X., Winkler, M. and Reske, S.N. (2003) Gene silencing by adenovirus-delivered siRNA. *FEBS Letters* 539, 111–114.

Siolas, D., Lerner, C., Burchard, J., Ge, W., Linsley, P.S., Paddison, P.J., Hannon, G.J. and Cleary, M.A. (2005) Synthetic shRNAs as potent RNAi triggers. *Nature Biotechnology* 23, 227–231.

Smith, K.R. (1999) Sperm cell mediated transgenesis: a review. *Animal Biotechnology* 10, 1–13.

Staeheli, P., Horisberger, M.A. and Haller, O. (1984) Mx-dependent resistance to influenza viruses is induced by mouse interferons alpha and beta but not gamma. *Virology* 132, 456–461.

Sullivan, C.S., Grundhoff, A.T., Tevethia, S., Pipas, J.M. and Ganem, D. (2005) Sv40-encoded microRNAs regulate viral gene expression and reduce susceptibility to cytotoxic T cells. *Nature* 435, 682–686.

Svoboda, P., Stein, P., Hayashi, H. and Schultz, R.M. (2000) Selective reduction of dormant maternal mRNAs in mouse oocytes by RNA interference. *Development* 127, 4147–4156.

Tabara, H., Sarkissian, M., Kelly, W.G., Fleenor, J., Grishok, A., Timmons, L., Fire, A. and Mello, C.C. (1999) The RDE-1 gene, RNA interference, and transposon silencing in *C. elegans*. *Cell* 99, 123–132.

Takigawa, Y., Nagano-Fujii, M., Deng, L., Hidajat, R., Tanaka, M., Mizuta, H. and Hotta, H. (2004) Suppression of hepatitis C virus replicon by RNA interference directed against the NS3 and NS5B regions of the viral genome. *Microbiology and Immunology* 48, 591–598.

Tijsterman, M., Ketting, R.F. and Plasterk, R.H. (2002) The genetics of RNA silencing. *Annual Review of Genetics* 36, 489–519.

Tiley, L.S., Hagen, M., Matthews, J.T. and Krystal, M. (1994) Sequence-specific binding of the influenza-virus RNA-polymerase to sequences located at the 5′ ends of the viral RNAs. *Journal of Virology* 68, 5108–5116.

Tiscornia, G., Tergaonkar, V., Galimi, F. and Verma, I.M. (2004) CRE recombinase-inducible RNA interference mediated by lentiviral vectors. *Birth Defects Research Communications* 101, 7347–7351.

Tompkins, S.M., Lo, C.Y., Tumpey, T.M. and Epstein, S.L. (2004) Protection against lethal influenza virus challenge by RNA interference in vivo. *Proceedings of the National Academy of Sciences of the United States of America* 101, 8682–8686.

Ui-Tei, K., Naito, Y., Takahashi, F., Haraguchi, T., Ohki-Hamazaki, H., Juni, A., Ueda, R. and Saigo, K. (2004) Guidelines for the selection of highly effective sirna sequences for mammalian and chick RNA interference. *Nucleic Acids Research* 32, 936–948.

van de Wetering, M., Oving, I., Muncan, V., Pon Fong, M.T., Brantjes, H., van Leenen, D., Holstege, F.C., Brummelkamp, T.R., Agami, R. and Clevers, H. (2003) Specific inhibition of gene expression using a stably integrated, inducible small-interfering-RNA vector. *EMBO Reports* 4, 609–615.

Van den Haute, C., Eggermont, K., Nuttin, B., Debyser, Z. and Baekelandt, V. (2003) Lentiviral vector-mediated delivery of short hairpin RNA results in persistent knockdown of gene expression in mouse brain. *Human Gene Therapy* 14, 1799–1807.

Ventura, A., Meissner, A., Dillon, C.P., McManus, M., Sharp, P.A., Van Parijs, L., Jaenisch, R. and Jacks, T. (2004) Cre-lox-regulated conditional RNA interference from transgenes. *Proceedings of the National Academy of Sciences of the United States of America* 101, 10380–10385.

Waterhouse, P.M., Graham, M.W. and Wang, M.B. (1998) Virus resistance and gene silencing in plants can be induced by simultaneous expression of sense and antisense RNA. *Proceedings of the National Academy of Sciences of the United States of America* 95, 13959–13964.

Whitelaw, C.B.A., Radcliffe, P.A., Ritchie, W.A., Carlisle, A., Ellard, F.M., Pena, R.N., Rowe, J., Clark, A.J., King, T.J. and Mitrophanous, K.A. (2004) Efficient generation of transgenic pigs using equine infectious anaemia virus (EIAV) derived vector. *Febs Letters* 571, 233–236.

Wightman, B., Ha, I. and Ruvkun, G. (1993) Posttranscriptional regulation of the heterochronic gene lin-14 by lin-4 mediates temporal pattern formation in *C. elegans*. *Cell* 75, 855–862.

Williams, B.R. (1997) Role of the double-stranded RNA-activated protein kinase (PKR) in cell regulation. *Biochemical Society Transactions* 25, 509–513.

Wilmut, I., Schnieke, A.E., Mcwhir, J., Kind, A.J. and Campbell, K.H.S. (1997) Viable offspring derived from fetal and adult mammalian cells. *Nature* 386, 810.

Wilmut, I., Beaujean, N., De Sousa, P.A., Dinnyes, A., King, T.J., Paterson, L.A., Wells, D.N. and Young, L.E. (2002) Somatic cell nuclear transfer. *Nature* 419, 583–586.

Wu, H.L., Huang, L.R., Huang, C.C., Lai, H.L., Liu, C.J., Huang, Y.T., Hsu, Y.W., Lu, C.Y., Chen, D.S. and Chen, P.Y. (2005a) RNA interference-mediated control of hepatitis B virus and emergence of resistant mutant. *Gastroenterology* 128, 708–716.

Wu, Y., Huang, A.L., Tang, N., Zhang, B.Q. and Lu, N.F. (2005b) Specific anti-viral effects of RNA interference on replication and expression of hepatitis B virus in mice. *Chinese Medical Journal* 118, 1351–1356.

Yang, D., Lu, H. and Erickson, J.W. (2000) Evidence that processed small dsRNAs may mediate sequence-specific mRNA degradation during RNAi in Drosophila embryos. *Current Biology* 10, 1191–1200.

Yokota, T., Sakamoto, N., Enomoto, N., Tanabe, Y., Miyagishi, M., Maekawa, S., Yi, L., Kurosaki, M., Taira, K., Watanabe, M. and Mizusawa, H. (2003) Inhibition of intracellular hepatitis C virus replication by synthetic and vector-derived small interfering RNAs. *EMBO Reports* 4, 602–608.

Yu, J.Y., DeRuiter, S.L. and Turner, D.L. (2002) RNA interference by expression of short-interfering RNAs and hairpin RNAs in mammalian cells. *Proceedings of the National Academy of Sciences of the United States of America* 99, 6047–6052.

Yu, J.Y., Taylor, J., Deruiter, S.L., Vojtek, A.B. and Turner, D.L. (2003) Simultaneous inhibition of GSK3 alpha and GSK3 beta using hairpin siRNA expression vectors. *Molecular Therapy* 7, 228–236.

Zamore, P.D. and Haley, B. (2005) Ribo-gnome: the big world of small RNAs. *Science* 309, 1519–1524.

Zamore, P.D., Tuschl, T., Sharp, P.A. and Bartel, D.P. (2000) RNAi: double-stranded RNA directs the ATP-dependent cleavage of mRNA at 21 to 23 nucleotide intervals. *Cell* 101, 25–33.

Zeng, Y., Wagner, E.J. and Cullen, B.R. (2002) Both natural and designed micro RNAs can inhibit the expression of cognate mRNAs when expressed in human cells. *Molecular Cell* 9, 1327–1333.

Zeng, Y., Cai, X. and Cullen, B.R. (2005) Use of RNA polymerase II to transcribe artificial microRNAs. *Methods in Enzymology* 392, 371–380.

Zhao, L.J., Jian, H. and Zhu, H. (2003) Specific gene inhibition by adenovirus-mediated expression of small interfering RNA. *Gene* 316, 137–141.

Zhou, H., Xia, X.G. and Xu, Z. (2005) An RNA polymerase II construct synthesizes short-hairpin RNA with a quantitative indicator and mediates highly efficient RNAi. *Nucleic Acids Research* 33, e62.

9 Towards an Integrated Approach to Stochastic Process-based Modelling: with Applications to Animal Behaviour and Spatio-temporal Spread

Glenn Marion,[1] David M. Walker,[2] Alex Cook,[3] David L. Swain[4] and Mike R. Hutchings[5]

[1]*Biomathematics and Statistics Scotland, Edinburgh, UK;* [2]*Department of Applied Mathematics, Hong Kong Polytechnic University, Hong Kong;* [3]*School of Mathematical and Computer Sciences, Heriot Watt University, Edinburgh, UK;* [4]*CSIRO Livestock Industries, Australia;* [5]*Scottish Agricultural College, Edinburgh, UK*

Abstract

Using example applications from our recent research we illustrate the development of an integrated approach to modelling biological processes based on stochastic modelling techniques. The goal of this programme of research is to provide a suite of mathematical and statistical methods to enable models to play a more central role in the development of scientific understanding of complex biological systems. The resulting framework should allow models to both inform, and be informed by data collection, and enable probabilistic risk assessments to reflect inherent variability and uncertainty in current knowledge of the system in question. We focus on discrete state–space Markov processes as they provide a general and flexible framework to both describe and infer the behaviour of a broad range of systems. Unfortunately the non-linearities required to model many real-world systems typically mean that such discrete state–space stochastic processes are intractable to analytic solution, necessitating the use of simulation and analytic approximations. We show how to formulate stochastic process-based models within this framework and discuss the representation of spatial and temporal heterogeneity. Simple population models are developed and used to illustrate these concepts. We describe how to simulate from such models, and compare them with their deterministic counterparts. In addition, we discuss two methods, closure schemes and linearization about steady-states, which can be used to obtain analytic insights into model behaviour. We outline how to conduct parameter estimation for such models when, as is typically the case for biological and agricultural systems, only partial observations are available. Having focused on familiar population-level models in introducing our integrated approach, its wider applicability is illustrated by two contrasting applications from our recent research. The first example combines the development and analysis of an agent-based model describing grazing in heterogeneous environments, with parameter inference based on data generated using a transponder system in a behavioural experiment on dairy cows. The second example makes use of large-scale data describing bio-geographical features of the landscape and the spatio-temporal spread of an alien plant to estimate the parameters of a stochastic model of dispersal and establishment.

Introduction

A key difficulty faced in the study of biological and agricultural systems is their immense complexity. This fact makes even the observation, recording and cataloguing of biological phenomena a formidable task and such problems are compounded when one seeks to understand, predict and manipulate the dynamic processes underlying this observed complexity. Mathematical modelling currently plays an important role in developing scientific understanding of complex biological processes, and model-based risk assessments make it relevant to policy makers and resource managers. However, the gathering pace of data acquisition and consequent advances in knowledge require the continuing development of both mathematical methods, to cope with increasing complexity, and statistical methods, to fully integrate data and models. Ideally such methods should allow mathematical models to both inform, and be informed by data collection, and enable probabilistic risk assessments to reflect inherent variability and uncertainty in current knowledge of the system in question. In this chapter, using examples from our research, we illustrate the development of such an integrated approach to modelling biological processes based on stochastic modelling techniques. However, the picture is still somewhat incomplete and we discuss areas where further research effort is required to fully integrate model development and empirical effort.

In the next few paragraphs we outline our approach, and in particular motivate our focus on stochastic models. We discuss: (i) the implementation, simulation and analytic treatment of stochastic models; (ii) the incorporation of individual variability and spatio-temporal heterogeneity, which are inherent properties of many biological systems; and (iii) statistically correct parameter inference and probabilistic model-based risk assessment, accounting for system variability and uncertainty in models and parameters. Advances in both data collection, such as radio-tracking, contact-logging and Global Positioning Systems (GPS), and data collation, such as the development of Geographic Information Systems (GIS) and the creation of large-scale species atlases, enable parameter inference across a widening range of applications. Despite increases in available computing power statistically correct inference is still a significant bottleneck for complex models. Moreover, a commonly encountered problem in modelling biological systems is the explosion of model complexity, leading to poorly understood behaviour and low predictive ability. Our approach therefore is to develop simple parsimonious stochastic process-based models which are more amenable to analysis and rigorous statistical treatment than complex mechanistic systems models.

An important feature of many biological and agricultural systems is that of heterogeneity in both space and time, and between individuals. It has been shown that such variability can quantitatively and qualitatively change model behaviour, and should therefore be considered. A particularly powerful approach is to model heterogeneity by introducing a stochastic, or random, element into models. The model then predicts variation, in time and space or between individuals, in the outcomes of a particular event. This has the advantage of representing heterogeneity parsimoniously with relatively few parameters and without increasing the number of variables needed to represent the state of the system. Rand and Wilson (1991) partition such stochastic effects for epidemics into two types: (i) demographic fluctuations arising from the stochastic nature of contacts and infection events; and (ii) randomness in the environment and thereby in the parameters affecting the epidemic. This demographic and environmental stochasticity can produce model behaviour quite distinct from deterministic implementation (see, for example, Gurney and Middleton, 1996; Kokko and Ebenhard, 1996; Wilson and Hassell, 1997; Marion *et al.*, 1998). As a first step the stochastic approach models heterogeneity at the population level and there is no need to model individuals explicitly. However, the representation of heterogeneity can be enhanced by combining such stochasticity with an explicit representation of space, and/or of individuals. In the simplest case, individuals and locations are treated identically, but none the less variation will be observed when comparing across individuals or locations for a given model run. Such spatial and temporal heterogeneities have been shown to be critical in understanding epidemiological and ecological processes (for a selection of papers see, e.g., Tilman and Karieva, 1997). In some circumstances it may be possible to partially attribute differences between locations or individuals to variations in some measurable factors or covariates, for example land-use category or animal status (lactating/non-lactating). In such cases the response will remain stochastic but the probability of different outcomes will depend on the covariates and thus vary between individuals or locations. For example, in studying the spread of an invasive plant, Cook *et al.* (2007) define the suitability of local habitats in terms of land-use classes, temperature and altitude.

A useful framework for developing stochastic process-based models is that of discrete state–space Markov processes (Cox and Miller, 1965). The theory provides a general and flexible framework to both describe and infer behaviour. It is a powerful tool for developing models and provides a framework to parameterize such models from data. Crucially this framework enables the discrete nature of populations to be modelled in contrast to models based on deterministic ordinary differential equations, or diffusion-like approximations such as stochastic differential equations (see below). The Markovian assumption is extremely widely used, for example being implicit in many ordinary differential equation models. However, in some cases a non-Markovian description of the system might be more parsimonious, but in principle any system can be expressed as a Markov process by a suitable expansion of the state–space. Unfortunately the non-linearities required to model many real-world systems typically mean that such discrete state–space stochastic processes are intractable to analytic solution (Isham, 1991; Bolker and Pacala, 1997; Filipe and Gibson, 1998; Matis *et al.*, 1998; Keeling, 2000a,b); however, simulation is usually straightforward (Renshaw, 1991) unless the expected number of events is extremely large. The stochastic approach can also point the way towards better deterministic process models by accounting for variability and spatial heterogeneity using suitable limiting processes and approximations (Whittle, 1957; Isham, 1991; Bolker and Pacala, 1997; Matis *et al.*, 1998; Keeling, 2000a,b; Holmes *et al.*, 2004; Marion *et al.*, 2005). Such approaches are typically based on so-called closure approximations, which we discuss below, and they offer analytic insights into system behaviour in addition to a computationally efficient alternative to, and check on, stochastic simulation. A widely applicable alternative approximation procedure, based on expansion around the steady state (Bailey, 1963), re-casts the model in terms of linearized stochastic differential equations from which fluctuation characteristics may be obtained (Nisbet and Gurney, 1982; Marion *et al.*, 2000).

An important aspect of model development is the handling of uncertainty and the integration with data, and these are key difficulties for process-based models. Statistical approaches are data driven, naturally incorporate uncertainty, and a large toolkit is available for parameter estimation and assessing model performance. On the other hand such methodology is poorly developed for process-based models.

Typically uncertainty can be broken down into uncertainty about the parameter values for specific models, and model uncertainty. If parameters are jointly estimated from data in a statistical fashion then the effect of the resulting parameter uncertainty can be accounted for in model output by repeatedly drawing parameter combinations from their joint distribution and running the model. If a distribution over a set of stochastic models can also be inferred then predictions can reflect the inherent variability of the system and uncertainty in both models (Draper, 1995) and parameter values. It is easy to see that a proper accounting of variability and uncertainty is crucial in any model-based risk assessment as it changes the nature of the advice given to resource managers or policy makers from unequivocal recommendation to probabilistic (e.g. a given course of action will result in the desired outcome with a certain probability). This distinction is especially important for phenomena, such as epidemics, which exhibit threshold behaviour.

Fitting stochastic dynamical models directly to observations allows parameter uncertainty to be treated more completely since the model itself defines the error distribution and implicitly accounts for correlations in the data. In contrast estimation based on least-squares, as often used for deterministic models, typically makes the additional assumptions that errors are uncorrelated and Gaussian. However, a full analytic treatment of parameter estimation for dynamical stochastic systems is rarely feasible since observations of biological processes from practical experiments or field studies typically record only a subset of the information that defines the evolution of the system. In such cases we must 'integrate out' the missing information, which typically leads to analytically intractable high dimensional integrals. Recent advances in computing power mean that sampling methods and in particular Markov chain Monte Carlo, or MCMC (Metropolis *et al.*, 1953; Hastings, 1970; Gelfand and Smith, 1990; Smith and Roberts, 1993; Besag and Green, 1993), are flexible enough to be used to make inferences about missing data and unknown parameters by providing robust approximations to such difficult integrals. The methods are based on Gibbs sampling, Metropolis–Hastings algorithms and the methodological advance of reversible-jump MCMC, which is specifically tailored to explore state spaces of varying dimension (Gelman *et al.*, 1995; Green, 1995; Gilks *et al.*, 1996; Gamerman, 1997). The need to sample from state spaces of varying dimension arises here because the observed data does not determine the numbers of all event types. Therefore sampling from the range of plausible reconstructions of the missing data implies sampling over different numbers of reconstructed events. It should be noted that this approach is limited to relatively small numbers of missing events although later we will show an example requiring the estimation of $\sim 10^4$ missing events.

The joint estimation of parameters and missing data (also referred to as nuisance parameters) is typically conducted within the framework of Bayesian estimation (Lee, 2004) in which explicit quantification of uncertainty in model parameters (and indeed the missing data) is given by their posterior distributions with respect to the observed data and of course the model. A requirement, which should be mentioned, is the need for the selection of subjective prior distribution of parameters in the Bayesian methodology. A potential advantage of this approach is that the shape of the prior can be chosen to quantify information gained from previous studies. In many cases, however, little prior information is available and the prior distribution is often chosen to be uniform, perhaps over some range of parameters determined from the literature. In either case, prior influence lessens with more information, and large observation samples mean that posterior distributions are determined largely by the data. In addition the robustness of results to prior assumptions can be checked. The Bayesian approach coupled with MCMC techniques has been applied in recent years to infer the parameters of stochastic epidemic models (Gibson, 1997; Gibson and Renshaw, 1998, 2001a,b; Renshaw and Gibson, 1998; O'Neill and Roberts, 1999). In principle the Bayesian approach can be extended to enable model uncertainty (among a defined set of models) to be accounted for. However, such methods are computationally expensive and this approach has been applied rarely (Gibson and Renshaw, 2001b).

In the next section we show how to formulate stochastic process-based models and discuss how they may be used to represent spatial and temporal heterogeneity. A simple population model, the immigration–death process is developed and used to introduce discrete state–space Markov processes. Subsequently this model is extended to include a simple disease dynamic, and then used to illustrate how spatial heterogeneity can be handled within this stochastic framework. We describe how to simulate from such models, and compare them with their deterministic counterparts. In addition, we discuss two methods, closure schemes and linearization about steady-states, which can be used to obtain analytic insights in to model behaviour. This culminates in a description of a general discrete state–space Markov process, and in the following section we outline how to conduct parameter estimation for such models when, as is typically the case for biological and agricultural systems, only partial observations are available. Having focused on familiar population-level models in introducing our integrated approach to stochastic modelling, we then illustrate its wider applicability by discussing two contrasting applications. First, we describe in some detail the formulation and analysis of a model of grazing behaviour, for which we infer parameter values from partial data gathered in a behavioural experiment on dairy cows. Finally, we consider the spread of an alien plant species across Britain, and describe the development of a model which incorporates the effect of spatial covariates on the suitability for colonization. Bayesian estimation of parameters enables uncertainty in parameter values to be incorporated into model outputs predicting the risk of future colonizations.

Formulation and Analysis of Stochastic Models

In this section we illustrate the formulation and analysis of stochastic process-based models using a simple toy example. We start with perhaps the simplest non-spatial population model imaginable, namely the immigration–death process. Subsequently we introduce some non-linearity by modelling the spread of an infection in this population, and finally consider the impact of spatial heterogeneity on the system. We use this example to demonstrate how to formulate discrete state–space Markov processes, and contrast stochastic model behaviour with that of deterministic counterparts. The solution of the linear immigration–death model is discussed and the complications introduced by adding epidemiology used to illustrate the difficulties posed by the analysis of non-linear stochastic processes. The relative merits of several approximate approaches are discussed in this context. Finally we consider the effect of spatial heterogeneity and the utility of approximations to such complex stochastic processes.

Model formulation and stochastic simulation for a simple example

To begin, consider a population which is subject only to the twin effects of immigration at some constant rate ν, and death at per-capita rate μ. Then a standard deterministic continuous time model would describe the size of this population at time t by the positive real valued variable $X(t)$ whose rate of change with time is given by the ordinary differential equation:

$$\frac{dX(t)}{dt} = \nu - \mu X(t) \tag{9.1}$$

Where immigration balances mortality the population reaches the steady state where the rate of change $dX(t)/dt = 0$. This steady state condition is equivalent to $\nu - \mu X(t) = 0$, which is satisfied once the population size reaches $X = \nu/\mu$.

Demographic stochasticity

Reformulating this model as a discrete state–space Markov process introduces fluctuations in immigrations and deaths, which represent demographic stochasticity that crudely accounts for individual variation. To do so the population size is now represented, more realistically, as an integer-valued stochastic process $n(t)$ which increases by 1 when an immigration event occurs and decreases by 1 if a death event occurs. The occurrence of each event is governed by the rates defined above for the deterministic model as follows. The probability of an immigration event occurring during a sufficiently small time interval from t to $t + \delta t$, which is written as $(t,t + \delta t)$, is proportional to the immigration rate ν and the length δt of the time interval:

$$P(immigraton : n \rightarrow n+1) = \nu\delta t$$

Similarly, the probability that a death occurs in the short time interval $(t,t+\delta t)$, given that the population is of size $n(t)$ at time t, is proportional to the death rate $\mu\, n(t)$

$$P(death: n \to n-1) = \mu n(t)\delta t$$

When the system is in state $n(t)$ the time to the next event is exponentially distributed with the total event rate $R(n; \nu,\mu) = \nu + \mu\, n(t)$ (see, e.g., Cox and Miller, 1965; Renshaw, 1991). This means that the time to the next event can be generated by calculating $\tau = -\ln(y)/R(n; \nu,\mu)$ where $y \sim U(0,1)$ is a random number drawn uniformly between zero and 1 (Renshaw, 1991). Note that uniform random number generators are available as standard in many programming languages, numerical libraries and software applications. The time is then advanced to $t + \tau$, and an immigration event occurs with probability $\nu/R(n; \nu,\mu)$ or else a death event occurs. To simulate this step draw a second random number $y_2 \sim U(0,1)$, then an immigration event occurs if $y_2 \leq \nu/R(n; \nu,\mu)$ otherwise choose a death event. The population size is then adjusted according to the event type chosen (e.g. reduced by 1 for a death and increased by 1 for an immigration event) and the process can be simulated to arbitrary time in the future by iterating this procedure. For simple linear processes, such as immigration–death, simulation is not always required as analytic solutions can often be obtained (see below).

The model underlying both the deterministic representation (1) and the stochastic formulation can be summarized in terms of the definition of the state–space and the specification of events and associated mean rates, as shown in Table 9.1. The stochastic model then follows from specifying that the event times are exponentially distributed, or equivalently that if rate R_i is associated with event i the probability of the occurrence of event i in a small time interval $(t,t+\delta t)$ is given by $R_i\, \delta t$.

Environmental fluctuations

Within this framework it is also relatively straightforward to account for temporal fluctuations in the environment by making the parameters of the model vary in time. The immigration rate $\nu(t)$ and per capita death rate $\mu(t)$ can be deterministic functions of time (e.g. sinusoidal, reflecting diurnal or seasonal variation), stochastic processes representing random fluctuations or a combination of both. Although the exact simulation algorithm described earlier can be adapted for deterministically varying parameters, the easiest way to simulate the process is to adopt the following approximate algorithm. First, choose $\delta t = \min(1/R(n;\nu(t),\mu(t)), \delta t_{\min})$ where $\delta t_{\min}$ is chosen so that changes in the time-varying parameters can be ignored in the interval $(t,t+\delta t)$, and update time to $t+\delta t$. Second, generate $y \sim U(0,1)$ then if $y<\nu(t)\delta t$ choose an immigration event, else if $y<\nu(t)\delta t + \mu(t)n(t)\delta t$ then a death occurs, otherwise no event occurs.

Marion *et al.* (2000) explore the modelling of environmental fluctuations within stochastic population models. They suggest using some simple continuous-valued stochastic processes to model random variation in rate parameters. In particular they consider transformations $B(Z(t))$ of the auto-correlated Gaussian mean-reverting Uhlenbeck-Ornstein process $Z(t)$. Two transformations, $B(Z)=Z^2$ and $B(Z)=e^Z$, are employed to ensure that the rate parameters (e.g. $\nu(t) = B(Z(t))$) remain

Table 9.1. Definition of the linear immigration–death process where the population size $n(t)$ is governed by immigration at rate ν, and death at rate μ per capita.

Event description	Change in state space δn	Event rate at time t
Immigration	+1	ν
Death	−1	$\mu\, n(t)$

non-negative. $Z(t)$ can be generated by iterating the following difference equation:

$$Z(t+\delta t) = Z(t) + b(\overline{Z} - Z(t))\delta t + \sigma\sqrt{\delta t}\, dB(t)$$

where $dB(0)$, $dB(\delta t)$, $dB(2\delta t)$,..., $dB(n\delta t)$,..., are uncorrelated Normal random variables with zero mean and unit variance. In equilibrium the mean and variance of $Z(t)$ are $\overline{Z}$ and $\sigma^2 / 2b$ respectively. A third characteristic of this noise process is its auto-correlation, or colour and this has an important effect in determining how the system responds to environmental perturbations. It is well known that the periodicity of deterministic perturbations strongly determines size of the resulting population fluctuations. For example, Roberts and Grenfell (1992) analyse the effect of seasonal fluctuations on the dynamics of a generic model of nematode infection in ruminants. Based on linearization around a steady state it can be shown that each system has a resonant frequency and any deterministic environmental perturbation at or near this will drive large fluctuations compared with environmental perturbations of a similar magnitude but of a different frequency. A similar effect is seen with stochastic perturbations where environmental noise with a colour (auto-correlation) close to the resonant frequency will produce relatively larger fluctuations.

Analysis of stochastic models

Marion *et al.* (2000) solve the stochastic immigration–death process, showing for example that if the population size $n(0) = 0$ then the population has a Poisson distribution with mean $(1-e^{-\mu t})\nu / \mu$, which tends asymptotically to the deterministic steady state ν / μ. Indeed it is straightforward to show that in this linear model the expected population size $E[n(t)]$ obeys the deterministic equation (9.1). To do so consider the expected change, at time $t + \delta t$, in a population of size $n(t)$ at time t:

$$n(t+\delta t) = n(t) + (+1)\nu\delta t + (-1)\mu n(t)\delta t$$

in which the right-most terms are the event probabilities multiplied by the change associated with each. Rearranging this expression one obtains:

$$\frac{n(t+\delta t) - n(t)}{\delta t} = \nu - \mu n(t)$$

Finally, taking the expectation at time t and taking the limit $\delta t \to 0$ reveals that:

$$\frac{dE[n(t)]}{dt} = \nu - \mu E[n(t)]$$

which is seen to be equivalent to the deterministic equation (9.1) on identifying $X(t) = E[n(t)]$. We will see later, however, that when the event rates are non-linear functions of the state variables (population size in this case) the deterministic model does not describe the evolution of the mean. Indeed if one replaces the linear death rate $\mu n(t)$ in the preceding derivation with, say, $\mu n^2(t)$ we find that:

$$\frac{dE[n(t)]}{dt} = \nu - \mu E[n^2(t)]$$

That is the evolution of the first-order moment $E[n(t)]$ depends on the second-order moment $E[n^2(t)]$ which is related to the population variance (i.e. $Var[n(t)] = E[n^2(t)] - E[n(t)]^2$). It turns out that a similar equation can be obtained for the second-order moment, but that this depends on the third-order moment $E[n^3(t)]$, and in general the evolution of the kth order moment $E[n^k(t)]$ depends on the $(k+1)^{th}$ order moment $E[n^{k+1}(t)]$. This effect, which is indicative of a more general intractability of non-linear stochastic processes, is termed the problem of closure since any finite set of equations for the moments is not closed in the sense that it will depend on an additional variable not included in the set (e.g. in this case the $(k+1)^{th}$ order moment). The upshot of this is that any finite set of moment equations can't be solved, even numerically. Similar problems arise with statistics other than the raw moments considered here, for example central moments or cumulants (Kendall, 1994). Moreover, other methods of solution become problematic when non-linearities are introduced into stochastic processes (see, e.g., Renshaw, 1991) and in general it is not currently possible to obtain exact solutions for such systems. In such cases simulation remains the only generally applicable exact method for revealing model behaviour.

One approach to the development of approximations to the discrete state–space Markov process is to add a stochastic term representing noise on to the deterministic equation (9.1). Stochastic calculus provides a formal framework for doing this, but a more intuitive approach is to consider the stochastic difference equation of the form:

$$\delta X(t) = f(X(t), \nu, \mu)\delta t + \sum_{i=1}^{k} g_i(X(t), \nu, \mu) \sqrt{\delta t}\,\delta B_i(t) \qquad (9.2)$$

Where the functions f and g depend on state variable $X(t)$ and the model parameters ν and μ, and $\delta B_i(0), \delta B_i(\delta t), \delta B_i(2\delta t), \ldots, \delta B_i(n\delta t), \ldots$, are uncorrelated Normal random variables with zero mean and unit variance. In the limit $\delta t \to 0$, this can be expressed more formally as a stochastic differential equation (SDE, see, e.g., Mao, 1997). The uncorrelated nature of the noise terms means that, to first order in δt, the mean and variance of the update $\delta X(t)$ are given by:

$$E[\delta X(t)] = f(X(t), \nu, \mu)\delta t \text{ and}$$
$$E[\delta X^2(t)] = \sum_{i=1}^{k} g_i^2(X(t), \nu, \mu)\delta t$$

The corresponding moments of the updates for the discrete-state–space Markov process model are $E[\delta n(t)] = (\nu - \mu n(t))\delta t$ and $E[\delta n^2(t)] = (\nu + \mu n(t))\delta t$, which if there is a separate noise term for each event type suggests the following identities:

$$f(X(t), \nu, \mu) = \nu - \mu X(t), g_1(X(t), \nu, \mu) = \nu$$
$$\text{and } g_2(X(t), \nu, \mu) = \mu X(t)$$

Stochastic calculus gives mathematical meaning to the difference equation (9.2) in the limit $\delta t \to 0$, and provides tools for the analysis of the resulting SDE (Mao, 1997). However, as noted earlier it is straightforward to simulate the process by iterating equation (9.2), and in some cases it is possible to show that this is a numerical solution which converges to the corresponding SDE as $\delta t \to 0$ (Kloeden and Platen, 1992; Marion *et al.*, 2002a). It should be noted that this system is simply a continuous approximation to the integer-valued Markov process. One particularly general approach to the analysis of the SDE system is to linearize around the fixed points of the deterministic dynamics (i.e. X where $f(X(t), \nu, \mu) = 0$). The fluctuation characteristics (expected values, variances and time-lagged correlations) around this deterministic equilibrium can typically be obtained by spectral analysis (see, e.g., Nisbet and Gurney, 1982; Marion *et al.*, 2000). However, while this approach is quite general in its applicability, for example to demographic and environmental stochasticity, and can provide surprisingly accurate approximations to the discrete state–space Markovian system in equilibrium, it is not applicable to transient aspects of the process.

Closure methods (Whittle, 1957; Isham, 1991; Bolker and Pacala, 1997; Filipe and Gibson, 1998; Matis *et al.*, 1998; Keeling, 2000a,b; Holmes *et al.*, 2004; Marion *et al.*, 2005) on the other hand are applicable to both transient and equilibrium regimes, since they are based on equations describing the temporal evolution of quantities such as moments discussed above. As we saw above, closure methods are not necessary for linear systems, but non-linearity plays a central role in many biological systems and therefore such techniques are typically required for essentially all biologically plausible models. Moment-closure techniques are often based on the system of equations describing the evolution of moments up to some finite order k, which for non-linear models will depend on moments of order greater than k (in the case above the evolution of the kth order moments depended on the $(k+1)^{th}$ order moments). The system of equations is closed at order k, by making some assumption which enables moments of orders higher than k to be written in terms of moments of order less than or equal to k. This problem can be expressed in terms of raw moments $E[n^k(t)]$ as we have done here, or in terms of central moments $E\lfloor(n(t) - E[n(t)])^k\rfloor$, or so-called cumulants (the kth order cummulant is obtained by evaluating the kth derivative with respect to θ of $\ln(E\lfloor e^{\theta n(t)}\rfloor)$ at $\theta = 0$). The first-order cummulant corresponds to the expected value and the second-order to the variance (Kendall, 1994). The lowest-order closure schemes truncate the system of equations at $k = 1$, which corresponds to ignoring fluctuations and therefore reproduces the deterministic version of the model discussed above. Some stochastic features can be retained by assuming that cumulants (Matis and Kiffe, 1996; Matis *et al.*, 1998) or central moments (Bolker and Pacala, 1997) are zero above order $k > 2$. For example cummulant truncation at second-order is equivalent to assuming a Normal distribution (Whittle, 1957). Moment-closure techniques are based on generalizing this by making alternative distributional assumptions. Typically the parameters of the chosen distribution are determined from the set of moments up to order k, and higher-order moments can then be written in terms of these, thereby closing the system of equations. For example, Normal and (e.g. log) transformed Normal distributions are determined by first- and second-order moments, and provide expressions for third- and higher-order moments. Typically standard distributions are used (e.g. binomial, Poisson, Normal, log-Normal, etc.), but these may not always be suitable; for example when the coefficient of variation is large, a Normal distribution can offer significant support to negative values, which is an inappropriate description of population size. Krishnarajah *et al.* (2005) construct mixture distributions in a univariate setting in order to obtain better closure approximations in cases where standard distributions fail.

A more realistic example

In order to illustrate the development of more realistic stochastic models consider extending our immigration–death model to include susceptible–infected disease dynamics. The model must now account for the numbers of susceptible (n_S) and infected (n_I) individuals in the population. For simplicity assume only susceptible individuals are recruited into the population via immigration at rate ν, and that susceptible individuals become infected at rate $\beta n_S n_I$, and then remain so until death. Finally, assume that in addition to the population-wide per-capita death rate μ there is also excess disease-induced mortality at per-capita rate μ_I. The model is summarized in Table 9.2.

As noted earlier once the state–space is defined and possible events and event rates chosen, it is straightforward to translate these into either a deterministic or a stochastic model. For consistency with previous notation, which emphasized that the deterministic model represents population size using continuous-valued variables, we will write the number of susceptibles as X_S, and the number of infectives as X_I. The rates of change of the susceptible and infected populations are then written as:

$$\frac{dX_S}{dt} = \nu - \mu X_S - \beta X_S X_I$$
$$\frac{dX_I}{dt} = \beta X_S X_I - (\mu + \mu_I)X_I \qquad (9.3)$$

The steady-state is given by simultaneously setting the rates of change dX_S/dt and dX_I/dt to be zero, and solving for X_S and X_I. Doing so reveals two steady-state solutions, the first corresponding to epidemic extinction:

$$X_S^0 = \nu/\mu; X_I^0 = 0$$

and the second to endemic disease:

Table 9.2. Susceptible–infected disease dynamics with contact rate β, and demographic fluctuations induced by excess disease induced by background per-capita death rate μ, excess disease induced mortality at per-capita death rate μ_I, and immigration of susceptibles at rate ν.

Event description	Change in state space δn_S	Change in state space δn_I	Event rate at time t
Immigration of susceptible	+1	0	ν
Infection of susceptible	−1	+1	$\beta\, n_S(t)\, n_I(t)$
Death of susceptible	−1	0	$\mu\, n_I(t)$
Death of infective	0	−1	$(\mu + \mu_I) n_I(t)$

$$X_S^1 = (\mu + \mu_I)/\beta; X_I^1 = \frac{\nu\beta - (\mu + \mu_I)\mu}{(\mu + \mu_I)\beta}$$

At time t in the equivalent stochastic model, the time to the next event is exponentially distributed with a rate equal to the sum of event rates $R(n;\nu, \mu, \mu_I, \beta) = \nu + \mu\, n_S(t) + (\mu + \mu_I)\, n_I(t) + \beta\, n_S(t)n_I(t)$. And this event is an immigration, a susceptible death, an infective death or an infection with probabilities $\nu/R(n; \nu, \mu, \mu_I, \beta)$, $\mu\, n_S(t)/R(n; \nu, \mu, \mu_I, \beta)$, $(\mu + \mu_I)\, n_I(t)/R(n;\nu, \mu, \mu_I, \beta)$, and $\beta\, n_S(t)n_I(t)/R(n;\nu, \mu, \mu_I, \beta)$ respectively. Thus simulation of this process is straightforward and proceeds as described earlier for the immigration–death model.

Similarly, it is also possible to construct equations describing the evolution of the moments of this process as follows. Consider the expected number of susceptibles, at time $t + \delta t$, in population with $n_S(t)$ susceptible and $n_I(t)$ infected individuals at time t:

$$n_S(t+\delta t) = n_S(t) + (+1)\nu\delta t + (-1)\mu\, n_S(t)\delta t + (-1)\beta n_S(t) n_I(t)\delta t \qquad (9.4)$$

As before the right-most terms are the event probabilities (i.e. the product of the event rate and δt) multiplied by the change associated with each. Finally rearranging this expression, taking the expectation at time t and taking the limit $\delta t \to 0$, reveals that:

$$\frac{dE[n_S(t)]}{dt} = \nu - \mu E[n_S(t)] - \beta E[n_S(t) n_I(t)]$$

And similarly consideration of the change in the size of the infected population leads to:

$$\frac{dE[n_I(t)]}{dt} = \beta E[n_S(t) n_I(t)] - (\mu + \mu_I) E[n_S(t)]$$

Note that these equations for the expectations (first-order moments) are not closed since they depend on the second-order term $E[n_S(t)n_I(t)]$, and similarly the evolution equations for the second-order moments depend on third-order terms and so on. An important consequence of this lack of closure is that, unlike those in the linear-immigration–death model, the expected value of the stochastic model need not coincide with the solution of the deterministic dynamics. Indeed in this case simulation of the stochastic model reveals that even when the deterministic dynamics give rise to the endemic steady-state the corresponding stochastic system is unstable and the disease outbreak dies out (see Marion *et al.*, 2002b). Thus in this case it is extremely important to account for demographic fluctuations as they qualitatively change the behaviour of the model. Such stochastically induced instabilities are a common feature of a wide range of models in epidemiology and ecology. One of the most famous examples of this is the instability of predator–prey interactions, in which it is notoriously difficult to observe long-term persistence in non-spatial stochastic models or indeed in experimental systems (Renshaw, 1991). The standard solution is the introduction of spatial heterogeneities into the process, for example by allowing movement between discrete patches, which allows greater global persistence in the context of spatially asynchronous local extinctions and colonizations (Renshaw, 1991; Keeling, 2000a,b)

To better understand the role of spatial heterogeneities consider its introduction into the simple SI model with immigration, death and disease-induced mortality which was summarized in Table 9.2. Assume that there are now $(n_S)_i$ susceptible and $(n_I)_i$ infected individuals at patches i=1, ..., N. The dynamics within each patch are simply those discussed above, but now the patches are linked by the random movement of infected individuals between patches. Thus the rate of emigration of infectives from patch i is $(n_I)_i\lambda$, whilst their immigration rate into patch i is $\lambda\, \Sigma_{i=1}^{N}\, (n_I)_i/N$. This model can be simulated by extending the algorithm described previously (also see below for a generic simulation algorithm). It is also possible to derive equations for the moments of the numbers of susceptibles and infectives at each site, $(n_S)_i$ and $(n_I)_i$. However, it is both more convenient and instructive to analyse the system in terms of spatial averages:

$$< f > = \frac{1}{N} \sum_{i=1}^{N} f\left((n_S)_i, (n_I)_i\right)$$

By writing down equations for each patch analogous to (9.4), summing over all the sites, and dividing by N it is relatively straightforward to show that the first-order moments of the average numbers of susceptibles and infectives across all the sites,

$$< n_S > = \frac{1}{N}\sum_{i=1}^{N}(n_S)_i$$

and

$$< n_I > = \frac{1}{N}\sum_{i=1}^{N}(n_I)_i$$

respectively, obey the following equations,

$$\begin{aligned}\frac{dE[< n_S(t) >]}{dt} &= \nu - \mu E[< n_S(t) >] \\ &\quad - \beta E[< n_S(t) n_I(t) >] \\ \frac{dE[< n_I(t) >]}{dt} &= \beta E[< n_S(t) n_I(t) >] \\ &\quad - (\mu + \mu_I) E[< n_I(t) > \end{aligned} \quad (9.5)$$

Note that these equations are not closed in terms of spatial moments, as the evolution of (the expectation of) $<n_S(t)>$ and $<n_I(t)>$ depend on the (expectation of the) second-order spatial average $<n_I(t)n_S(t)>$. Therefore, to make progress we employ closure schemes. The simplest of which is to ignore the spatial covariance, $Cov(n_S(t),n_I(t))$, between numbers of infectives and susceptibles, and, since $<n_S(t)n_I(t)> = <n_S(t)><n_I(t)> + Cov(n_S(t),n_I(t))$. This recovers the non-spatial deterministic model; note that λ only enters equations (9.5) via the second order-terms. Furthermore it suggests that, for a sufficiently large number of patches, the deterministic model might be accurate in the limit of large λ as the resultantly high mixing rate would break down any spatial correlations. Marion *et al.* (2002b) analyse a closely related model applied to auto-catalytic reactions. They derive moment equations analogous to (9.5) for the second-order spatial statistics $<n_S(t)n_I(t)>$, $<n_S^2(t)>$, and $<n_I^2(t)>$, showing that these depend on the third-order spatial moments $<n_S^2(t)n_I(t)>$ and $<n_S(t)n_I^2(t)>$. It is possible to deduce a number of results for the model considered here from the approximate analysis conducted by Marion *et al.* (2002b), who, in order to close this system of moment equations, employ a log-normal approximation which enables third-order spatial moments to be written as a function of the second-order spatial moments. The approximation is completed by assuming that the spatial domain is infinite and that correlations between spatial averages can therefore be ignored, for example that:

$$E[< n_S(t) >< n_I(t) >] = E[< n_S(t) >]E[< n_I(t) >]$$

By solving for the steady-state solution of the resulting approximation one obtains the fixed point:

$$\begin{aligned} E[< n_S(t) >] &= E[< n_S^2(t) >] \\ &\quad - E[< n_S(t) >]^2 = \nu / \mu \\ E[< n_I(t) >] &= E[< n_I^2(t) >] \\ &\quad - E[< n_I(t) >]^2 = 0 \\ Cov(n_S(t), n_I(t)) &= 0 \end{aligned} \quad (9.6)$$

This shows disease extinction with the susceptible population governed by a Poisson distribution with mean ν/μ corresponding to the immigration–death process discussed earlier. In the limit of fast mixing, $\lambda \to \infty$, a second steady-state of the log-normal approximation can be identified to be:

$$\begin{aligned} E[< n_S(t) >] &= E[< n_I^2(t) >] - E[< n_I(t) >]^2 \\ &= (\mu + \mu_I) / \beta \\ E[< n_I(t) >] &= E[< n_I^2(t) >] - E[< n_I(t) >]^2 \\ &= \frac{\nu\beta - (\mu + \mu_I)\mu}{(\mu + \mu_I)\beta} \\ Cov(n_S(t), n_I(t)) &= 0 \end{aligned} \quad (9.7)$$

corresponding to a Poisson-like distribution about the endemic steady-state of the deterministic model. Thus when the system size is large and mixing is strong the deterministic model is a good approximation to the spatial stochastic dynamics. However, Fig. 9.1 shows that the stochastic and spatial system can have markedly different behaviour from the deterministic system. For small mixing rate λ the disease does not persist and steady-state (9.6) of the log-normal approximation is an accurate description of the spatial stochastic process. As λ increases the disease becomes persistent and the log-normal approximation captures this behaviour qualitatively, becomes increasingly accurate as the mixing rate increases. For very fast mixing the expected

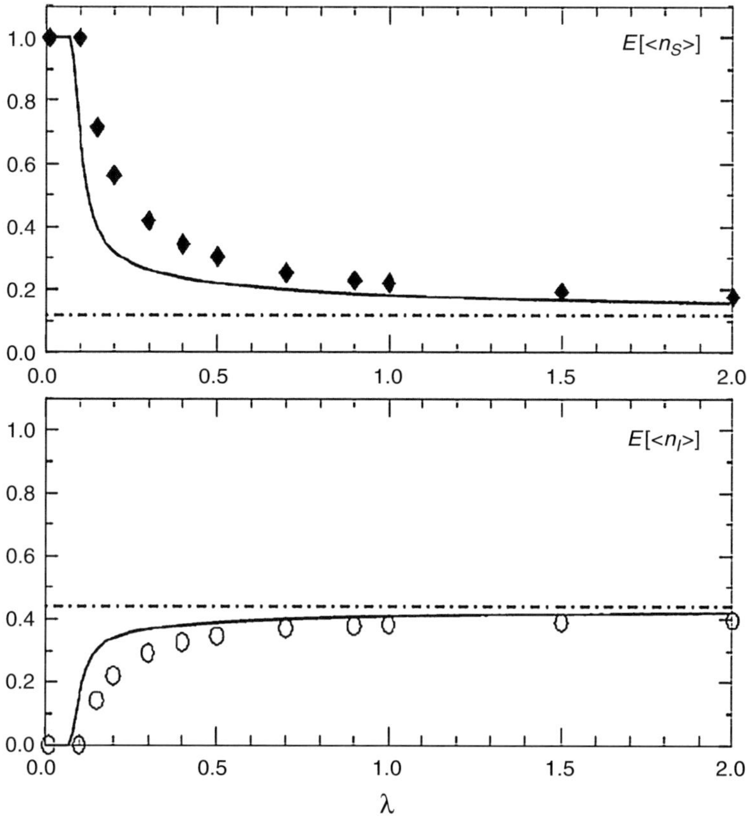

Fig. 9.1. Disease persistence. The upper graph shows the expected equilibrium (large time) value of the spatial average number of susceptible individuals $E[< n_s >]$ versus the movement rate of infectives λ, obtained from stochastic simulation (diamond symbols), the log-normal moment-closure approximation (solid line) and the non-spatial deterministic model (dot-dash line) described by (9.3). The lower graph shows the same for the equilibrium number of infective individuals $E[< n_I >]$. The parameter values used are $\beta = 1$, $\mu_I = \mu = \nu = 1/17$, and $N = 500$. The equilibrium values for the stochastic model were obtained from the output of ten simulations for $t = 500,\ldots,1000$, and the standard errors on the estimates are approximately equal to the size of the symbols.

value of the stochastic dynamics corresponds to the deterministic system, and the endemic steady-state of the log-normal approximation (9.7) provides an accurate representation of the first- and second-order spatial moments.

Model formulation and stochastic simulation in the general case

The formulation of the immigration–death process and the spatial SI model described above can be easily expanded to a general time-homogeneous Markov process defined in terms of the set of q allowed changes or event types $\{e_i: i = 1,\ldots,q\}$, where event type e_i induces a change δs_{ei} in the multi-dimensional state of the system, denoted $s(t)$ at time t. In other words if an event of type e_i occurs at time t the state of the system immediately afterwards is $s(t) + \delta s_{ei}$. The rate $r(e_i,s(t);a)$ at which event e_i occurs at time t is governed by the state of the system and the vector of model parameters a. The total event rate at time t is $R(s(t);a) = \Sigma_{i=1}^{q} r(e_i,s(t);a)$. As we saw above for the immigration–death model these rates may also be used to define deterministic or mean-field dynamics in terms of ordinary differential equations. The stochastic dynamics of the corresponding time-homogeneous Markov process are defined as follows: (i) the time τ to the next event (of any type) is drawn from an exponential distribution with rate $R(s(t);a)$; and (ii) the event type which occurs at time $t+\tau$ is chosen to be type e_i with probability $r(e_i,s(t);a)/ R(s(t);a)$ (Cox and Miller, 1965).

Then to simulate a realization of this stochastic process iterate the following procedure: (i) generated the time to the next event $\tau = -\ln(y)/R(n; v,\mu)$ where y~U(0,1); (ii) draw a second random variate y_2~U(0,1), and starting with $k = 1$ choose event k and update state–space accordingly, if $y_2 \geq \Sigma_{i=1}^{k} r(e_i,s(t);a)/ R(s(t);a)$, else increase k to $k = k + 1$ and repeat; and (iii) update time to $t + \tau$.

It is also relatively straightforward to account for temporal fluctuations in the environment by making the parameters of the model vary in time. In such cases the easiest way to simulate the process is to adopt the following approximate algorithm. First, choose $\delta t = \min(1/R(s(t);a(t)), \delta t_{\min})$ and update time to $t + \delta t$. Note $\delta t_{\min}$ is chosen so that changes in the time-varying parameters are negligible in the interval $(t,t+\delta t_{\min})$. Second, choose event e_i with probability $r(e_i,s(t);a(t))\delta t$. To do this generate a random variable y~U(0,1), calculate $y' = y\delta t$. Then starting with $k = 1$ choose event k if $y' \geq \Sigma_{i=1}^{k} r(e_i,s(t);a)$, else $k = k + 1$. Repeat until $k = q + 1$ unless an event is chosen, in which case update state–space appropriately. Repeat procedure for the next time step. Note that if $y' \geq \Sigma_{i=1}^{q} r(e_i,s(t);a)$ then the above algorithm will move on to the next time step and no event occurs in the interval $(t,t+\delta t)$.

Parameter Estimation in Stochastic Models: a General Formulation

Here we consider how to infer parameters from incomplete data for a general time-homogeneous Markov process defined above (see also Walker *et al.*, 2006). Recall that the total event rate at time t is $R(s(t);a) = \Sigma_{i=1}^{q} r(e_i,s(t);a)$ and the time τ to the next event (of any type) is drawn from an exponential distribution with rate $R(s(t);a)$; and the event type which occurs at time $t+\tau$ is chosen to be type e_i with probability $r(e_i,s(t);a)/R(s(t);a)$. This definition leads to a stochastic updating rule such that, conditional on the state of the system being $s(t)$ at time t, the probability density that an event of type e_i occurs before any other event type and does so at time $t + \tau$ is given by:

$$f\left(s(t+\tau) = s(t) + \delta s_{e_i} \,|\, s(t \leq y < t+\tau) = s(t)\right) = r(e_i, s(t);a)\, e^{-\tau R(s(t);a)} \quad (9.8)$$

Suppose the timings and nature of *all events* (e.g. all births, deaths, immigrations and infections, etc.) which occur in the interval $[t_0, t_n]$ are observed and recorded. Then let t_k be the time at which event k in the sequence occurs and denote its type by $E(k) \in \{e_i : i = 1,\ldots,q\}$. Suppose there are n events, then given an initial state $s(t_0)$ the finite and complete realization of the stochastic process:

$$S = \left\{s_{t_k}\right\}_{k=0}^{n}$$

can be generated from the set of events $\zeta = \{(E(k),t_k):k = 1,\ldots,n\}$. The *likelihood* of the complete data set, ζ, $L(a,\zeta) = P(\zeta | a, s(t_0))$, is the probability of observing the complete sequence of events ζ given the parameters a and the initial configuration $s(t_0)$ and is given by:

$$P(\zeta \mid a, s(t_0)) \propto \prod_{k-1}^{n} r(E(k), s(t_{k-1});a)\, e^{-(t_k - t_{k-1})R(s(t_{k-1});a)} \quad (9.9)$$

In general the final observation time T may not coincide with the occurrence of the final event at t_n. In such cases the likelihood (9.9) should be multiplied by an additional term $exp\{-(T-t_n)R(s(t_n);a)\}$ describing the probability that nothing happens between t_n and T. Given (9.9), if complete data are available likelihood methods (Edwards, 1992) can be used to estimate model parameters. The likelihood follows directly from the definition of the model via the stochastic update rule (9.8), and in general this is also true for non-Markovian stochastic processes, although the form of the likelihood will differ from that shown in (9.9). In the sequel we shall simply write the complete likelihood as $P(\zeta \mid a)$, dropping the explicit dependence on the initial condition $s(t_0)$, which may be either regarded as known and fixed or considered as an additional set of parameters to be estimated and thus incorporated into the vector a. Note that we have already suppressed the conditional dependence of the likelihood on the model since we will not formally compare different models.

In the case of incomplete data we observe a set of events D (the data), but there are also those hidden events H we do not observe. The complete realization is therefore characterized by the full set of events $\zeta = (D, H)$. We note that in this missing data context one could employ data augmentation within a likelihood framework; however, here we focus on a Bayesian treatment of this problem. Applying Bayes' rule:

$$P(A \mid B) = \frac{P(B \mid A)P(A)}{P(B)},$$

to $P(\zeta \mid a) = P(D, H \mid a)$ we obtain the joint *posterior* distribution for the parameters a and the unobserved events H:

$$P(a, H \mid D) = \frac{P(D, H \mid a)P(a)}{P(D)} \tag{9.10}$$

in terms of the likelihood for complete observations (9.9), the *parameter prior* $P(a)$ and the normalization constant $P(D)$. The prior distribution is typically chosen to reflect any knowledge about the parameters available before the data D were obtained. For example $P(a)$ may be derived from previous analysis, or simply be a uniform distribution over some plausible range of parameter values as ascertained from appropriate literature. In the absence of such information the prior is usually chosen to be some convenient form, for example for the rate parameters considered here, an (unnormalized) flat prior on the positive real line or a gamma distribution. In addition it is common to assume independence between the priors for each of the N components of the parameter vector, i.e.

$$P(a) = \prod_{k-1}^{N} P(a_k).$$

It is good practice to test the robustness of any analysis to prior specification.

Bayesian inference (see, e.g., Lee, 2004) is based on the posterior distribution (9.10), which for a given set of data is simply proportional to the likelihood and the prior. For example the distribution of parameters is given by:

$$P(a \mid D) = \int_H P(a, H \mid D)dH \tag{9.11}$$

which is just the joint posterior (10) marginalized over the hidden events. However, this integral is typically analytically intractable and the space of possible hidden events too large to allow evaluation by quadrature. Moreover, evaluation of the normalization constant $P(D)$ in (9.10) involves integrals of similar computational complexity. Fortunately, Markov chain Monte Carlo techniques allow parameter samples to be drawn directly from the posterior $P(a, H \mid D)$ without having to calculate the normalization constant $P(D)$. The Metropolis–Hastings algorithm and Gibbs sampling allow parameter samples to be drawn directly from the posterior, but since the number of unobserved events is in general unknown, in sampling over H, the Markov chain must explore spaces of varying dimension (corresponding to the numbers of events in a given realization) requiring application of reversible jump MCMC (Green, 1995). The samples generated from the posterior $P(a, H \mid D)$ using MCMC allow the calculation of essentially any statistic based on the parameters, and missing events. For example the marginal distribution of parameters described by (9.11) may be estimated by simply disregarding the sampled hidden events and forming a histogram of the sampled parameter values only. The marginal distribution of any single parameter (component of a) or the joint distribution of two or more may be obtained in a similar fashion. Such estimates improve as the number of samples generated from the Markov chain increases.

Reversible-jump Metropolis–Hastings algorithm

In order to implement the procedure described above samples of parameters and missing events must be generated from the posterior

(9.10). In order to do this we describe here two algorithms, the first samples hidden events H for fixed parameters a, and the second samples parameters for fixed H. Samples from the joint distribution are obtained by iteratively applying the first and then the second algorithm.

To generate samples of the missing events from the posterior $P(a, H|D)$ for a given set of parameters a we employ reversible jump MCMC (Green, 1995) based on the Metropolis–Hastings algorithm (Metropolis *et al.*, 1953; Hastings, 1970). Start with a set of hidden events H_0 which are consistent with the observations D. Let H_i denote the set of hidden events at the ith step and iterate the following procedure M times:

1. propose $H_i \rightarrow H'$ with probability $q(H_i, H')$

2. set $H_{i+1} = H'$ with probability:

$$\min\left\{1, \frac{P(D,H' \mid a)q(H',H_i)}{P(D,H_i \mid a)q(H_i,H')}\right\}$$

3. else $H_{i+1} = H_i$

Note that since in general the method only makes use of relative values of the posterior $P(a, H \mid D)$ the acceptance probability in step 2 is straightforward to calculate as the ratio of likelihoods of complete events (9.10) multiplied by a ratio of proposal probabilities (as shown). The proposal probabilities allow the exploration of the space of possible hidden events. In theory $q(,)$ can be any distribution, for example, uniform; however, selection of the proposal distribution determines how well the chain mixes, and thus convergence time (the number of samples that must be discarded as burn-in).

A general approach that enables a full exploration of the space of possible events, although it may not be optimal, is to allow the proposal of three basic changes to the current reconstructed realization: a birth step where a new event is added to the realization; a death step where an event is deleted from the realization; and a rearrange step which changes the time of an existing event in the realization.

In order to draw parameter samples from the posterior we can apply a variant on the above algorithm in which we keep the reconstructed events H fixed. If we draw parameter samples from the proposal distribution $q(a,a')$ then the acceptance probability becomes:

$$\min\left\{1, \frac{P(D,H \mid a')P(a')q(a',a)}{P(D,H \mid a)P(a)q(a,a')}\right\}$$

From which it is noted that if the proposal distribution were proportional to the posterior $q(a,a') \approx q(a') \propto P(D, H|a')P(a')$, then the acceptance probability would be 1. This is the basis of Gibbs sampling, which in principle is more efficient than using the Metropolis–Hastings algorithm as no samples are rejected. Of course the key difficulty is in drawing from a distribution proportional to the posterior. To see how this can be applied in the present context, suppose that $a_i \geq 0$, then we can assign a gamma prior distribution $P(a_i) \approx \mathrm{Ga}(\alpha,\beta) \propto a_i^{\alpha-1}\, e^{-\beta a_i}$. If the rates of the time-homogeneous Markov process (described earlier) are linear in a_i then, up to a constant of proportionality independent of a_i, the likelihood can be written:

$$P(D,H \mid a_i,a_{-i}) \propto a_i^{\alpha'(D,H,a_{-i})} e^{-a_i\beta'(D,H,a_{-i})}$$

where in general α' and β' depend on the data, the missing events and the other parameters a_{-i}. It follows that since the prior is a gamma distribution then so is the posterior:

$$\begin{aligned} P(a_i \mid D,H,a_{-i}) &= P(D,H \mid a_i,a_{-i})P(a_i) \\ &= Ga(\alpha+\alpha',\beta+\beta') \end{aligned}$$

It is therefore possible to sample values directly from this distribution. Moreover this is relatively efficient compared with the Metropolis–Hastings algorithm, since α' and β' are certainly no more computationally demanding to calculate than the likelihood itself.

To draw a set of samples $\{(a^i,H^i){:}i=1,\ldots N\}$ from the joint posterior $P(a, H|D)$ we must choose $\{(a^0,H^0)$ consistent with the data D and then iterate the following:

A. For fixed parameters a^i propose M changes to the hidden events using the reversible jump

Metropolis–Hastings algorithm described in steps 1–3 above. This generates H^{i+1}

B. Draw the next set of parameter samples a^{i+1} using Gibbs sampling from the univariate conditional posterior distributions.

This combined Metropolis–Hastings/Gibbs procedure implements a Markov chain (indexed by i) which asymptotically as $i \to \infty$ generates samples from the distribution $P(a, H \mid D)$ (Metropolis *et al.*, 1953; Hastings, 1970; Green, 1995). In other words if we run the chain for long enough (the burn-in period) it will settle down to an equilibrium in which each sample generated (post burn-in) is draw from the posterior distribution. In practice, the key problem is deciding on the burn-in period, that is, how many samples to discard before it is safe to assume that the Markov chain has converged to the desired distribution. There are a number of convergence diagnostics available (Gilks *et al.*, 1996) but none guarantee convergence in general, and by far the most common approach is visual inspection of the chain output. This is the method we rely on and we monitor time series of the parameter samples obtained by the chain after an initial burn-in (determined post-hoc by visual inspection of the output). The Markov chain tends to mix more efficiently if for each Gibbs sample of the parameters we typically perform many iterations $M > 1$ of the Metropolis–Hastings sampling of the missing events.

Applications

The examples in the section describing the formulation and analysis of stochastic models focused on population-level models with applications in ecology and epidemiology. Whilst there are many such applications of time-homogeneous Markov processes, these stochastic methods are remarkably flexible and can be employed across a wide spectrum of application. To illustrate this we now describe two contrasting applications drawn from our recent research. The first example combines the development and analysis of an agent-based model describing grazing in heterogeneous environments, with parameter inference based on data generated using contact logging in a behavioural experiment on dairy cows. The second example makes use of large-scale data describing bio-geographical features of the landscape and the spatio-temporal spread of an alien plant to estimate the parameters of a stochastic model of dispersal and establishment.

Modelling individual grazing behaviour

Marion *et al.* (2005) develop a simple stochastic agent-based model describing the grazing behaviour of herbivores in a spatially heterogeneous environment. The model reflects the biology in that decisions to move to a new location are based on visual assessment of the sward height (or some other proxy for nutritional value) in a surrounding neighbourhood, whilst the decision to graze the current location is based on the residual sward height and olfactory assessment of local faecal contamination. The model divides space into N discrete patches with c_i animals and sward height h_i in each patch $i = 1,\ldots,N$, and assumes that the agents (animals) either graze the current patch at rate $\beta c_i(h_i - h_0)$ or move to one of z neighbouring patches j at rate $\nu\ c_i\ h_j/z$. In addition, the sward growth in each patch, $I = 1,\ldots,N$, is assumed to be logistic $\gamma\, h_i(1 - h_i/h_{max})$. The model is summarized in Table 9.3.

As was shown above for the SI model with immigration and death it is possible to construct equations describing the spatial averages:

$$< f >= \frac{1}{N}\sum_{i=1}^{N} f(h_i, c_i)$$

of functions of the state variables. For example, note that since there are no births and deaths the animal density $< c(t) >$ is constant, and the equation describing the average sward height:

$$< h(t) >= \frac{1}{N}\sum_{i=1}^{N} h_i(t)$$

Table 9.3. Agent-based model of grazing behaviour defined in terms of the sward height h_i and the number of animals c_i at patch $i=1,\ldots,N$. The sward grows logistically at rate $\gamma h_i(1-h_i/h_{max})$, and the agents take bites from patch i at rate $\beta c_i(h_i - h_0)$ and move from patch i to j at rate $\nu c_i h_j/z$.

Event description	Change in state space			Event rate at time
	δh_i	δc_i	δc_j	t
Grass growth at patch i	+1	0	0	$\gamma h_i(1-h_i/h_{max})$
Animal bite at patch i	−1	0	0	$\beta c_i(h_i - h_0)$
Movement of animal from patch i to a neighbouring patch j	0	−1	+1	$\frac{\nu}{z} c_i h_j$

is written:

$$\frac{dE[<h(t)>]}{dt} = \gamma\left(E[<h(t)>] - \frac{E[<h(t)>]^2}{h_{\max}}\right) - \beta E[<c(t)>]\left(E[<h(t)>] - h_0\right) - \frac{\gamma}{h_{\max}} Var[h(t)] - \beta Cov[c(t), h(t)]$$

where the variance in sward height $Var[h(t)] = E[<h^2(t)>] - E[<h(t)>]^2$ measures the spatial heterogeneity and the covariances $Cov[c(t), h(t)] = E[<c(t)h(t)>] - E[<c(t)>]E[<h(t)>]$ measures the strength of association between tall swards and the grazing animals. It is worth noting that this equation involves no approximation and that the first line represents the equivalent non-spatial deterministic model. The variance and covariance terms in the second line therefore measure the importance of stochastic and spatial effects in the system; if both terms are close to zero then these effects are negligible. Evolution equations for the second-order quantities, $Var[c(t)]$, $Var[h(t)]$and $Cov[c(t), h(t)]$, depend on third-order spatial moments and also on correlations between nearest neighbours. Marion *et al.* (2005) show how to close these equations using a variant of the log-normal approximation discussed earlier in the section describing the formulation and analysis of stochastic models. As before this becomes more exact as the movement rate increases in a similar manner to the previous example.

Spatial heterogeneities play a significant role in this system, but the effects are typically quantitative as opposed to the qualitative differences between persistence and extinction we saw in our epidemiological example. For example, Fig. 9.2 plots the total intake rate against stocking density and shows that the optimal stocking density obtained from the spatial stochastic process is markedly different from that obtained from the non-spatial deterministic model. The maximum intake obtained is also lower when spatial heterogeneity is taken into account. The log-normal approximation partially captures this reduction in intake, but fails to predict the correct optimal stocking density. Marion *et al.* (2005) introduce additional features and also discuss various other aspects of the behaviour of the system including the relative efficiencies of random and directed searching, and in particular the impact of faecal avoidance. If the level of faecal contamination of patch i is described by the variable $f_i \geq 0$ avoidance behaviour can be accounted for by modifying the bite rate to be $\beta c_i(h_i - h_0)e^{-\mu f_i}$. Relative to the case of no avoidance the bite rate is progressively reduced as both the avoidance parameter $\mu \geq 0$ and the level of contamination increase. Friend *et al.* (2002) conducted an experiment at the Scottish Agricultural College (Dumfries, UK) to investigate avoidance behaviour in dairy cows. Prior to releasing animals into the outdoor experimental arena certain areas, amounting to approximately 5% of the total area of the paddock, were artificially contaminated with faeces. The four animals released into the paddock were fitted with faecal collection bags to prevent further contamination of the paddock during the experiment. Moreover, a data-

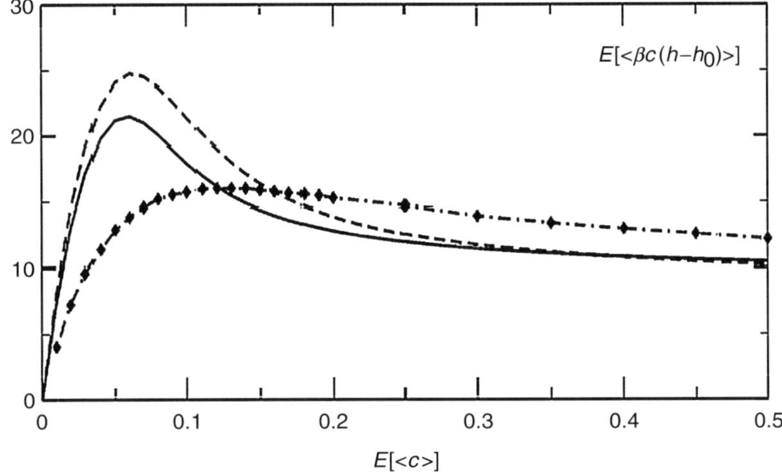

Fig. 9.2. The effect of fixed stocking density $E[<c>]$ on the total intake rate per patch across all animals at equilibrium $E[<\beta c(h - h_0>]$. The results show a marked difference in both the intake level and the optimal stocking rate obtained from the deterministic model (dashed line) and the stochastic spatial process (dot-dash with symbols). The fixed stocking density $E[<c>]$ was varied between zero and 0.5 as shown. The equilibrium values were obtained for $t \in [50,100]$ on a 10 by 10 lattice with $z = 4$ nearest neighbours, and initially uniform sward with $E[<h(t)>] = 2$ and randomly distributed animals. The other parameter values were $\beta = 1$, $\gamma = 0.1$, $v = 0.2$, $h_{max} = 10$, an initially uniform sward with $E[<h(t)>] = 2$ and randomly distributed animals.

logging system composed of transponders worn by the animals and aerials buried under the faecal contamination produced a record of every visit to the contaminated areas for each animal for the 4-day duration of the experiment. These data were supplemented by the daily measurement of the sward height in the contaminated zones and at a sample of points across the uncontaminated region.

If sward growth is discounted it is possible to regard the data produced by this protocol as a partial history (see above section on parameter estimation) of the stochastic model whose rates were defined in Table 9.3. The transponder data can be considered as direct observation of move events into the contaminated areas, although it should be noted that in some cases the transponder system logged multiple contacts in a short space of time and for the purpose of our analysis here these are regarded as a single visit, with the move event corresponding to the first contact. In addition the measured sward heights provide some information about bite events even though these are not recorded directly. In the general formulation of parameter estimation described above, the calculation of the likelihood was discussed in terms of events, but the sward height data are observations on the state–space of the system, rather than direct observations of events in the model. None the less it is straightforward to modify the event based likelihood (9.9) described above to account for such state–space observation by multiplying it with a noise model describing how state–space observations relate to the underlying state of the system. In this case assume that sward height measurements are subject to a Gaussian error with mean zero and standard deviation σ. Then if for a given event history the sward height of patch i at time t is $h_i(t)$ and the corresponding observation $h_{obs}(t)$ is available, then the likelihood will gain the factor:

$$\propto \exp\left(-\frac{\left(h_i(t) - h_{obs}(t)\right)^2}{2\sigma}\right)$$

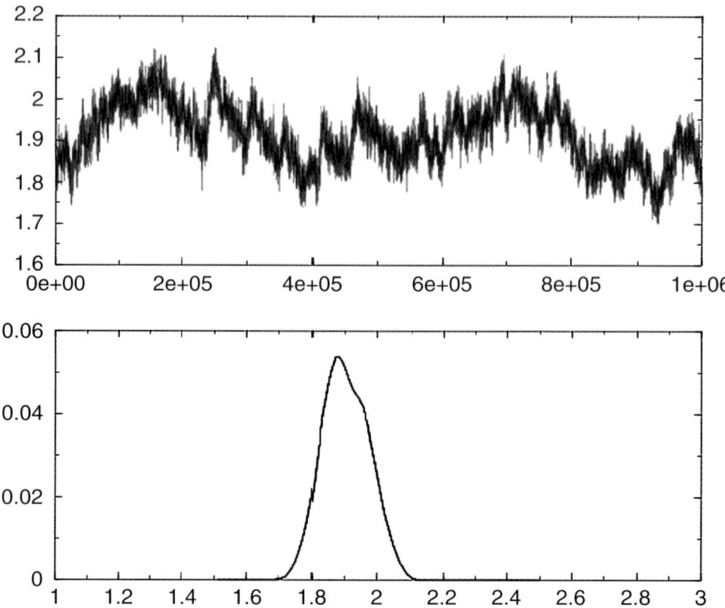

Fig. 9.3. Estimates of the movement rate v obtained by applying the Bayesian inference scheme described in the text to data from the SAC grazing experiment. The top graph shows samples of v generated from 1 million parameter samples of the Markov chain after an initial burn-in of 500,000. Note, for every proposed change to the parameters ten changes to the event history were proposed. Lower graph shows the histogram with 100 bins obtained from the post burn-in samples. The v samples have mean 1.90 and standard deviation 0.07.

Using this modification, the parameter inference techniques described earlier have been applied to a variant of the with-avoidance foraging model described above. In particular, the paddock was divided into 20 patches of equal size, one of which was considered to be the 5% contaminated area. The level of faecal contamination was described by $f_i = 1$ in the contaminated patch and $f_i = 0$ in the clean areas. Sward growth was ignored during the 4-day period of the experiment and for the movement the neighbourhood size was taken to be the entire paddock. In addition h_0 was assumed zero and $\sigma = 1$. The data provided all the move events into the contaminated patch and sward height measurements in centimetres for clean and contaminated patches initially and at subsequent daily intervals. The initial values of h_i for each patch were set equal to the initial sward height measurements. Flat Gamma priors (see e.g. O'Neill and Roberts, 1999) were chosen for each parameter β, v, and μ, and a combination of Gibbs sampling of the parameters and reversible-jump MCMC sampling of the (reconstructed) missing move and bite events was used to generate samples from the associated posterior as described above in the section on parameter estimation in stochastic models.

The estimates obtained from the Markov chain sampler are shown in Figs 9.3–9.5 for the parameters, while Fig. 9.6. shows the numbers of reconstructed missing events. The key difficulty in running the MCMC algorithm was obtaining a representative sample of the large number of missing movement events. The data recorded ~450 moves into the contaminated areas whereas Fig. 9.6. shows that the total number of move events (into all patches) is estimated to be ~8000. With the missing bite events, this requires a large number of nuisance parameters (event times) to be estimated alongside the more interesting model parameters β, v, and μ and a large number of samples are required to ensure adequate mixing of the Markov chain.

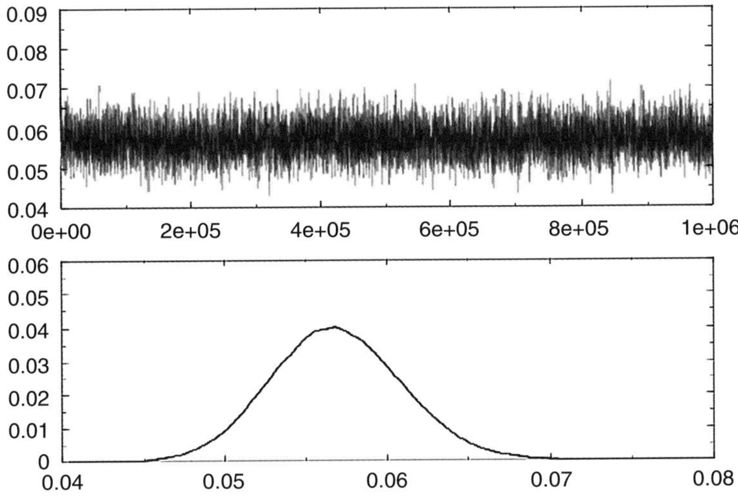

Fig. 9.4. As for Fig. 9.3 but for the bite rate β. The β samples have mean 0.06 and standard deviation 0.004.

These difficulties are reflected in the time-series plots of the three parameters. While those of β and μ fluctuate rapidly around their mean value, the samples of the movement rate v vary relatively slowly. None the less all three parameters and the numbers of missing events seem to have converged to steady states, indicating that the Markov chain has itself converged and the values obtained are representative of the posterior distribution. Moreover, numerical studies (not shown) based on simulated data mirroring those obtained in the Dumfries grazing experiment suggest that this procedure can reliably estimate model parameters. Although it might be possible to improve

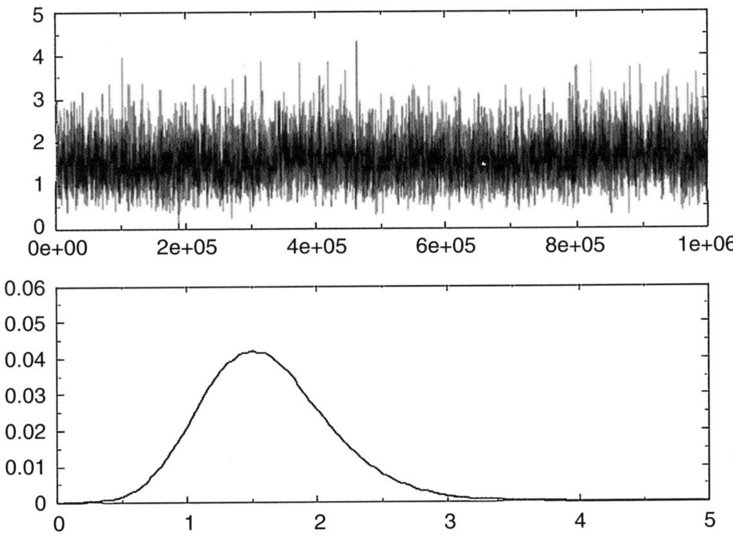

Fig. 9.5. As for Fig. 9.3 but for the avoidance parameter μ. The μ samples have mean 1.59 and standard deviation 0.5.

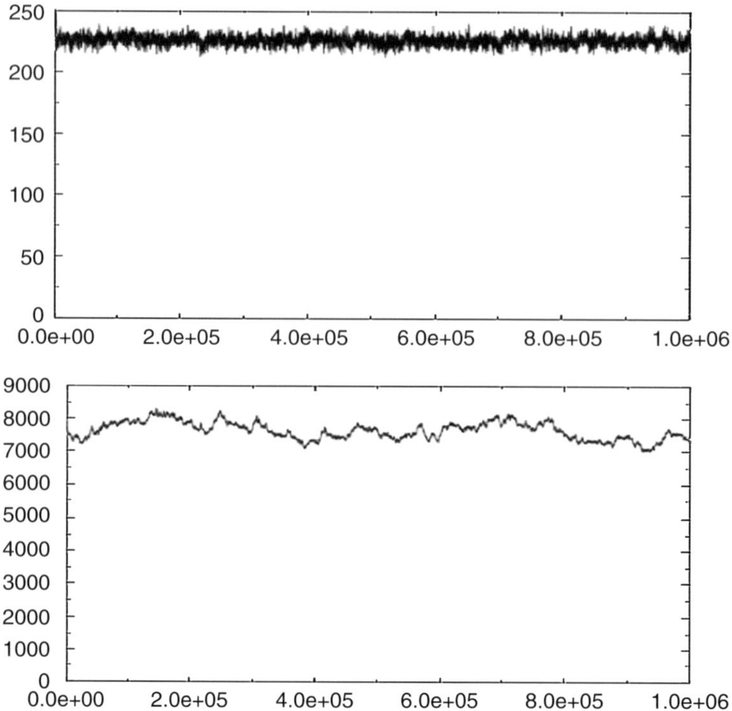

Fig. 9.6. The total number of move (upper graph) and bite (lower graph) events in the reconstructed event history obtained by applying the Bayesian inference scheme described in the text to data from the SAC grazing experiment. The graph shows the total number of each event type generated for 1 million parameter samples of the Markov chain after an initial burn-in of 500,000. Note, for every proposed change to the parameters ten changes to the event history were proposed.

mixing of the Markov chain by some alternative parameterization with respect to animal movement, it is difficult to see how this could be achieved without radically changing the model and whilst remaining faithful to the description of the underlying process. In addition, here we seek to develop a statistical treatment of an existing process-based model and demonstrate that progress can be made even with large amounts of missing data. However, it is clear that additional information on animal movements would greatly facilitate this analysis.

The detection of avoidance behaviour was of key interest in terms of the original experiment (Friend *et al.*, 2002), and the estimates of the avoidance parameter shown in Fig. 9.5 provide strong evidence that the animals are avoiding faecal contamination, with bite rates on contaminated patches being, on average, 20% of those on clean patches. Another interesting issue is how to design a new experiment, and it is possible to use the model and estimation framework outlined here to address such questions. First, it should be noted that if data describing all movements were available, for example via GPS tracking, then, as discussed above, far fewer nuisance parameters would be required and the estimation procedure would be much easier to implement. Second, the numerical experiments (discussed above) in which parameter estimates are obtained from simulated data suggest that the frequency of sward height measurements could be significantly reduced without serious effect on the parameter estimates obtained. Thus these methods can suggest how best to use novel technologies and point the way towards making better use of scarce resources. Moreover, confronting models with data in this manner often suggests ways in which they can be improved. For example, in the present study it may be more appropriate to estimate separate avoidance parameters for each animal. Thus, the techniques described here are potentially useful tools in linking empirical and theoretical developments more tightly.

Modelling the spread of an invasive species across a landscape

Parameter estimation for fully spatio-temporal stochastic models is an area of ongoing development which forms a bridge between complex systems models and spatial statistical regression. As discussed above, the increasing capabilities of computational statistics in recent years, most notably the development of MCMC techniques (see Gilks *et al.*, 1996, for a review), and their application to population processes such as epidemics (Gibson, 1997; Marion *et al.*, 2003; Hohle *et al.*, 2005; Gibson *et al.*, 2006), means that it is now possible to fit ever more realistic models to observational data in a statistically rigorous fashion. Moreover, estimation in stochastic spatio-temporal models has the advantage over more traditional statistical techniques, such as logistic and autologistic regression (Beerling, 1993; Huffer and Wu, 1998; Collingham *et al.*, 2000), of explicitly modelling change over time rather than assuming that the species' distribution (or other spatial pattern) has already reached equilibrium. However, one area in which the statistical treatment of spatio-temporal models is currently deficient is in its handling of covariate information. To date most applications have made little or no use of such data, largely focusing on estimating dispersal kernels and other key processes. On the other hand, regressive techniques excel in their ability to estimate the effects of large numbers of covariates. Moreover, Hastings *et al.* (2005) suggest that the realistic incorporation of environmental spatial heterogeneity into models of spatio-temporal spread is a key challenge.

Such issues are of practical significance to the spread of a disease or an alien species across a landscape. Non-indigenous species can cause substantial economical losses, and are recognized as one of the largest threats to native biodiversity and ecosystem functioning (Sala *et al.*, 2000; Kolar and Lodge, 2001), and increasing globalization has promoted the intentional and accidental spread of species through their natural dispersal barriers. Considerable research effort has therefore focused on this issue, and mathematical and statistical approaches have contributed to our understanding of the ecological processes underlying biological invasions (Kolar and Lodge, 2001). Cook *et al.* (2007) attempt to bridge the gap between complex systems models and spatial regressions by developing a time-homogeneous Markov process to model the spread of an invasive species across a landscape in terms of a contact process, describing both natural dispersal and human interventions such as transportation (deliberate and accidental), and the variations in bio-geographical features described a range of covariates such as habitat type and climate which cause spatial heterogeneity in local suitability for colonization. This approach necessarily balances model parsimony, computational tractability and the desire for realism, for example by describing varied and complex transport processes by means of a single dispersal kernel. Within a Bayesian framework MCMC techniques are applied to estimate the parameters of this model from data describing the spread of the riparian weed *Heracleum mantegazzianum* (giant hogweed) in Britain in the 20th century. A key advantage of the Bayesian approach is that model predictions can reflect both inherent variability and the estimated parameter uncertainty. In this case the risk of future spread is calculated by repeatedly re-running the model for a large number of parameter samples from the posterior distribution.

The model describing the spread of an alien species across the landscape is defined as follows. First, the landscape is divided into N patches and the presence of the alien species in patch i is defined by $\sigma_i = 1$ ($\sigma_i = 0$ denotes an absence). The *H. mantegazzianum* data were recorded on 10 × 10 km grid squares (hectads), which sets the patch scale, requiring N~3000 to model the UK. The rate of colonization of any empty patch j by a colonized patch i is given by $\beta_j \lambda^2 d_{ij}^{-2\lambda}$, where β_j represents the suitability of site j for colonization, d_{ij} the Euclidean distance between sites i and j, and λ is a parameter characterizing the decay of the power-law dispersal kernel which determines the range of spread. For computational reasons the spread was limited to patches j for which $d_{ij} \leq 150$ km. Spatial heterogeneity in the local environment is assumed to affect only the colonization suitability β_j via covariates describing the altitude A_j, the temperature T_j, and the proportion of land occupied by habitat class k in patch j, $H_{j,k}$ as follows:

$$\beta_j = \exp\left(-aA_j\right)\exp\left(\tau(T_j - \overline{T})\right)\sum_{k-0}^{9} b_k H_{j,k}.$$

The parameter a measures how colonization suitability decays with altitude, $\overline{T}$ measures the mean temperature across Britain 1920–1999, the sign of τ determines a preference for high or low temperature and the magnitude of τ defines the sensitivity to deviations from $\overline{T}$, and $b_k \in [0, \infty)$ measures the suitability of habitat type k.

All the covariate information were obtained from the Countryside Information System (www.cis-web.org.uk © Crown copyright 2006). The land-use classification and data were from the United Kingdom, Department for Environment, Food and Rural Affairs and Natural Environment Research Council (NERC) Land Cover Map 2000 GBv7 data set © NERC (2006). This comprises the proportion of each hectad covered by each of ten land-cover types: sea, coastal, arable, broadleaf forest, built-up, conifer forest, improved grassland, open water (rivers, lochs, etc.), semi-natural and upland. The temperature used was taken to be the annual temperature per hectad averaged over the period 1920–1999, derived from the Met Office's UK 5 × 5 km gridded monthly data set © Crown copyright (2006), (published by the Met Office in association with UK Climate Impact Programme and the Department for Environment, Food and Rural Affairs), and the mean altitude per hectad, from the Ordnance Survey Altitude and Slope Data: 1995 version 2 © Crown copyright (2006).

The *H. mantegazzianum* distribution data were obtained from the UK National biodiversity network (www.nbn.org.uk) and represent a collation of data from a number of sources including surveys and reported presences. The data used records presences in 10 × 10 km grid squares (there are approximately N~3000 such *hectads* in the UK) along with a time window during which the observation was made. The minimum length of a time window was a year, but many were considerably longer. Some sites had multiple observations (i.e. more than one time window), and since the weed is extremely persistent we make the assumption that once a site is colonized it remains colonized forever; therefore, for the model considered here, subsequent observations provide no further information about the colonization process. In this example we also make the simplifying assumption that the colonization of a site occurs at the earliest end-point of all of the observation windows recorded at that site, since this is the first time we can be certain the plant has reached that site. In reality there is uncertainty in the colonization time, which could in principle be accounted for in the general inference scheme described above by specifying an observation model and treating colonizations as unobserved events. However, this may lead to problems of parameter identifiability and in any case will dramatically increase the computationally complexity of the problem. This is a reasonable compromise since the focus here is the treatment of spatially explicit covariate information.

Estimation of the model parameters under these assumptions reveals that the habitat with the highest posterior mean colonizability is broadleaf. This is in contrast to reports (Clapham *et al.*, 1985; Tiley *et al.*, 1996; Stace, 1997) that the weed's primary habitats include waste-lands, and rivers and canals, which are included in the land-use classes built-up (estimated to be second most colonizable), and open-waters (fourth, just behind improved grasslands). There is evidence that hectads with a lower mean altitude are more suitable to *H. mantegazzianum* than higher areas. Interestingly, however, this is countervailed by the weed's preference for lower temperatures (evinced by posterior support for negative values of a).

This is especially surprising since altitude is negatively correlated with temperature, and indeed we might have expected confounding of these two effects. Both results are, however, supported to some extent by the literature: Pysek *et al.* (1998) found that *H. mantegazzianum* was represented less in areas of the Czech Republic with warm winters than cold; Willis and Hulme (2002) found experimentally that survival and biomass were lower for *H. mantegazzianum* at high altitudes.

We can also combine the effects of all covariates in the full model to evaluate which areas are more suitable for the weed. Figure 9.7. shows the posterior mean colonizability β_j for each hectad in Britain, with darker areas being more suited to the plant. There are considerable differences across sites, with the areas that have been colonized (most of England, the lowlands of Scotland) for the most part having high β_j. To evaluate where future colonizations are likely to occur 3000 simulations were performed, using parameter values sampled from the posterior, for the time period 2000–2025 (the data extend to the end of 1999). Figure 9.7 maps the probability of sites being colonized by 2010, obtained by recording against time the proportion of simulations resulting in the colonization of sites uncolonized at the start of 2000. The model predicts that upland areas are likely to remain uncolonized over this timescale. Cook *et al.* (2006) show that

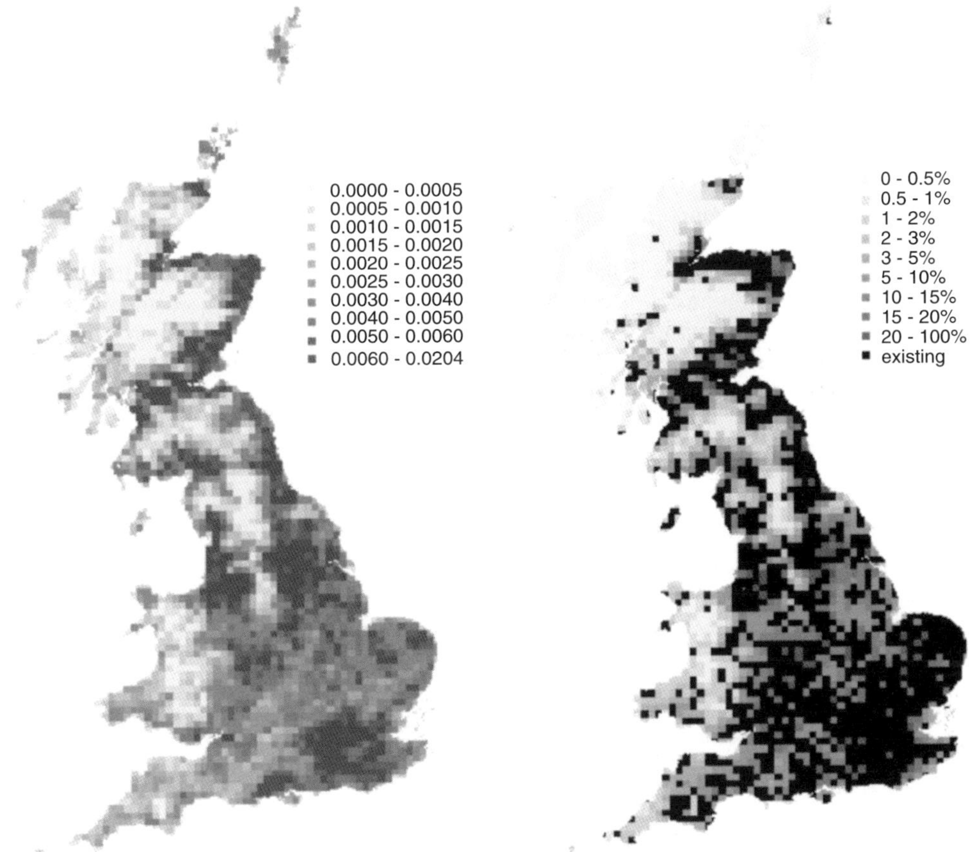

Fig. 9.7. Risk maps for *H. mantegazzianum*. The left-hand map shows the posterior mean value of the potential colonizability $\beta_j : j = 1,\ldots N$ of every hectad in Britain. The right-hand map shows hectads colonized by 1999 in black whilst the other colours indicate the probability of colonization by 2010 as described in the text.

the exclusion of covariates describing local suitability not only produces biased estimates of the dispersal kernel, but also leads to the unlikely conclusion that upland areas will become colonized over this period. The inclusion of land-use and temperature covariates means that the model described here could be used to produce similar risk maps under future land-use and climatic scenarios.

Conclusion

In this chapter we have presented a range of tools which enable closer integration of model development and data collection for complex biological systems. The methods presented relate to dynamic stochastic process-based models, and in particular we have focused on time-homogeneous Markov processes. The framework can account for spatio-temporal heterogeneities and individual variability, which are often crucial in understanding the properties of biological systems. Unfortunately, the non-linear nature of such phenomena typically leads to models which are analytically intractable. However, a number of analytic approximations are available which help to verify simulation results and provide valuable insights into model behaviour. The applications considered revealed that such approximations often perform well in some parameter regions, but rather poorly in others. Therefore the development of improved approximations and analytic approaches to such models is an area of ongoing research. Computational statistical methods such as MCMC can be applied to infer model parameters even when observations are incomplete. Moreover, in a Bayesian framework this enables uncertainty in parameters, and in principle even in models, to be estimated. We described two applications in which parameter estimation was carried out using MCMC methods. In the first, incomplete observations of animal behaviour were used to estimate the parameters of an agent-based model, and the framework used to suggest improvements to the design of any future experiment. In the second, covariate effects on local establishment and parameters describing spatial spread were jointly estimated, and the resulting parameter uncertainty incorporated into projections of the future spread of an alien plant species across Britain. In practice such statistically correct parameter inference can be too computationally demanding to apply to complex models, or when there are too many missing observations. Therefore computationally efficient, and possibly inexact, methods for parameter inference are required. Moreover, whilst there is a wide literature base on model selection *per se,* model selection with respect to stochastic process-based models is poorly understood and further research is required. None the less the applications discussed demonstrate that currently available methods are already useful tools in developing understanding of complex biological phenomena.

References

Bailey, N.T.J. (1963) *The Elements of Stochastic Processes.* John Wiley, New York.

Beerling, D.J. (1993) The impact of temperature on the northern distribution limits of the introduced species *Fallopia japonica* and *Impatiens glandulifera* in north-west Europe. *Journal of Biogeography* 20, 45–53.

Besag, J. and Green, P.J. (1993) Spatial statistics and Bayesian computation. *Journal of the Royal Statistical Society B* 55, 25–37.

Bolker, B. and Pacala, S.W. (1997) Using moment equations to understand stochastically driven spatial pattern formation in ecological systems. *Theoretical Population Biology* 52, 179–197.

Clapham, A.R., Tutin, T.G. and Moore, D.M. (1985) *Flora of the British Isles,* 3rd edn. Cambridge University Press, Cambridge.

Collingham, Y.C., Wadsworth, R.A., Huntley, B. and Hulme, P.E. (2000) Predicting the spatial distribution of non-indigenous riparian weeds: issues of spatial scale and extent. *Journal of Applied Ecology* 37 (Suppl. 1), 13–27.

Cook, A., Marion, G., Butler, A. and Gibson, G. (2007) Bayesian inference for the spatio-temporal invasion of alien species. *Bulletin of Mathematical Biology* (in press).

Cox, D.R. and Miller, H.D. (1965) *The Theory of Stochastic Processes.* Chapman & Hall, London.

Draper, D. (1995) Assessment and propagation of model uncertainty (with discussion). *Journal of the Royal Statistical. Society B* 57, 45–97.

Edwards, A.W.F. (1992) *Likelihood.* Johns Hopkins University Press, Baltimore, Maryland.

Filipe, J.A.N. and Gibson, G.J. (1998) Studying and approximating spatio-temporal models for epidemic spread and control. *Philosophical Transactions of the Royal Society London B* 353, 2153–2162.

Friend, M.A., Swain, D.L., Mayes, R.W. and Hutchings, M.R. (2002) Combining an active transponder system with sprayed n-alkanes to monitor spatial and temporal grazing patterns of contaminated pastures. In: Sveinbjörnsson, J. (ed.) *Techniques for Investigating Intake and Ingestive Behaviour by Farmed Animals: Proceedings of the Xth European Intake Workshop*. Agricultural Research Institute, Iceland, pp. 1–5.

Gamerman, D. (1997) *Markov Chain Monte Carlo.* Chapman & Hall, London.

Gelfand, A.E. and Smith, A.F.M. (1990) Sampling-based approaches to calculating marginal densities. *Journal of the American Statistical Association* 85, 398–409.

Gelman, A., Carlin, J.B., Stern, H.S. and Rubin, D.B. (1995) *Bayesian Data Analysis*. Chapman & Hall/CRC, Boca Raton, Florida.

Gibson, G.C., Otten, W., Filipe, J.A.N., Cook, A., Marion, G. and Gilligan, C.A. (2006) Bayesian estimation for percolation models of disease spread in plant communities. *Statistics and Computing* 16(4), 391–402.

Gibson, G.J. (1997) Markov chain Monte Carlo methods for fitting spatio-temporal stochastic models in plant epidemiology. *Applied Statistics* 46, 215–233.

Gibson, G.J. and Renshaw, E. (1998) Estimating parameters in stochastic compartmental models using Markov chain methods. *IMA Journal of Mathematics Applied in Medicine and Biology* 15, 19–40.

Gibson, G.J. and Renshaw, E. (2001a) Inference for immigration–death processes with single and paired immigrants. *Inverse Problems* 17, 455–466.

Gibson, G.J. and Renshaw, E. (2001b) Likelihood estimation for stochastic compartmental models using Markov chain methods. *Statistics and Computing* 11, 347–358.

Gilks, W.R., Richardson, S. and Spiegelhalter, D.J. (1996). Introducing Markov chain Monte Carlo. In: Gilks, W.R., Richardson, S. and Spiegelhalter, D.J. (eds) *Markov Chain Monte Carlo in Practice*. Chapman & Hall, London, pp. 1–19.

Green, P.J. (1995) Reversible jump Markov chain Monte Carlo computation and Bayesian model determination. *Biometrika* 82, 711–732.

Gurney, W.S.C. and Middleton, A.J. (1996) Optimal resource allocation in a randomly varying environment. *Functional Ecology* 10, 602–612.

Hastings, A., Cuddington, K., Davies, K.F., Dugaw, C.J., Elmendon, S., Freestone, A., Harrison, S., Holland, M., Lambrinos, J., Malvadkar, U., Melbourne, B.A., Moore, K., Taylor, C. and Thomson, D. (2005) The spatial spread of invasions: new developments in theory and evidence. *Ecology Letters* 8, 91–101.

Hastings, W.K. (1970) Monte Carlo sampling methods using Markov chains and their applications. *Biometrika* 57, 97–109.

Hohle, M., Jurgensen, E. and O'Neill, P.D. (2005) Inference in disease transmission experiments by using stochastic epidemic models. *Applied Statistics* 54, 349–366.

Holmes, P., Brown, E., Moehlis, J., Boyacz, R., Gao, J., Hu, P., Aston-Jones, G., Clayton, E., Rajkowski, J. and Cohen, J.D. (2004) Optimal decisions: from neural spikes, through stochastic differential equations, to behaviour. In: *Proceedings 2004 International Symposium on Nonlinear Theory and its Applications (NOLTA2004)* NOLTA, Fukuoka, Japan, pp.15–18.

Huffer, F.W. and Wu, H. (1998) Markov chain Monte Carlo for autologistic regression models with application to the distribution of plant species. *Biometrics* 54, 509–524.

Isham, V. (1991) Assessing the variability of stochastic epidemics. *Mathematical Biosciences* 107, 209–224.

Keeling, M.J. (2000a) Metapopulation moments: coupling, stochasticity and persistence. *Journal of Animal Ecology* 69, 725–736.

Keeling, M.J. (2000b) Multiplicative moments and measures of persistence in ecology. *Journal of Theoretical Biology* 205, 269–281.

Kendall, M.G. (1994) *Kendall's Advanced Theory of Statistics: Distribution Theory*. 6th edn. Stuart, A. and Ord, J.K. (eds), Edward Arnold, London.

Kloeden, P.E. and Platen, E. (1992) *Numerical Solution of Stochastic Differential Equations*. Springer-Verlag, Berlin.

Kokko, H. and Ebenhard, T. (1996) Measuring the strength of demographic stochasticity. *Journal of Theoretical Biology* 183, 169–178.

Kolar, C. and Lodge, D. (2001) Progress in invasion biology: predicting invaders. *Trends in Ecology and Evolution* 16, 199–204.

Krishnarajah, I., Cook, A., Marion, G. and Gibson, G. (2005) Novel moment closure approximations in stochastic epidemics. *Bulletin of Mathematical Biology* 67, 855–873.

Lee, P.M. (2004) *Bayesian Statistics: an Introduction*. Arnold, London.

Mao, X. (1997) *Stochastic Differential Equations and Applications.* Horwood, Chichester.

Marion, G., Renshaw, E. and Gibson, G. (1998) Stochastic effects in a model of nematode infection in ruminants. *IMA Journal of Mathematics Applied in Medicine and Biology* 15, 97–116.

Marion, G., Renshaw, E. and Gibson, G.J. (2000) Stochastic modelling of environmental variation for biological populations. *Theoretical Population Biology* 57, 197–217.

Marion, G., Mao, X. and Renshaw, E. (2002a) Convergence of the Euler scheme for a class of stochastic differential. *International Mathematical Journal* 1, 9–22.

Marion, G., Mao, X., Renshaw, E. and Liu, J. (2002b) Spatial heterogeneity and the stability of reaction states in autocatalysis. *Physical Review E* 66, 051915-1– 051915-9.

Marion, G., Gibson, G.J. and Renshaw, E. (2003) Estimating likelihoods for spatio-temporal models using importance sampling. *Statistics and Computing* 13, 111–119.

Marion, G., Swain, D.L. and Hutchings, M.R. (2005) Understanding foraging behaviour in spatially heterogeneous environments. *Journal of Theoretical Biology* 232, 127–142.

Matis, H.J. and Kiffe, T.R. (1996) On approximating the moments of the equilibrium distribution of a stochastic logistic model. *Biometrics* 52, 980–991.

Matis, J.H., Kiffe, T.R. and Parthasarathy, P.R. (1998) On the cumulants of population size for the stochastic power law logistic model. *Theoretical Population. Biology* 53, 16–29.

Metropolis, M., Rosenbluth, A.W., Rosenbluth, M.N., Teller, A.H. and Teller, E. (1953) Equation of state calculations by fast computing machines. *Journal of Chemical Physics* 21, 1087–1092.

Nisbet, R.M. and Gurney, W.S.C. (1982) *Modelling Fluctuating Populations.* Wiley, Chichester.

O'Neill, P.D. and Roberts, G.O. (1999) Bayesian inference for partially observed stochastic epidemics. *Journal of the Royal Statistical Society A* 162, 121–129.

Pyšek, P., Kopecký, M., Jarošík, V. and Kotková, P. (1998) The role of human density and climate in the spread of *Heracleum mantegazzianum* in the Central European landscape. *Diversity and Distributions* 4, 9–16.

Rand, D.A. and Wilson, H.B. (1991) Chaotic stochasticity: a ubiquitous source of unpredictability in epidemics. *Proceedings of the Royal Society of London B* 246, 179–184.

Renshaw, E. (1991) *Modelling Biological Populations in Space and Time.* Cambridge University Press, Cambridge.

Renshaw, E. and Gibson, G.J. (1998) Can Markov chain Monte Carlo be usefully applied to stochastic processes with hidden birth times? *Inverse Problems* 14, 1581–1606.

Roberts, M.G. and Grenfell, B.T. (1992) The population dynamics of nematode infections of ruminants: the effect of seasonality in the free living stages. *IMA Journal of Mathematics Applied in Medicine and Biology* 9, 29–41.

Sala, O.E., Chapin, F.S., Armesto, J.J., Berlow, E., Bloomfield, J., Dirzo, R., Huber-Sanwald, E., Huenneke, L.F., Jackson, R., Kinzig, A., Leemans, R., Lodge, D., Mooney, H.A., Oesterheld, M., Poff, N.L., Sykes, M.T., Walker, B.H., Walker, M. and Wall, D.H. (2000) Global biodiversity scenarios for the year 2100. *Science* 287, 1770–1774.

Smith, A.F.M. and Roberts, G.O. (1993) Bayesian computation via the Gibbs sampler and related Markov chain Monte Carlo methods. *Journal of the Royal Statistical Society B* 55, 3–23.

Stace, C. (1997) *New Flora of the British Isles.* Cambridge University Press, Cambridge.

Tiley, G.E.D., Dodd, F.S. and Wade, P.M. (1996) Biological fauna of the British Isles: *Heracleum mantegazzianum* Sommier & Levier. *Journal of Ecology* 84, 297–319.

Tilman, D. and Karieva, P. (1997) *Spatial Ecology: the Role of Space in Population Dynamics and Intraspecific Interactions.* Princeton University Press, Princeton, New Jersey.

Walker, D.M., Perez-Barberia, F.J. and Marion, G. (2006). Stochastic modelling of ecological processes using hybrid Gibbs samplers. *Ecological Modelling* 198, 40–52.

Whittle, P. (1957) On the use of the normal approximation in the treatment of stochastic processes. *Journal of the Royal Statistical Society B* 19, 266–281.

Willis, S.G. and Hulme, P.E. (2002). Does temperature limit the invasion of *Impatiens glandulifera* and *Heracleum mantegazzianum* in the UK? *Functional Ecology* 16, 530–539.

Wilson, H.B. and Hassell, M.P. (1997) Host–parasitoid spatial models: the interplay of demographic stochasticity and dynamics. *Proceedings of the Royal Society of London B* 264, 1189–1195.

10 Reef Safe Beef: Environmentally Sensitive Livestock Management for the Grazing Lands of the Great Barrier Reef Catchments

Iain J. Gordon[1] and Brigid Nelson[2]

[1]*Sustainable Ecosystems, CSIRO – Davies Laboratory, Queensland, Australia;*
[2]*Department of Primary Industries & Fisheries, Queensland, Australia*

Abstract

Extensive beef production is one of the major land uses of the Great Barrier Reef (GBR) catchments. Across Queensland it brings in over A$3 billion to the national economy annually and employs nearly 9000 people, many of them in rural communities. As well as being the major industry supporting economic activities and urban centres in remote rural Queensland, over 80% of terrestrial sediments and nutrients deposited in the GBR World Heritage Area, affecting the health of vulnerable reef ecosystems, originate from the extensive grazing lands of Queensland's interior. Recent research indicates that the quantity of sediments and nutrients lost from these grazing lands is strongly dependent upon grazing management practices which lead to degradation of soil and water resources, reduced infiltration and vegetation production. This has led to growing public concern about the environmental performance of the beef industry and increasing pressures on graziers to change their management practices to decrease off-property impacts. There is now an opportunity to assess options scientifically and provide guidance for the development of new on-property actions. These would be designed to allow an increase in profit from adopting sustainable grazing practices while reducing the off-property impacts of sediment and nutrient loss. In this chapter we argue that improvements in the quality of water draining into the GBR lagoon can best be achieved by demonstrating the productivity and economic benefits of science-based improved grazing management practices for graziers, leading to 'Win–Win' outcomes for all concerned.

Background

Queensland is renowned for its beef industry, which brings in over A$3.7 billion per annum (2004/05) to the national economy (North Australia Beef Research Council, 2006; Meat and Livestock Australia, 2006). In the last agricultural census held in 2001 there were approximately 13,000 farm businesses with beef cattle in Queensland, employing approximately 9000 people in North Queensland, many of them in rural communities, making the beef sector the State's largest primary industry in terms of both value and number of enterprises (Australian Bureau of Statistics, 2003; Productivity Commission, 2003).

Grazing lands cover over 75% of the Great Barrier Reef catchments (Fig. 10.1)

(eds D. Swain, E. Charmley, J. Steel and S. Coffey)

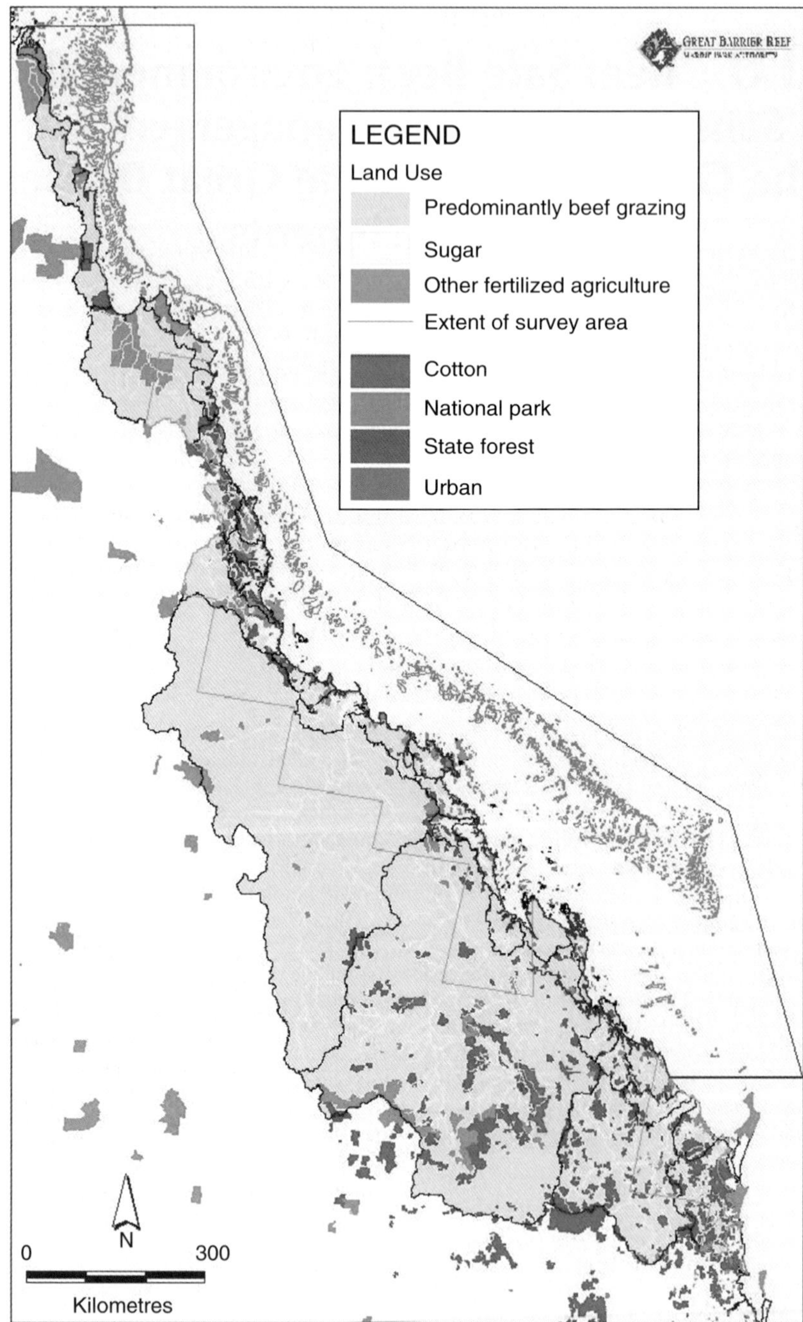

Fig. 10.1. Land use in the catchments on the eastern seaboard of Queensland that drain into the lagoon of the Great Barrier Reef.

and most beef is produced under extensive systems on large properties that have relatively low levels of infrastructure (dams and fencing). Stocking rates are generally in the range of one animal unit per 5–10ha (Quirk *et al.*, 2004; Bortolussi *et al.*, 2005). Semi-extensive beef production and dairy systems also occur but these are located on the relatively higher rainfall areas closer to the coast. Within the extensive systems livestock productivity is primarily determined by rainfall and its interaction with soil characteristics (nutrients and hydrological properties).

The Great Barrier Reef (GBR) is the largest system of coral reefs in the world. It includes approximately 3000 reefs and covers an area of approximately 350,000 km^2 on the north-eastern Australian continental shelf (Craik, 1992; Wachenfeld *et al.*, 1998). The reef ecosystems have a complex interdependent relationship with the adjacent coastal river catchments. Some 30 major rivers and hundreds of small streams drain into the GBR lagoon. The linkages of the catchment to reef continuum are not simply downstream, but also involve migrations of many species, such as fish, between coastal marine habitats and inland waterways and wetlands. The water quality of the coastal zone of the GBR is adversely impacted by increasing sediment, nutrient and other pollutants and by significant alterations to the hydrodynamic regime of the floodplain (freshwater, estuarine, and marine) (Haynes and Michalek-Wagner, 2000). Tourism is a major employer in Queensland and with the flow-on effects associated with the use of the reef, it underpins a significant portion of Queensland's regional economy (Productivity Commission, 2003). Most urban areas in Queensland's coastal zone are also experiencing significant growth (Anon., 1999); as a consequence, local governments along the coast face the challenge of balancing the demands of economic development associated with changes in land use, shifts in agricultural activity and urban and industrial expansion with maintenance of healthy coastal and reef ecosystems. An integral component of healthy coastal and reef ecosystems is protection of local water quality and maintenance of aquatic habitats. There are also indirect values provided by the GBR, such as ecosystem services (e.g. shoreline protection, maintenance of biological diversity, waste assimilation and reception, visual amenity, and lifestyle values), existence (moral requirement to protect natural ecosystems) and bequest (still around for future generations) values.

The GBR catchments have been extensively modified since European settlement through forestry, urbanization and agriculture (Lukas *et al.*, 1997; Furnas, 2003). At the end of the 19th century a much smaller proportion of the rangelands were grazed compared with today, mainly because of the constraints on animal movement by the position of watering points and rivers. However, with the sinking of bores and the replacement of the European breeds (*Bos taurus*) with hardier, drought- and tick-resistant (Frisch and O'Neill, 1998) Brahman breeds (*Bos indicus*), the proportion of the landscape that remains ungrazed is now small as most water-points are now within 10 km of each other (Abbott and McAllister, 2004). Consequently beef cattle numbers in the catchments of the GBR are approximately 4,500,000, with the highest stock numbers occurring in the Fitzroy catchment.

While grazing is the largest single land use in the catchment, cropping, mainly of sugarcane, and urban/residential development are located on the sensitive coastal floodplain (Gilbert *et al.*, 2003) and greatly influence the GBR. Hydrological modification of the coastal floodplain has resulted in loss or degradation of wetland systems in most of the GBR catchments (e.g. Tait, 1994). Loss of riparian vegetation in both rangelands and cropping lands and agricultural expansion into areas of acid sulphate soils that drain into the reef has also been extensive (Anon., 1999). The sugarcane cultivation area has increased steadily over the past 100 years with a total of 390,000 hectares reached by 1997 (1% of GBR catchment area). Sugarcane cultivation uses a lot of fertilizer, so with increasing sugarcane cultivation there has been a rapid increase in fertilizer use (Mitchell *et al.*, 2001). In addition to sugarcane, both the cotton and horticultural industries (particularly banana) have undergone considerable expansion in the coastal catchments (Gilbert *et al.*, 2003). The use of herbicides and pesticides is also significant in areas of crop cultivation (Hamilton and Haydon, 1996) with negative effects on corals and marine vertebrates (Hutchings and Haynes, 2000).

But it is the grazing industry that accounts for the majority (over 80%) of terrestrial sediments and nutrients deposited in the GBR (Gilbert *et al.,* 2003). The Burdekin catchment, where the primary land use is grazing, delivers on average 3.8 million tonnes of fine sediments, 8600 tonnes of nitrogen and 1300 tonnes of phosphorous per annum into the lagoon of the reef. The figures for the Fitzroy are 2.9 million tonnes of sediment per year exported to the coast, in addition to 8000 tonnes of N and 2000 tonnes of P (Brodie *et al.*, 2003). Present-day export of sediments and phosphorus from these catchments is six times, and nitrogen four times, greater than pre-European levels (Brodie *et al.*, 2003).

The quantity of sediments and nutrients lost from these grazing lands depends strongly on grazing management practices such as stocking rate, tree clearing and positioning of watering and supplementation points (McIvor, 2002). Poor grazing management practices are often exacerbated by drought, which leads to degradation of soil and water resources. These include impaired soil biological function, depletion of organic matter and reduced nutrient-holding capacity, soil acidification, erosion, compaction and rising water tables and salinization (Gifford, 1985). Ultimately, excessive sediments and nutrients flow into the coastal waters of the lagoon of the reef and have a negative impact on corals reefs (particularly near-shore reefs) and seagrass beds, with consequences for the fish assemblages (Fabricius, 2005).

Because of these environmental effects, graziers have been under pressure to change their management practices to decrease off-property environmental impacts (see Anon., 2003). Soil and vegetation loss through unsustainable management practices is also likely to lead to reduced livestock production, since vegetation acts as a barrier to the overland flow of water, thereby increasing ground infiltration (Ludwig *et al.*, 2005).

The variable and unpredictable climate of the east coast of Queensland (Fig. 10.2) further exacerbates the problems faced by pastoralists. A consequence is that graziers will tend to be over-optimistic about the probability of rainfall and will maintain stock levels above those which can be carried over the dry season, using supplementation to buffer against reduced pasture resource levels. This strategy, exacerbated by government drought relief policies, leads to pasture degradation, reducing the long-term carrying capacity of the land.

The preceding background shows that there are challenges to the grazing lands of the catchments emptying into the GBR lagoon. However, recent research demonstrates that graziers can change their management practices to decrease the negative effects they have on the quality of water leaving these properties. This involves maintaining high levels of ground cover and increasing the amount of water, sediments and organic matter captured by vegetation. In the long term, this could lead to an increase in the productivity of the vegetation and, therefore, the profitability of the enterprise, through reduced inputs of supplements and less variation in stock numbers over time.

Grazing Ecology

Beef production in the extensive rangelands of the GBR catchments primarily relies upon native vegetation, with some introduced species of legumes and productive grasses, to supply nutrition to the animal for most of the year (Bortolussi *et al.*, 2005). Most enterprise management is based upon stocking rate and the timing of purchase and sale of animals. The extensive beef system typical of northern Queensland has limited technological intervention compared, for example, with the dairy industry, other than the delineation of paddocks and the positioning of waterholes (Bortolussi *et al.*, 2005; Stokes *et al.*, 2006) and the use of nutrient supplements that are provided to overcome protein

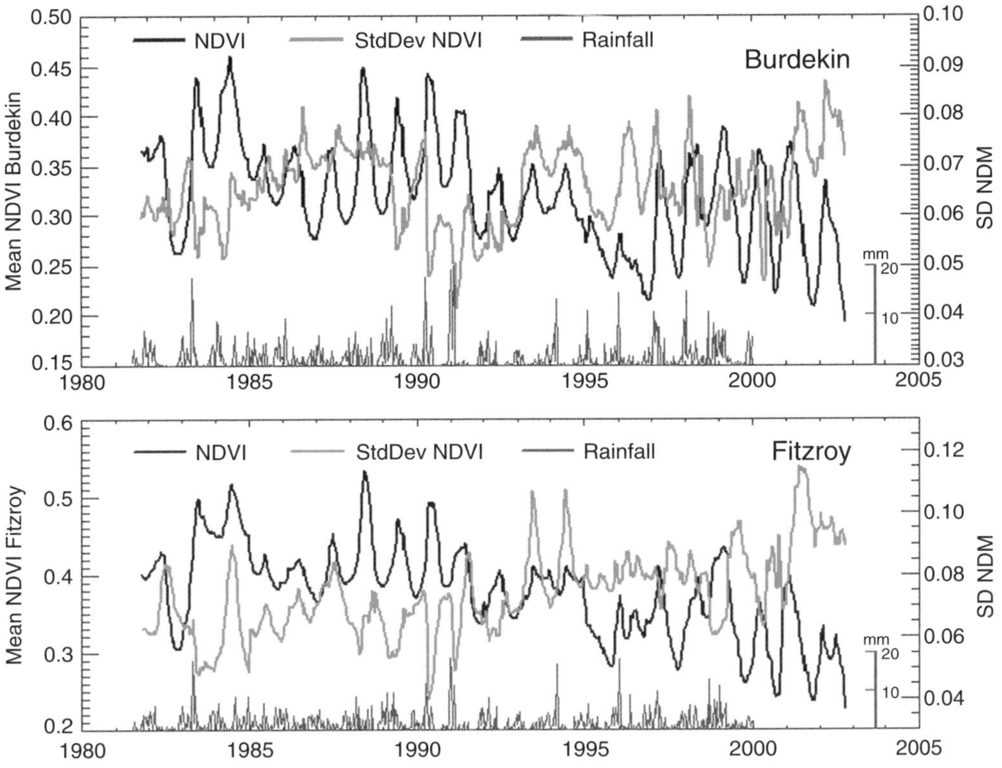

Fig. 10.2. Rainfall pattern and vegetation responses (Normalized Difference Vegetation Index; NDVI) in two main catchments of eastern seaboard of Queensland.

deficiencies in the dry season (McLennan *et al.*, 1981) or phosphorus deficiency year-round (Miller *et al.*, 1990).

The dominant native vegetation in the rangelands of the catchments of the GBR is the box–ironbark–bloodwood (Eucalyptus–Corymbia complex) communities, which are characterized by an open woodland layer over an understorey of perennial tussock grasses such as *Heteropogon contortus* (black spear grass), *Triodia* spp. (spinifex), *Bothriochloa* spp., *Dichanthium* spp. and mixtures of *Chrysopogon* spp., *Aristida* spp. and *Themeda* spp. with a few herbs (Tothill and Gillies, 1992). These native pastures evolved under limited grazing pressure until the arrival of livestock with the Europeans in the 19th century. Heavy livestock grazing pressure quickly leads to a depletion of the perennial grass root reserves (Walker *et al.*, 1999) with eventual loss of perennial grass plants and replacement by annuals, forbs and/or exotic perennial grasses (e.g. Indian couch, *Bothriochloa pertusa*) (Bisset, 1980; McIvor and Scanlon, 1994). This change in vegetation composition has dramatic effects on the system's ability to sustain livestock during the dry season, especially in droughts. Annual grasses have a very short growing season and senesce rapidly (Ash and McIvor, 1998) and stoloniferous grasses (*B. pertusa*) may have higher yields (Jones, 1997) but are vulnerable to drought (J. Corfield, Townsville, 2005, personal communication); this results in vegetation of low abundance and sometimes nutritive value, relative to a perennial grass dominated system (Ash and McIvor, 1998). Overall intake requirements for the herd often exceed pasture yield, leading to a requirement for supplementation, which causes further reductions in ground cover and pasture degradation (McIvor, 2002).

Surveys in the 1990s and early 21st century have found that nearly 50% of the pastures in the region are degraded and that pastoralists recognize the impact of degradation on their properties and for their livelihoods (Tothill and Gillies, 1992; McIvor, 2002; Bortolussi *et al.*, 2005).

Rehabilitation of degraded rangelands involves the restoration of the perennial grasses and herbs (Ash, 2004). This requires a reduction in grazing pressure, particularly when perennial grasses are most vulnerable to defoliation i.e. the early wet season (Ash *et al.*, 2001). Controlled grazing experiments have shown that wet season spelling (removal of livestock from a paddock during the wet season) leads to an improvement in the proportion of perennial grasses in the paddock (Ash *et al.*, 2001; Ash, 2004). The time it takes to achieve pasture rehabilitation will depend upon a range of factors, including whether there is a seed bank or seed source for the perennial grasses, the degree of scalding of the soil surface layer and the rainfall.

Hydrology

The rangelands of northern Queensland are classed as semi-arid, with less than 650 mm of rain per annum and a high coefficient of variation of annual rainfall (Ash *et al.*, 2001). There is a marked wet and dry season, usually December through April for the former, and year to year variations in mean rainfall, with extended drought periods (Fig. 10.2). Most of the rain that falls in the wet season comes in a few events, where > 50mm can fall in an hour. These patterns of rainfall mean that the system is highly pulsed, with large amounts of water at certain periods and none at others. This has profound impacts on the hydrology, ecology and agronomic potential of the system.

While the GBR catchments are not huge, relative to those in other continents, the water that falls there as rain discharges into the lagoon of the Great Barrier Reef. There has always been sediment and nutrient runoff from the rangelands of the catchments bordering the reef, and modelling work now suggests that there has been a threefold increase in the rate of delivery of sediments since the introduction of livestock grazing in the 19th century (Brodie *et al.*, 2003) and a much greater increase in the amount of nutrients delivered into the lagoon (Wooldridge *et al.*, 2006). While still controversial (McIvor *et al.*, 1995), there is evidence that much of this has been caused by tree clearing in, for example, the Fitzroy catchment and by reductions in ground vegetation cover associated with grazing (McIvor, 2002). There is still a great deal of controversy over the consequences of the source of sediments (Stark, 2005) relative to those already in the lagoon; however, for marine ecosystems, it is clear that there are major impacts of sediment and nutrient plumes on the seagrass and near-shore coral reef communities (Fabricius, 2005). If these impacts continue then the health of the reef may suffer, with consequences for reef-based tourism (Kragt *et al.*, 2006). There will also be reductions in the productivity potential of beef enterprises because of loss of soil structure and fertility (Nelson and Roth, 2004).

Effectively, rainfall is the sole water source for vegetation in the rangelands. Precipitation leaves the system through one of four routes: evaporation, transpiration, infiltration or run-off into streams and rivers. The proportion of rainfall that leaves through these various routes is determined by a range of factors, including soil type, solar radiation, heat and, most important, vegetation cover. Vegetation reduces soil surface temperature, reducing evaporation. It also acts as a barrier that impedes the flow of water and increases infiltration rates. These processes increase the soil organic matter and moisture content, both of which increase plant productivity in these semi-arid systems (Fig. 10.3).

As discussed in the section above ('Grazing ecology'), livestock grazing can have a dramatic effect on vegetation composition, cover and distribution. Since all of these impacts increase the ratio of runoff to infiltration, water/nutrients are lost from the system, which reduces plant productivity and leads to a spiralling cycle of rangeland degradation (McKeon *et al.*, 1990).

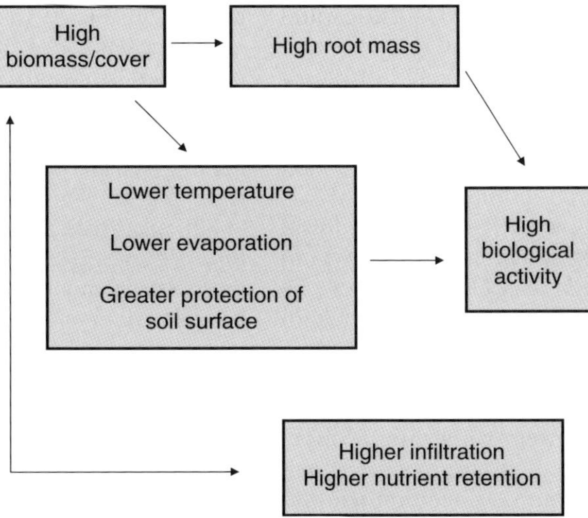

Fig. 10.3. Relationship between vegetation biomass/cover and ecological function and processes in Australian savannahs.

Also, the increased runoff carries with it sediments and nutrients that can impact on the GBR. In the only published study to date O'Reagain *et al.* (2005) found no differences in run-off of sediments (3–20kg/ha per event), N (10–1900g/ha per event) or P (1–71g/ha per event) across different grazing treatments; all of the treatments yielded water of medium quality. However, these authors suggested that, at this early stage of the experiment, the impacts of the grazing treatments might not yet have had sufficient impacts on vegetation cover to show a difference and we should beware of extrapolating from relatively small experimental plots to the whole of the Burdekin catchment.

Grazing Land Management

Grazing land management research, development and extension (R, D and E) programmes are being challenged by industry to account for the complexity of beef enterprise decisions, to build the capacity of people and partnerships, and to achieve measurable improvements in profit, environment, beef production and lifestyle. The philosophy is that improving land managers' knowledge, skills, and enterprise viability will facilitate change and enable adoption of practices more conducive to enhanced resource protection.

Grazing Land Management R, D and E has undergone major shifts in focus since the development era of the 1960s. Innovative technologies were adopted that increased overall stocking rates by lowering mortality rates and spreading grazing pressure away from natural waterways (Shepherd and Chilcott, 2006). In addition, livestock productivity was increased through introducing drought- and tick-resistant Brahman cattle in the 1960s (McCullough, 2004). In combination with the introduction of new breeds, tropically adapted pasture plants such as stylo (*Stylosanthes hamata*, *S. scabra*) became widespread (Miller and Stockwell, 1991) and the use of molasses–urea supplementation increased productivity and reduced mortality (Frisch and Vercoe, 1977).

Not until the 1980s were the limitations of the rangeland natural resources acknowledged in the quest for continued increases in production (Gardener *et al.*, 1990; McIvor *et al.*, 1995). In June 1988, beef producers in the region organized to form a graziers' group to address grazing land sustainability. Since then, and with the support of government agencies, groups across the region have been a catalyst for increasing awareness of better land management practices and participatory on-ground activities to improve the natural resources of the rangelands.

In the 1990s with increasing interest from the grazing industry, research agencies started investigating how the grazing ecosystem worked, defining what was meant by 'sustainable production' and quantifying degradation processes. Between 1990 and 2000 considerable research was carried out in the rangelands of the GBR catchment (see McCullough and Musso, 2004); however, in the late 1990s it was recognized that industry could be getting greater benefit from the information and outputs of past research and development programmes, particularly those with an ecological approach to grazing land management. A development and extension phase consequently began.

Education packages that would enhance management of grazing lands in northern Australia by transfer of the best R&D information to graziers were commissioned by industry in early 2000. The Grazing Land Management (GLM) Workshop (Chilcott *et al.*, 2003) was developed in 2003 for the Burdekin, Burnett, Victoria River Downs (Northern Territory), and Mitchell Grasslands of Queensland. Expansion of the GLM Workshop to all regions within Queensland, Central Australia, and other regions within the Northern Territory were completed in late 2006.

The State government through its Department of Primary Industries and Fisheries (DPI & F) led a consortium of organizations in the development of the Grazing Land Management (GLM) education package, in partnership with Meat and Livestock Australia. Development of the education package has been guided by market research which identified that:

- Producers wanted to address both the landscape health and productivity of their properties;
- There was a strong need for on-going technical support to ensure implementation and on-going exchange of information;
- There was a strong preference for locally derived and calibrated information and decision tools.

The GLM education package has assembled the best local R&D information accessible to industry in a structured learning environment. The education package is delivered through focused workshops which include education modules and participatory activities to understand the grazing ecosystem, and to manage grazing, fire, the tree–grass balance, weeds, and sown pastures. As new research information becomes available from local trials and demonstration sites each workshop module is updated.

During the workshop the R&D information is applied to a case study property and then to the participants' situation to ensure relevance and learning by doing. One of the main outcomes of the workshop, is for participants to start a grazing land management plan for their property. Participants are asked, throughout the workshop, to think about how the R&D information can be used to improve land condition and productivity within their beef business. To this end, information relating to ground cover thresholds, how to improve land condition, and how to reduce sediment and nutrient leaving their property is used to plan activities and to design sustainable grazing systems. Participants are asked to implement part of their grazing land management plan within 3 months after the workshop and the group will then meet to discuss each other's progress and to refresh key concepts, such as the ABCD Land Condition Framework (Fig. 10.4).

The Land Condition Framework was developed for the GLM Workshop decision-support tool kit so that producers could understand the characteristics and visual attributes of land in good to very poor condition. As land condition characteristics, such as the variety of native pasture species, weed species, tree density and soil erosive potential, are dependent on land type, the ABCD condition classes were developed for a broad range of land types throughout Queensland. The framework has proven to be very successful, with graziers finding the classification scheme easy to relate to what they are seeing in the paddock. It is has also allowed participants to benchmark their current management practices against land type and land condition classes and to use this information to monitor

Good or **'A' condition** has the following features:

- Good coverage of perennial grasses dominated by those species considered to be 3P grasses for that land type; little bare ground (<30%);
- Few weeds and no significant infestations
- Good soil condition: no erosion, good surface condition;
- No sign, or early signs, of woodland thickening.

Fair or **'B' condition** has at least one or more of the following features, otherwise similar to A condition:

- Some decline of 3P grasses; increase in other species (less favoured grasses, weeds) and/or bare ground (>30% but <60%);
- Some decline in soil condition; some signs of previous erosion and/or current susceptibility to erosion is a concern;
- Some thickening in density of woody plants.

Poor or **'C' condition** has one or more of the following features, otherwise similar to B condition:

- General decline in 3P grasses; large amounts of less favoured species and/or bare ground (>60%);
- Obvious signs of past erosion and/or susceptibility currently high;
- General thickening in density of woody plants.

Very poor or **'D' condition** has one or more of the following features:

- General lack of perennial grasses or forbs;
- Severe erosion or scalding, resulting in hostile environment for plant growth;
- Thickets of woody plants cover most of the area.

Fig. 10.4. Land condition classes developed by the Department of Primary Industries and Fisheries for the dry tropical rangelands in northern Queensland.

trends in condition after implementing better grazing land management practices. The simplicity of the framework is its strength.

The GLM Workshop is an excellent example of programmed learning extension methodology. The immediate challenge for R, D and E service providers is to develop programmes that use a mix of extension methods including technological development, team facilitation, information access, programmed learning and the consultant model (Coutts *et al.*, 2004). Grazing land management research often used a 'technology transfer' approach to extension, forgetting that the end-user needs to test and adapt technology before it is useful to their business.

Producer-driven, participative processes that provide tangible evidence of the impact of adoption of new and best technology, strategies and practices, on beef enterprise profit and sustainability that will be applicable to the northern beef cattle industry are gaining momentum and will create new and exciting directions for research, development and extension activities for grazing land management well into the future.

The Future

Typically graziers attempt to use technology to increase the productive capacity of the land they are managing (see review of technologies in Quirk *et al.*, 2004). Over-sowing native pasture with leguminous species or exotic grass species can lead to short-term increases in livestock production (Jones, 1997). However, continuous heavy stocking can lead to a decline in the proportion of native species in the pasture (Ash *et al.*, 2001) or degradation in soil health (Holt *et al.*, 1996), which reduces the capacity of the land to supply forage in the long-run (McIvor and Gardener, 1995). It is likely that future technologies (e.g. Virtual Fencing; http://www.ars.usda.gov/Main/docs.htm?docid=5564) that alter animal distribution in relation to pasture availability or the sensitivity of parts of the landscape to degradation will come to the fore.

Setting the stocking rates to match herbage supply at a level that allows the vegetation resource base to recover from past degradation, or the maintenance of vegetation cover levels, requires the grazier to make decisions about how much stock the enterprise can carry. As mentioned above, wet season spelling allows pastures to attain their maximum yield potentials; however, it is really over the dry season that the grazier will need to assess their purchasing and selling strategies, based on the vegetation biomass held at the end of the wet. This can be done on the ground, using rapid survey techniques, from modelled predictions derived from past stocking rates and rainfall (GRASP see Howden *et al.*, 1999), the Southern Oscillation Index (e.g. Australian RAINMAN, Clewett *et al.*, 1996) or from remotely sensed information (Pastures from Space, http://www.pasturefromspace.csiro.au/).

The interventions referred to above will allow the livestock manager to develop grazing practices to match forage removal with forage production. However, for sustainable grazing practices to be adopted, the triple bottom line outcomes i.e. economic, environmental and social, need to be assessed. In the first instance this will include predicting the relationship between grazing management practices and their economic and environmental consequences (MacLeod and McIvor, 2006). The grazier, or society, will then have to decide the optimum trade-off point between economic and environmental outcomes. Given the relationship between cover, water retention, pasture growth and possibly biodiversity benefits, it may be possible to develop 'Win–Win' grazing strategies that achieve profit for the grazier while securing environmental gain. This may be a utopian view but one we believe is worth striving for.

Currently nearly 45% of the land within the catchments flowing into the GBR lagoon is freehold, compared with 42% leasehold. The proportion of freehold to leasehold varies across the region, with the Fitzroy Basin having a greater proportion of freehold land and the Burdekin Basin having mostly leasehold. Policy changes relating to leasehold land tenure and tree clearing has increased the focus on management practices and land stewardship (Neldner, 2006). However, one of the major problems facing the beef industry in northern Queensland is that it may take a number of years for the restoration of perennial grasses and herbs in degraded rangelands (Ash *et al.*, 2001). This means that the economic benefits associated with the changes in management practice could take some time. This is likely to reduce the enthusiasm with which graziers adopt and implement changes in grazing management practice. There are various options to overcome this; first, changes in management practice could be achieved through legislation, for example in contracts associated with the renewal of leaseholds, or there may be incentives such as a subsidy/payment provided to bridge the gap between a change in grazing management practice and response of the system in ecological and economic terms (Fig. 10.5). One of the options for allowing graziers direct benefit from adoption

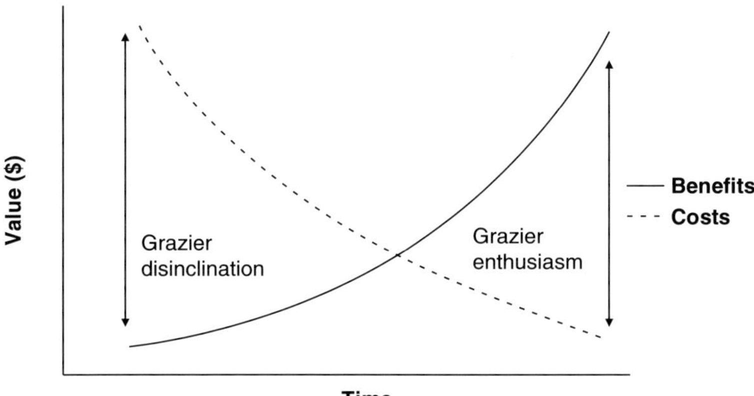

Fig. 10.5. Bridging the gap between change in grazing management practice and improved ecological and economic outcomes for rangelands.

of practices that reduce sediment and nutrient loss from paddock and thereby protect the reef is to link the management practice with the product being sold. For example, throughout the developed world local livestock products are being marketed so as to differentiate them from cheaper imports. This is often on the basis of an appeal to nationalism (i.e. 'Australian Owned and Grown') or because of some distinctive feature associated with the product's place of origin (e.g. http://www.levelsbest.co.uk/saltmarshlamb.html). There may be an opportunity to add economic value to beef reared in a sustainable fashion in northern Queensland (for example, Reef Safe Beef) as long as it can be demonstrated that the grazing management does not have detrimental effects off-farm.

Sustainable grazing practices for the rangelands of northern Queensland need to be based upon management of vegetation composition, cover and distribution, resulting in improved livestock performance and reduced off-farm impacts (Ash, 2004). It is our view that the knowledge exists to achieve this objective with the beef industry in northern Queensland showing the way to the rest of the grazing industry in the tropics as to how livestock production can achieve its goals while also maintaining landscapes and habitats of conservation importance.

Acknowledgements

Many of the ideas and much of the research presented in this chapter results from the long-term funding that has been provided by Meat and Livestock Australia to both CSIRO and Queensland DPI&F. We are grateful for that support.

References

Abbott, B.N. and McAllister, R.R.J. (2004) Using GIS and satellite imagery to estimate the historical expansion of grazing country in the Dalrymple Shire. In: Bastin, G.N., Walsh, D. and Nicolson, S. (eds) *Conference Papers: Australian Rangeland Society 13th Biennial Conference*. Australian Rangeland Society, Adelaide, pp. 405–406.

Anon. (1999) *State of the Environment Queensland 1999*. Queensland Environment Protection Agency, Brisbane.

Anon. (2003) *Reef Water Quality Protection Plan: for catchments adjacent to the Great Barrier Reef World Heritage Area*. www.thepremier.qld.gov.au/reefwater (accessed July 2006).

Ash, A.J. (2004) Dynamics of the grass layer. In: McCullough, M. and Musso, B. (eds) *Healthy Rangelands. Principles for Sustainable Systems. Focus on Australia's Burdekin Rangelands*. Tropical Savannas CRC, Charles Darwin University, Darwin, NT, pp. 69–95.

Ash, A.J. and McIvor, J.G. (1998) Forage quality and feed intake responses of cattle to improved pastures, tree killing and stocking rate in open eucalypt woodlands of north-eastern Australia. *Journal of Agricultural Science* 131, 211–219.

Ash, A.J., Corfield, J. and Ksiksi, T. (2001) *The Ecograze Project – Developing Guidelines To Better Manage Grazing Country*. CSIRO, Townsville.

Australian Bureau of Statistics (2003) *Queensland Agricultural Census 2000–2001*. Australian Bureau of Statistics, Canberra.

Bisset, W.J. (1980) Indian bluegrass has special uses. *Queensland Agricultural Journal* 106, 507–517.

Bortolussi, G., McIvor, J.G., Hodgkinson, J.J., Coffey, S.G. and Holmes, C.R. (2005) The Northern Australian beef industry, a snapshot. 1. Regional enterprise activity and structure. *Australian Journal of Experimental Agriculture* 45, 1057–1073.

Brodie, J., McKergow, L.A., Prosser, I.P., Furnas, M., Hughes, A.O. and Hunter H. (2003) *Sources of Sediment and Nutrient Exports to the Great Barrier Reef World Heritage Area*. Australian Centre for Tropical Freshwater Research, James Cook University, Townsville.

Chilcott, C.R., McCallum, B.S., Quirk, M.F. and Paton, C.J. (2003) *Grazing Land Management Education Package Workshop Notes – Burdekin*. Meat and Livestock Australia, Sydney.

Clewett, J.F., Clarkson, N.M. and Owens, D.T. (1996) Australian Rainman – a package of rainfall information for better pasture management. *Tropical Grasslands* 30, 172.

Coutts, J., Roberts, K., Frost, F. and Coutts, A. (2004) *A Review of Extension in Australia in 2001–2003 and its Implications for Developing Capacity into the Future. A RIRDC report for the Cooperative Venture for Capacity Building*. RIRDC, Barton, ACT.

Craik, W. (1992) Great Barrier Reef Marine Park: its establishment, development and current status. *Marine Pollution Bulletin* 25, 122–133.

Fabricius, K.E. (2005) Effects of terrestrial runoff on the ecology of corals and coral reefs: review and synthesis. *Marine Pollution Bulletin* 50, 125–146.

Frisch, J.E. and O'Neill, C.J. (1998) Comparative evaluation of beef cattle breeds of African, European and Indian origins. 2. Resistance to cattle ticks and gastrointestinal nematodes. *Animal Science* 67, 39–48.

Frisch, J.E. and Vercoe, J.E. (1977) Food intake, eating rate, weight gains, metabolic rate and efficiency of feed utilization in *Bos taurus* and *Bos indicus* crossbred cattle. *Animal Production* 25, 343–358.

Furnas, M. (2003) *Catchments and Corals: Terrestrial Runoff to the Great Barrier Reef*. Australian Institute of Marine Science, Townsville.

Gardener, C.J., McIvor, J.G. and Williams, J. (1990) Dry tropical rangelands: solving one problem and creating another. *Proceedings of the Ecological Society of Australia* 16, 279–286.

Gifford, G.F. (1985) Cover allocation in rangeland watershed management (a review). In: Jones, E.B. and Ward, T.J. (eds) *Watershed Management in the Eighties*. American Society of Civil Engineers, New York, pp. 23–31.

Gilbert, M., Waterhouse, J., Ramsay, M. and Brodie, J. (2003) *Population and Major Land Use in the Great Barrier Reef Catchment Area: Spatial and Temporal Trends*. Great Barrier Reef Marine Park Authority, Townsville.

Hamilton, D. and Haydon, G. (1996) *Pesticides and Fertilisers in the Queensland Sugar Industry – Estimates of Usage and Likely Environmental Fates*. Department of Primary Industries, Queensland.

Haynes, D. and Michalek-Wagner, K. (2000) Water quality in the Great Barrier Reef World Heritage Area: past perspectives, current issues and new research directions. *Marine Pollution Bulletin* 41, 428–434.

Holt, J.A., Bristow, K.L. and McIvor, J.G. (1996) The effects of grazing pressure on soil animals and hydraulic properties of two soils in semi-arid tropical Queensland. *Australian Journal of Soil Research* 34, 69–79.

Howden, S.M., McKeon, G.M., Walker, L., Carter, J.O., Conroy, J.P., Day, K.A., Hall, W.B., Ash, A.J. and Ghannoum, O. (1999) Global change impacts on native pastures in south-east Queensland, Australia. *Environmental Modelling and Software* 14, 307–316.

Hutchings, P. and Haynes, D. (2000) Sources, fates and consequences of pollutants in the Great Barrier Reef. *Marine Pollution Bulletin* 41, 265–266.

Jones, R.J. (1997) Steer gains, pasture yield and pasture composition on native pasture and on native pasture oversewn with Indian couch (*Bothriochloa pertusa*) at three stocking rates. *Australian Journal of Experimental Agriculture* 37, 755–765.

Kragt, M.E., Roebeling, P.C. and Ruijs, A. (2006) *Effects of Great Barrier Reef degradation on recreational demand: a contingent behavior approach*. Fondazione Eni Enrico Mattei (FEEM) working paper 45. 2006, pp. 19.

Ludwig, J.A., Wilcox, B.P., Breshears, D., Tongway, D.J. and Imeson, A.C. (2005) Vegetation patches and runoff-erosion as interacting ecohydrological processes in semiarid landscapes. *Ecology* 86, 288–297.

Lukas, P.H.C., Webb, T., Valentine, P.S. and Marsh, H. (1997) *The Outstanding Universal Value of the Great Barrier Reef World Heritage Area*. Great Barrier Reef Marine Park Authority, Townsville.

MacLeod, N.D. and McIvor, J.G. (2006) Reconciling economic and ecological conflicts for sustained management of grazing lands. *Ecological Economics* 56, 386–401.

McCullough, M. (2004) Snapshot of the upper Burdekin rangelands. In: McCullough, M. and Musso, B. (eds) *Healthy Rangelands. Principles for Sustainable Systems. Focus on Australia's Burdekin Rangelands*. Tropical Savannas CRC, Charles Darwin University, Darwin, NT, pp. 1–10.

McCullough, M. and Musso, B. (2004). *Healthy Rangelands. Principles for Sustainable Systems. Focus on Australia's Burdekin Rangelands*. Tropical Savannas CRC, Charles Darwin University, Darwin, NT.

McIvor, J.G. (2002) Pasture management in semi-arid tropical woodlands: effects on ground cover levels. *Tropical Grasslands* 36, 218–226.

McIvor, J.G. and Gardener, C.J. (1995) Pasture management in semi-arid tropical woodlands: effects on herbage yields and botanical composition. *Australian Journal of Experimental Agriculture* 35, 705–715.

McIvor, J.G. and Scanlan, J.C. (1994) State and transition models for rangelands. 8. A state and transition model for the northern speargrass zone. *Tropical Grasslands* 28, 256–259.

McIvor, J.G., Williams, J. and Gardener, C.J. (1995) Pasture management influences run-off and soil movement in the semi-arid tropics. *Australian Journal of Experimental Agriculture* 35, 55–65.

McKeon, G.M., Day, K.A., Howden, S.M., Mott, J.J., Orr, D.M., Scattini, W.J. and Weston, E.J. (1990) Northern Australian savannas – management for pastoral production. *Journal of Biogeography* 17, 355–372.

McLennan, S.R., Dunster, P.J., O'Rourke, P.K. and Murphy, G.M. (1981) Comparison of dry season supplements containing salt, sulphur or molasses for steers grazing native pasture in the dry tropics of northern Queensland. *Australian Journal of Experimental Agriculture and Animal Husbandry* 21, 457–463.

Meat and Livestock Australia (2006) *MLA Northern Beef Program – Livestock Production Research and Development Strategic Plan 2006–2011*. Meat and Livestock Australia, Sydney.

Miller, C.P. and Stockwell, T.G.H. (1991) Sustaining productive pastures in the tropics. 4. Augmenting native pastures with legumes. *Tropical Grasslands* 25, 98–103.

Miller, P.C., Winter, W.H., Coates, D.B. and Kerridge, P.C. (1990) Phosphorus and beef production in northern Australia 9. Strategies for phosphorus use. *Tropical Grasslands* 24, 239–249.

Mitchell, A.W., Reghenzani, J.R. and Furnas, M.J. (2001) Nitrogen levels in the Tully River – a long-term view. *Water Science and Technology* 43, 99–105.

Neldner, J. (2006) Why is vegetation condition important to government? A case study from Queensland. *Ecological Management & Restoration* 7, S5–S7.

Nelson, P. and Roth, C. (2004) Water and nutrient cycling processes. In: McCullough, M. and Musso, B. (eds) *Healthy Rangelands. Principles for Sustainable Systems. Focus on Australia's Burdekin Rangelands*. Tropical Savannas CRC, Charles Darwin University, Darwin, NT, pp. 23–67.

North Australia Beef Research Council (2006) North Australia Beef Research Council Homepage. www.nabrc.org.au (accessed July 2006).

O'Reagain, P.J., Brodie, J., Fraser, G., Bushell, J.J., Holloway, C.H., Faithful, J.W. and Haynes, D. (2005) Nutrient loss and water quality under extensive grazing in the upper Burdekin river catchment, North Queensland. *Marine Pollution Bulletin* 51, 37–50.

Productivity Commission (2003) *Industries, Land Use and Water Quality in the Great Barrier Reef Catchment. Research Report.* Productivity Commission, Canberra.

Quirk, M., Ash, A. and Holmes, W. (2004) Cattle production systems. In: McCullough, M. and Musso, B. (eds) *Healthy Rangelands. Principles for Sustainable Systems. Focus on Australia's Burdekin Rangelands*. Tropical Savannas CRC, Charles Darwin University, Darwin, NT, pp. 11–22.

Shepherd, R.N. and Chilcott, C. (2006) *Rangelands of the Burdekin Catchment; a History of Engagement in Grazing Land Management*. DPI&F, Brisbane.

Stark, W. (2005) 'Threats' to the Great Barrier Reef. *IPA Backgrounder* 17, 1–20.

Stokes, C.J., McAllister, R.R.J. and Ash, A.J. (2006) Fragmentation of Australian rangelands: processes, benefits and risks of changing patterns of land use. *The Rangeland Journal* 28, 83–96.

Tait, J. (1994) *Lowland Habitat Mapping and Management Recommendations Tully – Murray Catchments*. Queensland Department of Primary Industries, Brisbane.

Tothill, J.C. and Gillies, C. (1992) *The Pasture Lands of Northern Australia: Their Condition, Productivity and Sustainability*. Tropical Grassland Society of Australia, St Lucia, Queensland.

Wachenfeld, D.R., Oliver, J.K. and Morrisey, J.I. (1998) *State of the Great Barrier Reef World Heritage Area 1998*. Great Barrier Reef Marine Park Authority, Townsville.

Walker, L., Northup, B.J. and Brown, J. (1999) Developing techniques to quantifying root responses of tussock grasses to defoliation. In: Eldridge, D. and Freudenberger, D. (eds) *People and Rangelands: Building the Future. Proceedings of the VI International Rangeland Congress*. IRC, Townsville, pp. 193–197.

Wooldridge, S., Brodie, J. and Furnas, M. (2006) Exposure of inner-shelf reefs to nutrient enriched runoff entering the Great Barrier Reef Lagoon: Post-European changes and the design of water quality targets. *Marine Pollution Bulletin* 52, 1467–1479.

11 Meeting Ecological Restoration Targets in European Waters: a Challenge for Animal Agriculture

Penny J. Johnes
Aquatic Environments Research Centre, University of Reading, UK

Abstract

The EU Water Framework Directive requires the restoration of water bodies to good ecological status within a prescribed timetable. It applies in all member states, is mandatory, and requires classification of water quality as a function of the extent of ecological degradation from reference (unimpacted) conditions. Nutrient enrichment of water bodies is the single greatest pollution problem impacting on European waters, and diffuse nutrient export from agriculture is a primary source of this problem in the most intensively farmed regions of Europe. Here, the scale of the problem is illustrated with reference to the specific conditions in the UK, set within a wider European context. By way of example, the scale of nutrient enrichment in the waters of England and Wales is illustrated, and options for reduction of diffuse nutrient loading to these waters are analysed. Key trends identified in this analysis indicate the importance of animal agriculture as a contributor to the total diffuse agricultural nutrient loading on English and Welsh waters, and the overwhelming need to bring these sources under control if conditions suitable for sustaining 'Good Ecological Status' in these waters are to be generated. The analysis also highlights the likely impact of broad regional-scale scenarios on nutrient export rates across England and Wales. Key conclusions reached include the need to take some sensitive lands out of production, introduce ceilings on fertilizer use and stocking densities, and tight controls on agricultural practice in higher risk areas where intensive and inappropriate agriculture is combined with a low intrinsic nutrient-retention capacity in the landscape.

Introduction

In Europe there has been a pattern of post-Second World War intensification and expansion of agricultural production from upland and lowland areas. With the formation of the European Economic Community, new policies were introduced to support this process, providing economic price support mechanisms to encourage the adoption of new technologies, crop and livestock production practices. In arable areas, there have been changes in crop production including the introduction of new crop types, increases in fertilizer application rates, and the cultivation of highly sensitive and marginal areas of land not previously used for crop production. In livestock production areas, there have

 Redesigning Animal Agriculture (eds D. Swain, E. Charmley, J. Steel and S. Coffey)

been increases in stocking densities on grazing land, alterations to animal physiology and diet, a move of the majority of pig and poultry production to intensive rearing units, and changes in the perception of manures from fertilizer to waste product. The application of this 'one-size-fits-all' agronomic model for agricultural production across heterogeneous cultures and landscapes under the EU Common Agricultural Policy has led, in turn, to a reduction in the heterogeneity of agricultural landscapes, and a mismatch between the intrinsic nutrient-retention capacity of the landscape and the intensity of agricultural production. The result has been the development of pollutant transport hotspots, where inappropriate land use and management is practised in sensitive landscape areas. At the same time social, environmental and agricultural policy development within the European Union (EU) have been decoupled, with each resourced and controlled as a separate system. The intrinsic linkages and feedbacks between these systems have been ignored, resulting in a lack of cross-compliance in policies, with widespread degradation of terrestrial and aquatic environments across Europe and wholesale changes in rural social and economic systems. Recent recognition of the scale of damage to European waters, and public concerns over the state of many water bodies has led to the introduction of a new directive to tackle these problems: the EU Water Framework Directive (WFD).

The EU Water Framework Directive

The Water Framework Directive (2000/60/E) came into force on 22 December 2000. It establishes a new legal framework for the protection, improvement and sustainable use of all European water bodies. The WFD is the most substantial piece of water legislation ever produced by the European Commission, and will provide the major driver for achieving sustainable management of water in European member states for many years to come.

The WFD requires each member state to assess the current ecological status of all inland and coastal waters, defined as a function of deviation from an undamaged (reference) state. Each water body must be classified according to the extent to which its present ecological structure and function deviate from this reference state, with the resulting classification reported to the European Commission on a national basis. Five classes are available (Table 11.1), with waters showing little or no deviation from reference state being classed as being in high ecological status. Any waters showing

Table 11.1. Classification of ecological status under the EU Water Framework Directive (Environment Agency, 2005).

Class	Change from natural condition	Description
High	None or minimal	Biological, hydro-morphological and physico-chemical quality elements match reference conditions
Good	Slight	Biological and physico-chemical quality elements only slightly changed from reference conditions
Moderate	Moderate	Biological quality elements are moderately changed as a result of human activities
Poor	Major	Substantial changes to the reference biological communities through human activities
Bad	Severe	Large portions of the reference biological communities are absent through human activities

evidence of only slight degradation from reference conditions will be classed as being in good ecological status. Three further classes are then available for more severely degraded waters: moderate, poor and bad. The primary objectives of the WFD are that:

- All natural waters of Good or High Ecological Status at 2004 must be managed sustainably to maintain this status
- All natural waters of Moderate, Poor or Bad ecological status at 2004 must have achieved Good Ecological Status, not incurring disproportionate cost, by 2027
- All artificial waters, including canals, reservoirs and heavily modified water bodies must achieve 'Good Ecological Potential' by 2027.

The measures to be put in place to bring degraded waters up to good ecological status must form part of a river basin management plan (RBMP), introduced in each member state no later than 2008 (Table 11.2). The timetable for setting and meeting objectives for each river basin are highly prescribed under the WFD, which defines how the objectives should be achieved through the establishment of environmental objectives and ecological targets for both inland and coastal surface waters. Measures are also included in the WFD to ensure comparability in reporting standards across Europe, requiring inter-calibration of criteria for setting reference conditions and setting of class boundaries between member states, with distinct regional clusters set up to reflect the major geoclimatic regions of Europe. Compliance with the terms of the Water Framework Directive is mandatory in all member states, and requires control of a wide range of social, political and economic activities.

Table 11.2. Milestones to be met under the EU Water Framework Directive.

Date	Milestone
2003	WFD must be transposed to Member State Law. Members must identify: • 'River Basin Districts' for characterization and management to achieve GEC • 'Competent Authorities' responsible for practical compliance with the terms of the WFD
2004	Complete River Basin District Characterization of the risk of waters of not being in 'Good Ecological Status'; Complete first economic analysis of water use; Establish a register of protected areas in each RBD.
2006	Establish environmental monitoring programmes; Publish a work programme for producing first River Basin Management Plans (RBMPs).
2007	Publish interim overview of significant water management issues in each RBM district; Release for general consultation.
2008	Publish draft RBMPs for consultation; Identify measures to be undertaken to achieve GEC in all waters.
2009	Finalize and publish first RBMPs; Identify programme of measures to meet WFD objectives.
2012	Ensure all measures are fully operational; Publish timetable and work programme for second RBMPs.
2014	Publish second draft RBMPs.
2015	Achieve environmental objectives specified in first RBMPs; Finalize and publish second RBMP with revised programme of measures.
2021	Achieve environmental objectives specified in first RBMPs; Finalize and publish second RBMP with revised programme of measures.
2027	Achieve environmental objectives specified in first RBM; Finalize and publish second RBMP with revised programme of measures.

The WFD therefore represents a major shift in perceptions of water quality and its management in Europe. Traditionally, targets and class boundaries in national classification schemes have been based on chemical variables such as biological oxygen demand (BOD), ammonium and dissolved oxygen. Under the WFD the focus has shifted to targets based on the ecological structure and function of natural systems, set by reference to the ecological conditions expected in the absence or near absence of anthropological impacts. The natural variability in the ecology of waters is taken into account under the Directive, such that the definition of good ecological status will be, necessarily, water-body specific. As such, the environmental conditions required to meet these standards will also vary between water bodies and regions, and the management of anthropogenic impacts will also be, necessarily, catchment specific.

In terms of management under the WFD, a wide range of anthropogenic activities affecting the ecological structure and function of water bodies will need to be brought under control for degraded waters. This will include the control of hydrological or morphometric modification, direct damage to species composition and relative abundance through over-fishing, accidental or deliberate stocking with non-indigenous species, and modification of the hydrochemistry through direct and indirect discharge of polluting substances. Of these, nutrient enrichment through the discharge of N- and P-rich water from diffuse atmospheric, agricultural and septic tank sources and direct point-source discharges from sewage treatment works is a widespread problem in Europe. Such nutrient enrichment is widely recognized as having the capacity to alter both the structure and function of the biological communities in aquatic environments through the role of N and P as major nutrients limiting primary production.

The Scale of the Nutrient Enrichment Problem in Europe

A recent review paper highlights the importance of controlling point sources in the restoration of ecological status in lakes across Europe (Jeppesen *et al.*, 2005). However, with the introduction of a requirement for P stripping at major sewage treatment works (STWs), including those serving a population equivalent of over 10,000 people, under the EU Wastewater Treatment Directive (91/271/EEC) point sources are becoming less significant in the total P load delivered to surface waters in Europe. This serves to highlight the importance of continuing diffuse P export to waters from agricultural production systems. Similarly, point sources are rarely significant contributors to the total N loading delivered to waters where atmospheric deposition and agriculture are the primary sources (Johnes, 1996, 1999, 2001; Johnes and Butterfield, 2002). Recent studies have also highlighted the importance of agriculture as a major source of both P and N export to European waters (Moss *et al.*, 1996; Johnes, 1999; Leonard and Crouzet, 1999; McGuckin *et al.*, 1999; Meeuwig *et al.*, 2000; Johnes and Butterfield, 2002). Controlling the loss of both P and N from diffuse agricultural sources, together with improved removal of both N and P at sewage treatment works, is therefore a prerequisite for achieving good ecological status in many European freshwaters.

The problems facing European member states in complying with the WFD are greatest in the most densely populated areas where high population density is tied to intensive agricultural production to feed the people, and point-source discharges are a significant contributor to the nutrient load impacting on surface water and groundwater. This can be illustrated with reference to conditions in the UK, which has the fourth highest population density in Europe after Malta (1263 people per km^2), The Netherlands (465 people per km^2) and Belgium (335 people per km^2). The European average population density, based on 25 member states, is 110 people per km^2. Within the UK, England alone has a higher population density (383 people per km^2) than Belgium, though population densities are much lower in the upland areas of the UK with 142 people per km^2 in Wales, 125 people per km^2 in Northern Ireland, and only 65 people per km^2 in Scotland.

Throughout the UK, there are spatial variations in the nature of agricultural production, with livestock farming concentrated in the wetter western and upland areas, and arable farming predominant in the lowland areas of central and eastern England and the Scottish lowlands. There is little natural or semi-natural landscape remaining, with the greatest remnants found in the upland areas. Agriculture is the dominant land use, and the nutrient concentrations typically observed in a wide range of water bodies are, as a result, markedly higher than in other European nations.

This is illustrated in the data presented in Fig. 11.1 (rivers) showing riverine dissolved reactive P data for a range of European countries (extracted from the European Environment Agency database, after Foy, 2007). The countries are ranked along the x axis in ascending order based on the proportion of rivers with DRP above 50 μg Pl^{-1}. Concentrations for UK rivers are typically above 100 μg Pl^{-1} (Muscett and Withers, 1996) and the UK has the second highest ranking, exceeded only by Belgium, again highlighting the importance of population density in describing likely nutrient enrichment problems in European member states. The data highlight the scale of the problems to be addressed if UK waters are to attain good ecological status through the control of diffuse and point-source N and P discharges to its waters.

Within the UK, spatial variations in the distribution of population (highest in England) and the intensity of agricultural production are also reflected in regional differences in mean lake nutrient concentrations. This is illustrated in Fig. 11.2. The

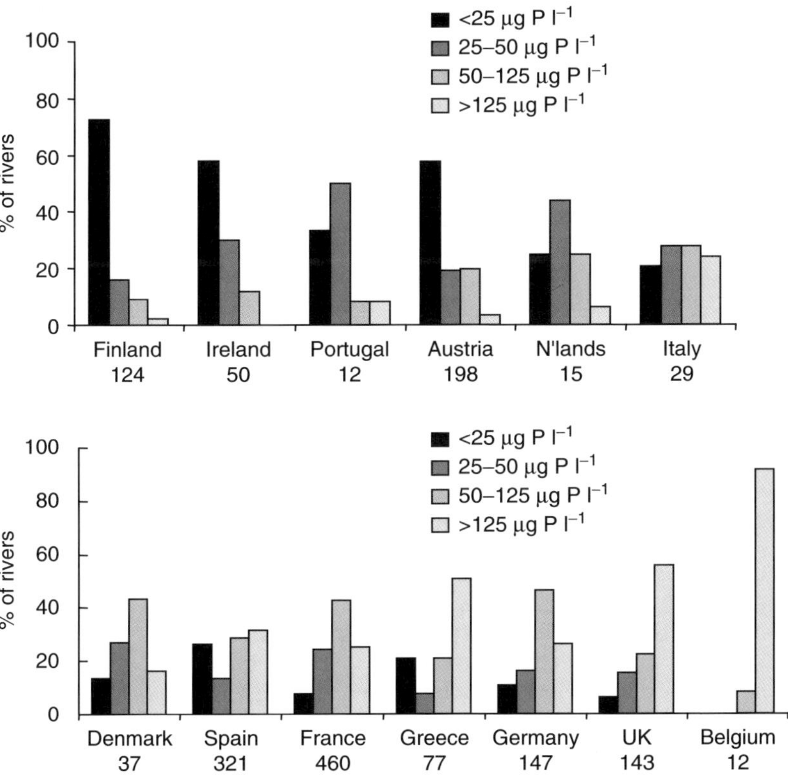

Fig. 11.1. Distribution of dissolved reactive phosphorus (DRP) concentrations in rivers of EC. Countries are ranked in ascending order by the % of rivers with DRP above 50 μg P l^{-1}. Numbers after country name refer to sample size. (Source of data: European Environment Agency; after Foy, 2007).

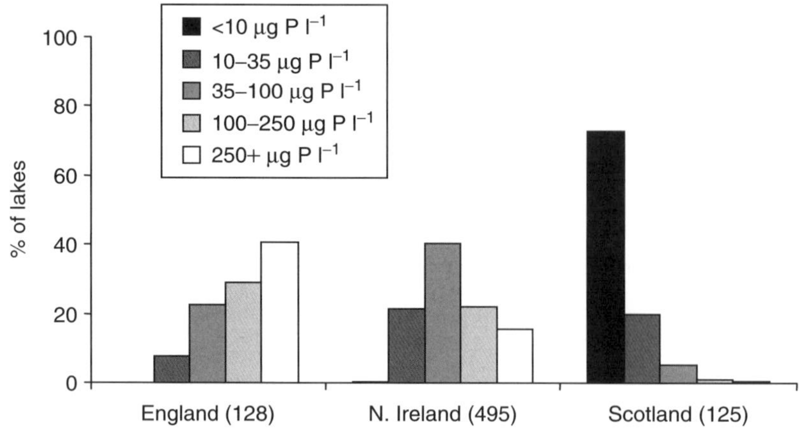

Fig. 11.2. Lake total phosphorus concentrations in the UK (Foy and Bailey-Watts, 1998). Numbers in parentheses refer to numbers of lakes.

coverage differs for each region: the Scottish data covers all large Scottish lakes but none of the smaller lakes, while the data from Northern Ireland covers all small lowland lakes between 1 ha and 50 ha in area (Ferrier *et al.*, 1996; Foy and Bailey-Watts, 1998). Few of these lakes were impacted by urban P inputs. The data from English lakes may be less representative as they were drawn from two published sources (Bennion, 1994; Carvahlo and Moss, 1995). These exclude upland lakes and are disproportionately weighted towards lakes that are statutorily designated (under domestic legislation) Sites of Special Scientific Interest.

Lakes that are oligotrophic (73%) dominate the Scottish data set, but all of these are located in the Highlands of Scotland, where extensive upland livestock production is the dominant land use, on open moorland and valley-bottom grasslands. This agricultural production system is common across all upland areas in the UK, and generates relatively low nutrient concentrations in upland waters, though these are nevertheless nutrient enriched relative to their catchment-specific reference conditions. Such lakes are absent from the data sets from England and Northern Ireland. For the English lakes the dominant class is for lakes with TP concentrations > 250 μg Pl^{-1} (41%) and 70% of lakes have concentrations above 100 μg Pl^{-1}, suggesting that these lakes could classed as hypertrophic. Only 8% of the English lakes have TP concentrations typical of mesotrophic conditions (<35 μg Pl^{-1}). These data highlight the geographic variation in the distribution of nutrient-rich lakes in the UK, with a very high proportion of lakes in England and lowland Wales falling into the eutrophic to hypertrophic classes, based on the OECD (1982) class bands. Notwithstanding the limitations of the data sets, however, it is clear that the distribution of trophic class varies between the nations of the UK according to population density and the proportion of land available for agricultural production. Thus less densely populated nations with a higher proportion of upland mountain and moorland, such as Scotland, have the majority of lakes with a lower trophic status than those located in the more densely populated and predominantly lowland agricultural landscape of England and Northern Ireland.

The scale of the problem, in terms of the degree to which current nutrient chemistry deviates from baseline or reference conditions, and thus the likely deviation of ecological structure and function from reference ecological state, is further illustrated where temporal trends in nutrient enrichment are examined. This is

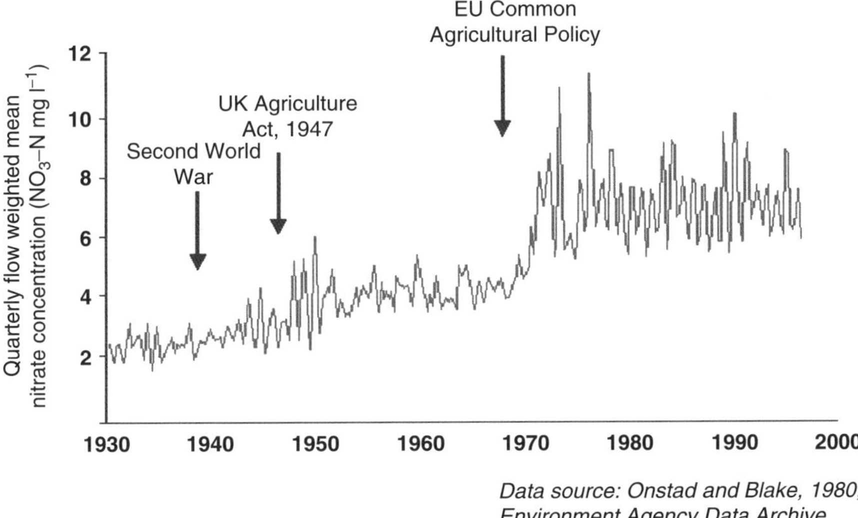

Fig. 11.3. Long-term trends in nitrate enrichment in the River Thames, 1930–2000.

illustrated using three long-term data sets for water bodies in England and Wales in Figs 11.3 and 11.4. In Fig. 11.3, long-term trends in nitrate concentrations in the River Thames at Walton on Thames are summarized (after Onstad and Blake, 1980; supplemented with more recent monthly data from the Environment Agency routine water quality monitoring programme). Key change points in the nitrate concentration trend occurred in conjunction with major phases of change in the social, economic and political systems. The first change point occurred during the Second World War, when much of the traditional grazing land was ploughed up, and many marginal lands were drained and cultivated in response food shortages and the UK Government 'Dig for Victory' campaign. In the post-war period, the 1947 Agriculture Act was enacted to support Government policy to develop self-sufficiency in food production at a time of severe food rationing following the collapse of food production and support in the post-war era. A third change point is apparent when the UK joined the EU in the early 1970s and benefited from the economic support structures introduced under the Common Agricultural Policy (CAP).

In Fig. 11.4, two long-term data sets collected by the Freshwater Biological Association at Windermere (now the Centre for Ecology and Hydrology, Lancaster) are presented, along with the output from two models reconstructing the likely P enrichment history for Windermere and Esthwaite Water in the English Lake District (after Bennion *et al.*, 2005). In these it is apparent that TP concentrations have more than doubled in both lakes since the mid-19th century, evident in both the observed P chemistry data, model simulations of nutrient export from catchment to lake, and in the response of the diatom communities in each lake. Key change points in both lakes can be tied to the opening of STWs on the shores of each lake, and the subsequent introduction of P stripping at these works in the late 20th century, but underlying these change points is a general trend of enrichment which is tied to increased diffuse P export from agricultural practices in the catchment of each lake (Bennion *et al.*, 2005).

The nature and extent of diffuse nutrient enrichment of waters is clearly a problem for the competent authorities charged with the implementation of the WFD in each member state, and will be a particular problem in those member states with intensive

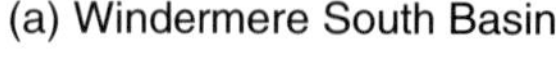

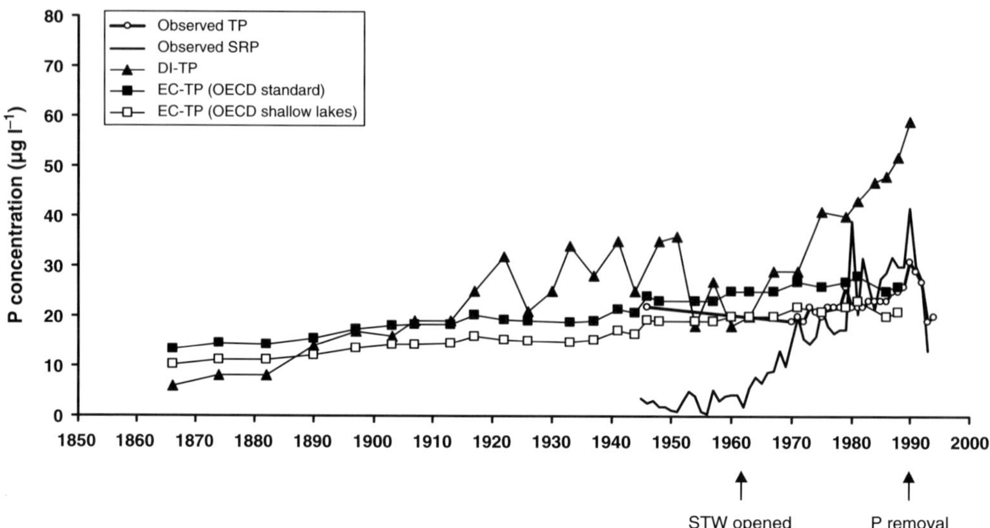

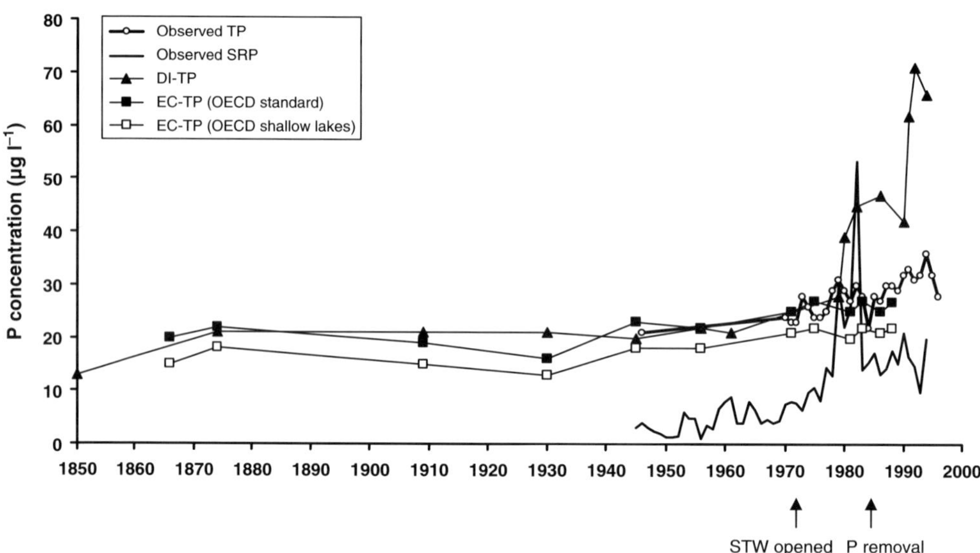

Fig. 11.4. Long-term trends in P enrichment (total P (TP), soluble reactive P (SRP), diatom transfer function (DI-TP) and export coefficient (EC_TP)) in the English Lake District (after Bennion *et al.*, 2005).

agricultural production systems. The scale of the challenge faced in England and Wales is shown in Fig. 11.5 (after Environment Agency, 2005). This shows some of the risk characterization maps submitted to the European Commission, as required under the WFD, by the Environment Agency as competent authority for England and Wales. The maps presented in Fig. 11.5 illustrate the calculated risk of waters not achieving Good Ecological Status by 2015 as a result of diffuse pollution pressures alone. Other maps illustrating the risk associated with other forms of chemical, biological or physical modification of waters are also available in Environment Agency (2005). Those presented in Fig. 11.5

suggest that the majority of inland and coastal waters in England and Wales are probably at risk of not achieving this target by 2015, and many of them are definitely at risk. Thus the problems facing the Environment Agency in implementing the WFD are substantial, and diffuse pollution pressures are acknowledged as being the single greatest challenge in England and Wales.

Key Contributors to the Diffuse Nutrient Pollution Problem

In meeting this challenge, the first step is to unpick the pollutant trend data to determine the likely causal factors to be addressed. In a recent paper Johnes and colleagues (2007) examined the likely scale of reduction in diffuse nutrient export needed to allow waters

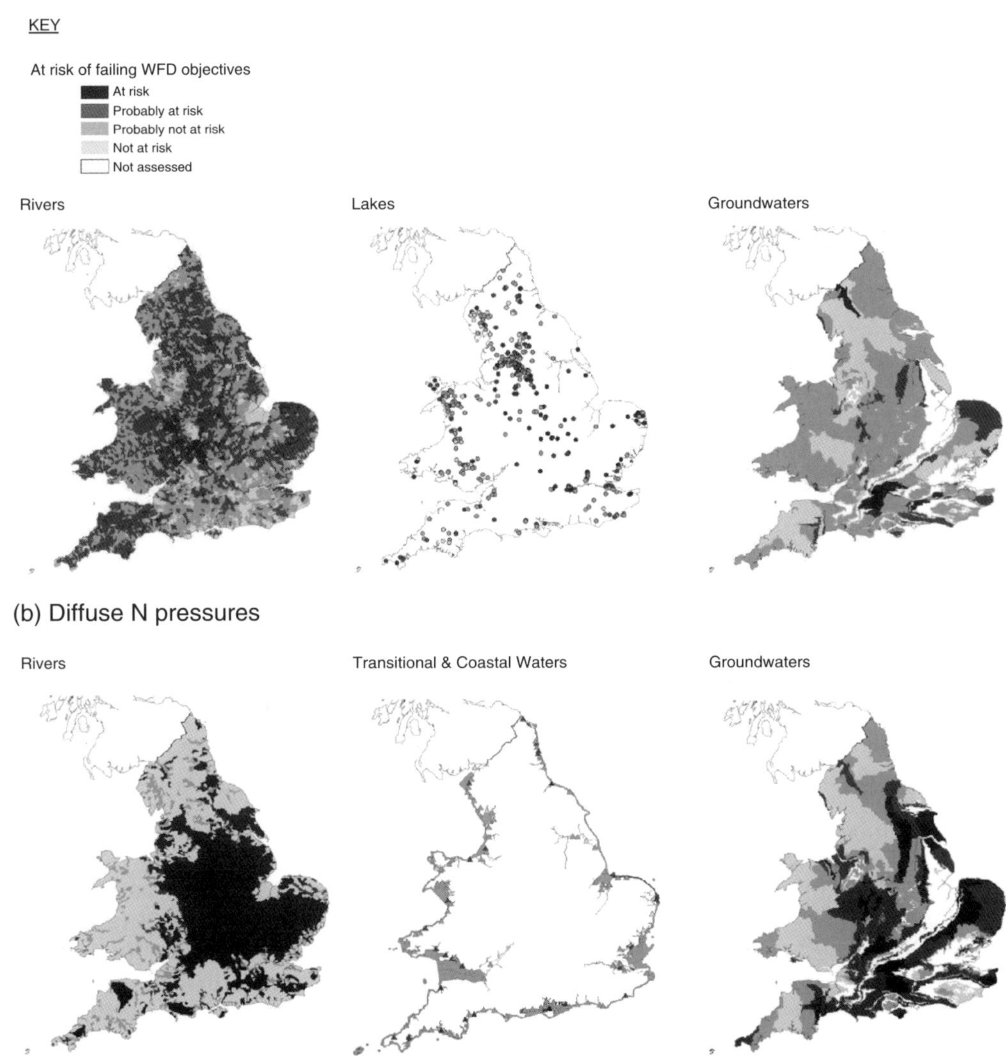

Fig. 11.5. Waters unlikely to meet European Union Water Framework Directive (EU-WFD) objectives by 2015 due to diffuse nutrient pressures (Environment Agency, 2005).

in England and Wales to achieve good ecological status under the WFD, as part of a programme of work for the Department of Food and Rural Affairs. The focus in that paper was to:

- Define the scale of nutrient reduction required for English and Welsh waters relative to catchment-specific nutrient reference conditions;
- Evaluate the potential for achieving good ecological status in English and Welsh waters through the application of national-scale strategies for nutrient and landscape management;
- Determine the areas where nutrient reduction will require control of agricultural production going beyond improvements to existing farm systems.

A summary of the key findings of that study is presented here, to illustrate the scale of the challenge facing animal agriculture in Europe if the ecological quality of European waters is to be restored. The evaluation was undertaken using the AERC Export Coefficient Modelling Approach (Johnes, 1996, 1999, 2001; Johnes *et al.*, 1996, 1998a,b; Johnes and Butterfield, 2002). This was developed as a tool capable of national simulation, at 1 km^2 grid scale, of nutrient export from diffuse agricultural, atmospheric and point sources, taking into account regional variations in catchment geology and climate. In this exercise, model outputs for 1931 and 1991 were compared, scenarios for nutrient reduction, tailored to regional variations in dominant land use and geoclimatic conditions, were evaluated in terms of their likely impact on nutrient export rates from land to water, and the likely impact of these scenarios in terms of the nutrient status of English and Welsh water bodies was discussed.

A summary of the current problems arising from agricultural production practice in England and Wales, as identified in this study, is presented in Table 11.3 (after Johnes *et al.*, 2007). The table is divided according to the major geoclimatic regions types (after Johnes and Butterfield, 2002), which represent quasi-homogenous landscape units with broadly similar climate, geology, soils and topography. Thus in the Upland Geoclimatic Region, there has been a reduction in the area of land used for fertilized grass, crop production, pig and poultry production in the period 1931–1991, but an increase in fertilized grass production on marginal (hillslope) lands, and an increase in fertilizer application rates to all crops and grass, and a substantial increase both in the scale of sheep and cattle production, and in the stocking densities on grazing land. This has led, in turn, to an increase in the size of the nutrient pool available for export from land to water, and a decrease in the intrinsic nutrient retention capacity of the land through trampling pressure, poaching and the resultant reduction in infiltration capacity of the soils in these regions. The net result is an average predicted increase in N export from land to stream of over 190%, and 104% increase in P export to waters. The major contributors to this trend are intensive sheep production on marginal lands and fertilizer application rates to crops and grass. By contrast, in permeable landscapes underlain by chalk or Jurassic limestone, typical of many lowland areas of the UK, mixed arable and livestock farming is the dominant practice. The greatest changes in agricultural production in these regions, 1931–1991, result from the introduction of new crops such as oilseed rape (canola), the intensity of fertilizer applications to these crops and grass, and the growth of the pig production industry on outdoor intensive pig-lots. The result has been 192% increase in N loading and 102% increase in P loading on waters in these regions. The most telling point, however, is that in each of the region types, livestock production, whether it is intensive beef and dairy cattle production on the heavier and wetter soils of Wales and western England, or pig and poultry production in intensive rearing units in eastern England, is one of the major contributors to the trend of enrichment. In livestock farming regions it is the key contributor, with nutrient loading on waters having doubled every decade, particularly in the uplands and marginal lands where the current stocking densities are environmentally unsustainable. Only in eastern England is fertilizer use the primary source

Table 11.3. Current problems in diffuse nutrient export by Geoclimatic Region type (after Johnes *et al.*, 2007).

	Upland regions	Lowland intensive dairying regions	Permeable lowland mixed farming regions	Impermeable lowland mixed farming regions	Intensive arable farming regions
Land use trend	Decrease in: • crop production area • area of fertilized grass • area of rough grazing on marginal lands • pig and poultry production	Decrease in: • area of permanent grass • area of rough grazing	Decrease in: • area of permanent and temporary grass • area of rough grazing	Decrease in: • area of all crops and grass • pig and poultry production	Decrease in: • area of all grazing land • cattle and sheep numbers
	Increase in: • fertilized grass production on marginal lands • fertilizer applications to all crops and grass • sheep and cattle production • stocking density on grazing land	Increase in: • area of all other crops and grass, particularly cereal crops and temporary grass • cattle, sheep and pig production • fertilizer applications to all crops and grass • manure applications to crops and grass in excess of crop requirements • stocking density on grazing land	Increase in: • area of all cereals and other arable crops • intensity of pig production in outdoor feedlots and sheep production • fertilizer applications to all crops and grass • manure applications to crops and grass • stocking density on grazing land	Increase in • fertilizer application rates to crops and grass • cattle production (+565%), sheep production (+145%) • manure applications to crops and grass • stocking density on grazing land	Increase in: • drainage of traditional grazing marshes for arable cultivation • area of arable crop production • fertilizer application rates to crops and grass • pig production on outdoor feedlots, poultry production in intensive indoor units • poultry slurry applications to crops and grass • stocking density on grazing land

Continued

Table 11.3. *Continued*

	Upland regions	Lowland intensive dairying regions	Permeable lowland mixed farming regions	Impermeable lowland mixed farming regions	Intensive arable farming regions
Impact	Increase in: • nutrient store available for export from animal manures, slurry, fertilizer applications and soil P reserves • nutrient and sediment export potential on grazing land, resulting from soil compaction and vegetation removal in heavily grazed areas	Increase in: • nutrient store available for export from animal manures, slurry, fertilizer applications and soil P reserves • nutrient and sediment export potential on grazing land, resulting from soil compaction and vegetation removal in heavily grazed areas	Increase in: • nutrient store available for export from animal manures, slurry, fertilizer applications and soil P reserves • nutrient and sediment export potential on arable land, left bare in winter • nitrate leaching to aquifers • nutrient and sediment export from pig lots	Increase in: • nutrient store available for export from animal manures, slurry, fertilizer applications and soil P reserves • nutrient and sediment export potential on grazing land resulting from soil compaction and vegetation removal on grazing lands • nutrient and sediment export potential on arable land	Increase in: • nutrient store available for export from animal manures, slurry, fertilizer applications and soil P reserves • nutrient and sediment export potential on arable land left bare in winter • nitrate leaching to aquifers • nutrient and sediment export from pig lots
Model estimate of changes in nutrient export rate,1931–1991	N = + 190% P = + 104%	N = + 157% P = + 83.3%	N = + 192% P = + 102%	N = + 196% P = + 98.0%	N = + 79.3% P = + 96.2%
Major contributors	Intensive sheep production on marginal lands, fertilizer applications to crops and grass	Cattle production intensity, fertilizer and manure applications to temporary grass and cereal crops	Cereal and other arable crop production, pig production on outdoor lots	Cattle and sheep production on grazing land, fertilizer applications to crops and grass	Fertilizer applications to arable crops and grass, poultry and pig production in intensive lots

of diffuse nutrient loading on waters. Major causes for concern arise from the following production systems in the UK:

- Dairy and beef cattle production in lowland Wales and western England;
- Sheep production in sensitive upland areas of Wales and northern England;
- Pig production on intensive outdoor pig-lots in lowland permeable areas;
- Cattle and sheep production through strip grazing in lowland permeable catchments;
- The disposal of poultry litter from intensive rearing units in eastern England.

The impact of these trends on nutrient export to waters is spatially variable across England and Wales. This is illustrated in Fig. 11.6, where the spatial variations in the rates of N and P export per hectare of agricultural land are shown for 1931 and 1991 (after Johnes *et al.*, 2007). The highest rates of both N and P export from agriculture are occurring in the lowland intensive dairying regions, resulting from the intensity of livestock production in these areas. The lowest rates of N and P export occur in the upland regions and also in flatter arable lands of East Anglia. The low rates of export in the upland regions result from relatively low intensity agricultural production in the national context. The low rates of export in East Anglia result from the low intrinsic vulnerability of this flat and dry landscape to water-borne nutrient transport, again in the UK context, despite the intensive agricultural production practised in this region, both in the present day and also in the baseline year.

These trends are interesting from the perspective of illustrating national patterns of nutrient flux from agriculture. However, this does not mean that the waters in the uplands and East Anglia are of better quality or ecological status than waters in the remainder of England and Wales. What is of greater ecological significance is not the absolute rate of nutrient flux and the resulting N and P concentrations instream, but the degree of deviation (pollution) that is potentially experienced by the biota living in the water body. Although water bodies in upland regions have typically low nutrient concentrations they are nevertheless under considerable pressure from diffuse agricultural sources, with the greatest rates of increase in nutrient loading in the UK predicted for the uplands of Wales, Cumbria and south-west England. The rivers in eastern England, by contrast, whilst having typically moderate to high nutrient concentrations, have experienced, in many instances, a decrease or only a modest increase in nutrient loading over the period 1931–1991, with marked spatial heterogeneity evident in the trends within each region. In order to comply with the definition of 'Good Ecological Status' under the Water Framework Directive, it is necessary to develop management strategies which will generate nutrient concentrations capable of sustaining a healthy ecology in terms of both structure and function in these impacted waters, regardless of absolute nutrient concentration per se. The challenge is to determine which management strategies are most appropriate in each region, given the different starting points and pressures for each water body, and to assess whether broad regional strategies can deliver an appropriate degree of nutrient export control for individual water bodies within each region.

Identifying the Nature and Scale of Appropriate Management of Diffuse Pollution Under the WFD

There are a wide range of options available for reduction of nutrient export from agricultural sources, including modifications to the rates and timing of fertilizer applications to crops and grass, genetic modifications to crop varieties to improve nutrient uptake efficiency, and for animal agriculture a wider range including:

- Manipulation of digestive physiology through animal selection and genomics;
- Modification of the nutrient content and availability in animal feeds;
- Modification of the timing, method and rate of manure application to crops and grass;

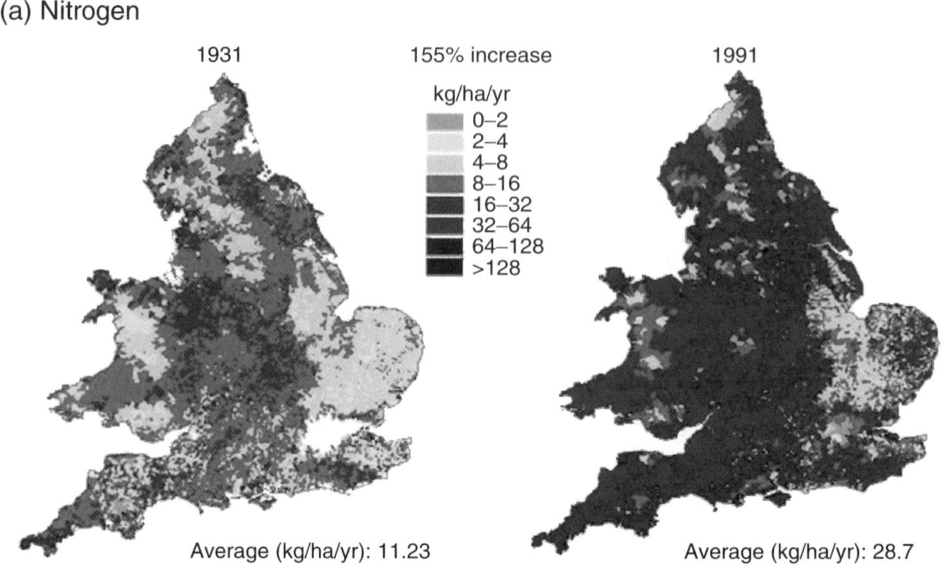

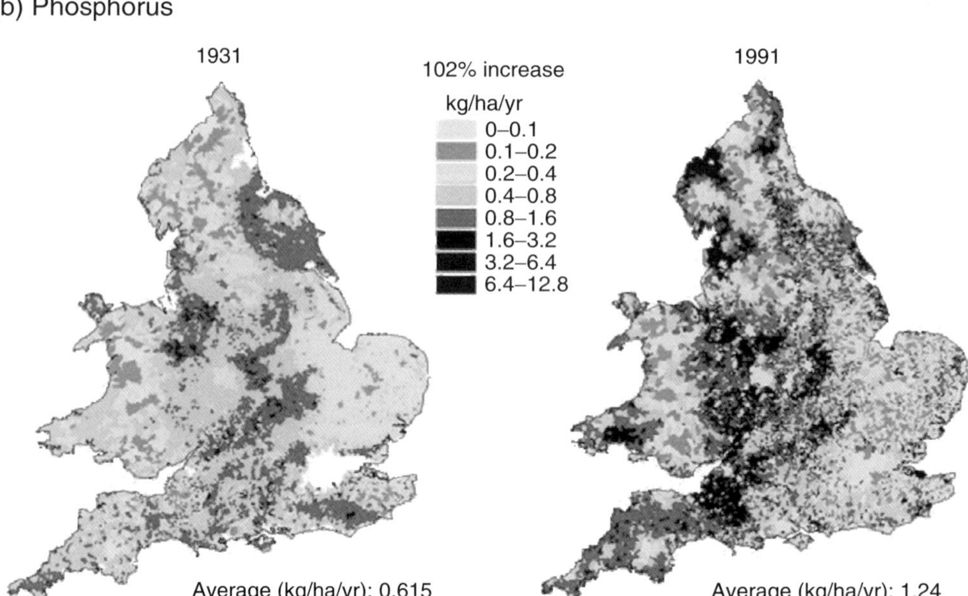

Fig. 11.6. Nutrient export from land to water in England and Wales, 1931–1991 (after Johnes *et al.*, 2007).

- Introducing full nutrient budgeting in farm management plans: seeing manure as a fertilizer, not as a waste product;
- Reduction in stocking densities on grazing land (feedlots, fodder cropping, strip-grazing) with implications for the crop:grass imbalance in European agriculture;
- Interrupting pollutant transport pathways from land to water, by introducing measures such as:
  - moving farm gates
  - replanting hedgerows
  - removing drainage;
- Reduction in stock numbers in sensitive areas, with implications for livestock production and related industries.

In the work undertaken by Johnes and colleagues (2007) only those measures which had quantified impacts on nutrient export in the full range of geoclimatic regions were evaluated. A two-tiered approach was adopted with separate scenarios evaluated for each geoclimatic region. The purpose of this exercise was not to determine the 'best' scenario for the future of agriculture in England and Wales but to undertake a theoretical exploration of the scale of changes which might be necessary to bring about a change in nutrient loading sufficient to sustain good ecological status in the water bodies of each geoclimatic region. In each case, management strategies were run through the model, targeting key sources of increases in nutrient flux in each region 1931–1991, as identified in Table 11.3. Scenario 1 was designed to achieve a moderate reduction in sectoral nutrient export for each of the key nutrient sources in each geoclimatic region. Scenario 2 was designed to introduce additional measures over and above those introduced under scenario 1, in order to generate more substantial reductions in sectoral nutrient export in each region. The scenarios run through the models are summarized in Table 11.4.

The impact of running these scenarios through the model is illustrated in Fig. 11.7, where the model output at parish scale for 1991 is plotted next to the model output

Table 11.4. Scenario runs for geoclimatic regions in England and Wales (after Johnes *et al.*, 2007).

Upland extensive livestock grazing regions	Lowland intensive dairy and beef cattle farming regions	Mixed farming regions underlain by permeable bedrock	Mixed farming regions underlain by impermeable bedrock	Intensive arable farming regions in E. England
Key problems:				
Sheep and cattle production on marginal lands	Intensity of cattle production, and P saturation of soils	Intensity of arable crop and outdoor pig production	Conversion of grazing land to arable land	Intensity of arable production
Scenario 1				
Reduce sheep nutrient export by 50%; Reduce fertilizer application rates to all crops and grass by 50%	Reduce N fertilizer applications by 50% and P by 80% to all crops and grass; Reduce cattle nutrient export by 25%	Convert 10% cereals to PG; Reduce fertilizer applications to all crops and grass by 50%	Convert 50% cereals to fertilized grass and 20% of fertilized grass to rough grazing; Reduce sheep and cattle nutrient export by 25%	Reduce fertilizer applications to all crops and grass by 50%
Scenario 2				
As scenario 1 AND reduce the cattle nutrient export by 25%	As scenario 1 AND reduce sheep and pig nutrient export by 25%	As scenario 1 AND reduce sheep and pig nutrient export by 25%	As scenario 1 AND reduce fertilizer application rates by 50%	As scenario 1 AND reduce pig and poultry nutrient export by 25%

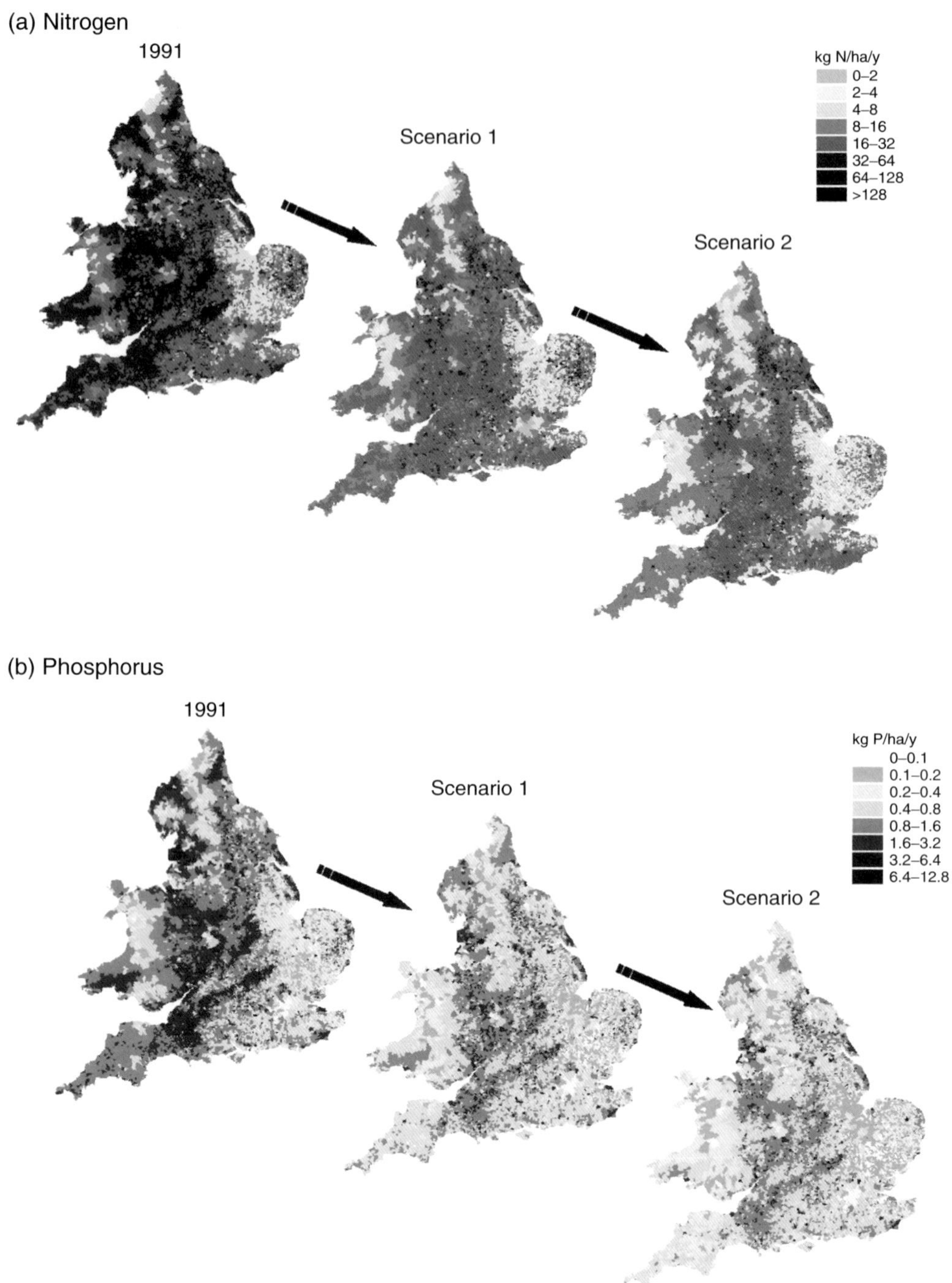

Fig.11.7. Impact of scenarios on rates of nutrient export from agriculture to water (after Johnes *et al.*, 2007).

following scenario 1, and scenario 2 for (a) N and for (b) P. In each of these it is apparent that there is a substantial reduction in the predicted rate of N and P export across England and Wales. There is a much greater reduction in N and particularly P flux when the additional measures in scenario 2 are applied to each region. However, it is important to measure the relative success of these measures as region-wide management strategies. It is likely that the management of diffuse pollution from agriculture, given the wide range of variation in current rates of N and P flux both within and between regions, will require at least river basin-scale or even farm-scale management.

The modelling of land-use scenarios presented in this earlier paper indicates the scale of change required in sectoral exports of N and P from animal and arable agriculture, in order to generate conditions with the potential to support good ecological status in English and Welsh water bodies. The scale of reduction is substantial, and some reductions might be generated by the application of broad regional-scale policies across each of the geoclimatic region types. This analysis allowed Johnes and colleagues to draw some important conclusions regarding the scale of approach needed to reduce diffuse nutrient export from agriculture to a level which has the potential to support Good Ecological Status in English and Welsh waters. They concluded that:

- Generating 'Good Ecological Status' in all waters will require:
  - substantial changes in agricultural land use and management;
  - taking sensitive lands out of production;
  - ceilings on fertilizer use and stocking densities;
  - tight controls on agricultural practice in high risk areas, particularly livestock.
- Control of both N and P is likely to be needed to generate Good Ecological Status in some waters, particularly shallow lakes.
- Additional measures may be required in resilient systems, particularly in shallow lakes.
- Management at regional scale is a blunt approach and would lead to ineffective management, which would be costly and fail to achieve targets in some areas.
- Targeted management at river-basin scale would allow for appropriate management at the farm scale, and redistribution of farming activities within river basins to take account of the varying risk of different activities and the varying vulnerability of the landscape to nutrient export.
- Sectoral reductions in both N and P export of 25–50% will be required from livestock production systems if European waters are to meet WFD objectives and have the potential to achieve 'Good Ecological Status' within the prescribed timetable.

The message for animal agriculture from this analysis is clear, and represents a major challenge for the future.

Conclusion

In order to rise to this challenge, those involved in animal agriculture need to resolve the clash between the drive for increases in livestock production and the need to curb and (possibly) reverse the extent of environmental degradation. This might involve a number of approaches including the development of new technologies and feedstuffs, modifications to digestive physiology to improve feed efficiencies, and further research into genomics and trait selection. However, as a first step, it will be important to re-think the role of animal agriculture as part of the wider socio-economic and environmental systems. Without compliance in social, economic and agricultural policies, it is unlikely that the necessary changes will be seen as desirable and sustainable. With this in mind, it is also important that policy makers and farmers accept an environmental truth: not all landscapes are the same, and some landscapes cannot sustain current production levels, no matter what might be done to modify production efficiency. Some stock reductions are inevitable, particularly in sensitive marginal lands, if ecological targets are to be attained and the progressive degradation of the natural environment is to be halted. In Europe, under the WFD, reductions in animal production are inevitable, and there is not an option, in the present policy context, to argue to maintain the status quo. The challenge for animal agriculture, therefore, is to develop solutions which reduce the impact of mandatory environmental legislation on the industry. Failure to rise to this challenge would be to sanction the continued degradation of the global ecosystem to the point at which it would be irretrievable in any meaningful sense.

References

Bennion, H. (1994) A diatom-phosphorus transfer function for shallow, eutrophic ponds in southeast England. *Hydrobiologia* 275/276, 391–410.

Bennion, H., Johnes, P.J., Ferrier, R., Phillips, G.L. and Haworth, E. (2005) A comparison of diatom phosphorus transfer functions and export coefficient models as tools for reconstructing lake nutrient histories. *Freshwater Biology* 50, 1651–1670.

Carvahlo, L. and Moss, B. (1995) The current status of a sample of English Sites of Special Scientific Interest subject to eutrophication. *Aquatic Conservation: Marine and Freshwater Ecosystems* 5, 191–204.

Environment Agency (2005) Assessing risks to the water environment – 2005 results. Environment Agency, Bristol, UK.

Ferrier, R.C., Malcolm, A., McAlister, E., Edwards, A.C., Morrice, J. and Owen, R. (1996) *Hindcasting of In-loch Phosphorus Concentrations Based on a Land Cover Classification. Macaulay Land Use Research Institute. Report to the Scottish and Northern Ireland Forum for Environmental Research, Aberdeen*. FER, Aberdeen.

Foy, R.H. and Bailey-Watts, A.E. (1998) Observations on the spatial and temporal variation in the phosphorus status of lakes in the British Isles. *Soil Use and Management* 14, 131–139.

Foy R.H. (2007) Variation in the reactive phosphorus concentrations in rivers of northwest Europe with respect to their potential to cause eutrophication. *Soil Use and Management,* in press.

Jeppesen, E. Søndergaard, M., Jensen, J.P., Havens, K.E., Anneville, O., Carvalho, L., Coveney, M.F., Deneke, R., Dokulil, M.T., Foy, R.H., Gerdeaux, D., Hampton, S.E., Hilt, S., Kangur, K., Kohler, J., Lammens, E., Lauridsen, T.L., Manca, M., Miracle, M.R., Moss, B., Noges, P., Persson, G., Phillips, G.L., Portielje, R., Romo, S., Schelske, C.L., Straile, D., Tatrai, I., Willen, E. and Winder, M. (2005) Lake responses to reduced nutrient loading – an analysis of contemporary long-term data from 35 case studies. *Freshwater Biology* 50, 1747–1771.

Johnes, P.J. (1996) Evaluation and management of the impact of land use change on the nitrogen and phosphorus load delivered to surface waters: the export coefficient modelling approach. *Journal of Hydrology* 183, 323–349.

Johnes, P.J. (1999) Understanding catchment history as a tool for integrated lake and catchment management. *Hydrobiologia* 395/396, 41–60.

Johnes, P.J. (2001) Quantifying the non-point source contribution to nutrient loading on freshwaters in 32 UK catchments. *Verhandlungen Internationale Vereinigung für Limnologie* 27, 1306–1309.

Johnes, P.J. and Butterfield, D. (2002) Landscape, regional and global estimates of nitrogen flux from land to ocean: errors and uncertainties. *Biogeochemistry* 57/58, 429–476.

Johnes, P.J., Moss, B. and Phillips, G.L. (1996) The determination of water quality by land use, livestock numbers and population data – testing of a model for use in conservation and water quality management. *Freshwater Biology* 36, 451–473.

Johnes, P.J., Curtis, C., Moss, B., Whitehead, P., Bennion, H.B. and Patrick, S. (1998a) *Trial Classification of Lake Water Quality in England and Wales: a Proposed Approach*. R&D Technical Report E33, Environment Agency, Bristol.

Johnes, P.J., Bennion, H.B., Curtis, C., Moss, B., Whitehead, P. and Patrick, S. (1998b) *Trial Classification of Lake Water Quality in England and Wales: a Proposed Approach*. R&D Project Record E2-i721\5, Environment Agency, Bristol.

Johnes, P.J., Foy, R.H., Butterfield, D. and Haygarth, P.M. (2007) Land use scenarios for England and Wales: evaluation of management options to support 'good ecological status' in surface freshwaters. *Soil Use and Management,* in press.

Leonard, J. and Crouzet, P. (1999) *Lakes and Reservoirs in the EEA Area*. European Environment Agency, Copenhagen.

McGuckin, S.O., Jordan, C. and Smith, R.V. (1999) Deriving phosphorus export coefficients for CORINE land cover types. *Water Science and Technology* 12, 47–53.

Meeuwig, J.J., Kauppila, P. and Pitkanen, H. (2000) Predicting coastal eutrophication in the Baltic – a limnological approach. *Canadian Journal of Fisheries and Aquatic Sciences* 57, 844–855.

Moss, B., Johnes, P.J. and Phillips, G.L. (1996) The monitoring and classification of standing waters in temperate regions – discussion and proposal based on a worked scheme for British waters. *Biological Reviews* 71, 310–339.

Muscett, A.D. and Withers, P.J.A. (1996) The phosphorus content of rivers in England and Wales. *Water Research* 30, 1258–1268.

OECD (1982) *Eutrophication of waters, monitoring, assessment and control*. Organization for Economic Co-operation and Development, Paris.

Onstad, C. and Blake, J. (1980) Thames basin nitrate and agricultural relations. In: *Proceedings of the American Society of Civil Engineers Symposium on Watershed Management*. AMCE, Boise, Indiana, pp. 961–971.

12 Livestock – Fulfilling Social, Environmental and Economic Roles – is it Possible?

David L. Swain[1] and Margaret Gill[2]
[1]*CSIRO Livestock Industries, Australia;*
[2]*School of Geosciences, College of Physical Sciences, University of Aberdeen, UK*

Abstract

History has shown that when societies fail to consider the sustainable use of natural resources they eventually collapse. Science has enabled modern agriculture to increase productivity. In developed countries farmers have successfully provided an abundance of food and in so doing achieved economic success. There is a growing recognition that future livestock farming systems need to simultaneously address social, environmental and economic goals. With the advent of widespread environmental degradation, molecular technologies to redesign animals and a disconnected consumer, it appears more important than ever to question how we might derive resilient and sustainable agricultural systems.

Economic drivers are inevitably the most powerful incentives for change within both an interventionist and neo-liberal market framework. Environmental degradation is often a by-product of agricultural intensification resulting from farmers responding to consumer demands for cheaper food. Societies in developed countries have become disconnected from modern agriculture and thus consumer action may not always be able to target the perpetrators of the problems. Policies by their very nature take a broad approach and are thus insensitive tools for capturing the opportunities presented by the diversity of livestock production systems. Reconnecting farmers with consumers via mechanisms which encourage more direct engagement can provide a forum to address environmental concerns. In so doing this helps to minimize the conflicts between social, environmental and economic goals and builds knowledge and understanding.

Introduction

The first archaeological evidence of the domestication of animals by man dates from around 10,000 years ago. Over the intervening millennia, livestock and their husbandry have adapted to 'society's' changing needs and to changes in the local environments. It is relatively easy to understand how systems adapted to changes in the past. In the current millennium, the rate and complexity of change will require a more proactive approach to designing livestock systems which are fit for a future where sustainable resource use will be a priority objective. This proactive approach will provide an opportunity to consider, in advance, the contribution which livestock systems make on a sustainable society.

This chapter draws on two examples from recent history both to provide a context and to highlight some of the emerging issues that may direct future livestock farming systems. It also builds on some of the concepts presented in previous chapters of this monograph to attempt to provide a synthesis of where the challenges and opportunities may be for redesigning animal agriculture. It embraces sustainable development principles, exploring whether social, environmental and economic drivers can indeed shape livestock production systems which will sustain society into an uncertain future.

More recently, the importance of giving society a voice in defining the way in which food is produced has increasingly been recognized. Earlier chapters of this monograph have illustrated how advances in molecular biology, for example, could enable animals to be designed to meet the objectives of future livestock systems. History has taught us that the wide-scale use and acceptance of emerging technologies will only succeed if they are applied appropriately in the context of the broader social, environmental and economic drivers. Can science use the lessons of the past to increase the likelihood of acceptance? Have we learned from these lessons?

Redesigning Livestock Farming Systems

Scientists, by the very nature of their discipline, often spend more time looking into the future than learning the lessons of the past. There are visionaries, however, who do explore the past and thereby enable others to incorporate the lessons of the past into consideration of the future. Diamond (2005) considered failed societies of the past and identified factors that led to their eventual collapse. He highlighted the dire consequences for some societies of failing to anticipate major impacts before or even when they happened.

Sir Kenneth Blaxter was another visionary who analysed the past and called the progress made in the 50-year period from 1936 to 1986 the 'modern agricultural revolution' (Blaxter and Robertson, 1995). Science was 'properly harnessed to the improvement in agricultural productivity', but Blaxter could also see that the scientific community was slow in recognizing the point at which agricultural productivity alone was no longer the sole challenge in developed countries. Improvements in productivity were often achieved at the cost of environmental degradation as the public (and policy makers) maintained the pressure on farmers to provide cheap food. In the 1990s the necessity to consider the impact of livestock systems on the environment received increased attention.

The potential for designing animals to suit future livestock production systems

As modern agriculture has moved through the industrial revolution and the chemical revolution we are now at the start of a molecular revolution. Both marker-assisted selection and genetic engineering have the potential to initiate large-scale changes in the livestock industry. However, in many societies there is widespread suspicion of genetic engineering and the challenge will be to build knowledge and trust between scientists, farmers and the broader community.

Prayaga and Reverter (Chapter 5 this volume) discussed precision breeding and the relationship with quantitative genetics. Traditional quantitative genetics relies on measuring how animals perform within a given system. Measuring the performance of a desired characteristic is done within a 'systems' framework. Molecular techniques provide the opportunity to link genes to physiological characteristics or traits and thus provide a powerful new tool for precision breeding through linking gene(s) function with phenotype. Molecular selection techniques are able to provide tools to quantify traits which are difficult to measure.

Genetic engineering is the process of manipulating genetic information to create a more desirable phenotypic outcome. There are a number of potential applications for techniques associated with genetic engineering. In earlier chapters, Doran and Lambeth (Chapter 8 this volume) discussed using RNAi to create transgenic chickens that are resistant to avian influenza and Wells and Laible (Chapter 7 this volume) discussed the use of cloning and transgenesis as an effective tool for disseminating superior genotypes.

The ability to effectively 'design' animals for specific systems could potentially bring very significant economic benefits to producers and both environmental and societal benefits to the general public. Our hypothesis is that the principles of sustainable development can provide a suitable framework within which to explore the opportunities and application of new molecular technologies to design more resilient livestock production systems that meet the needs of future societies.

Application of the principles of sustainable development to the livestock industry

Sustainability is a word that has multiple definitions and meaning (Tilman *et al*., 2002). It is easy to say that a sustainable approach to livestock development embraces economic, environmental and social outcomes. Can these tenets be given equal weighting or are they even independent of one another? Multinational businesses have learnt that they are not. For example, Shell saw the impact of public opinion on their 'bottom line' when they tried to dump the Brent Spar oilrig at sea. A quote from Shell captures the importance of listening and engaging with the broader community.

> Today we have learnt a serious lesson... We must take the views of the public into account, even where they have been influenced by a single-issue organisation... behaving entirely irrationally.

This is but one of many examples that could be quoted and global corporate industries are responding. Agriculture, however, is in the main a different type of industry and livestock industries are comprised of multiple small and medium-sized enterprises. There is only a loose connection between the individual farmer and the broader community (Pierce, 1994, Smithers *et al*., 2005). It can be difficult to recognize and respond individually or corporately to social and environmental pressures (Burton and Wilson, 2006). TV images of farmers in Latin America chopping down rainforests for beef production can have an impact on meat sales in any location, even if individual farmers locally are following best environmental practice. So how do we apply sustainable development principles to such a fragmented industry?

Farmers are in the business of food production and need to see an economic return, while consumers want to continue to pay low prices for food, but at the same time to dictate what is, or is not, an acceptable social and environmental cost (Ignatow, 2006). If it is the scientist who initiates and develops knowledge-based tools then it is the farmer that chooses to use them and the consumer that decides whether they accept the way in which the products have been produced. Previous chapters have demonstrated the technologies that are available for redesigning animal agriculture, but the application of sustainable development principles supports the proposition that the integration of the emerging technologies within farm systems needs to be a partnership involving the broader community, as suggested by Todt (2004). This partnership should not, however, be considered a linear process. There needs to be active dialogue between scientists, farmers, consumers and policy makers, to ensure that scientific advances are acceptable to society as a whole and so can be adopted to achieve better economic, environmental and social outcomes.

To some extent this is already happening. In the last 15 years there has been a shift from 'productivism' to 'consumerism' with a change in emphasis from maximizing food production to delivering a broader range of social and environmental benefits. Farm systems are now expected to simultaneously meet both production and environmental goals in a socially acceptable manner

(Firbank, 2005; Potter and Tilzey, 2005) and successful farmers are increasingly recognizing the need to build social capital with the non-farm community (Sharp and Smith, 2003). Global trading and agricultural intensification has resulted in geographical separation between where food is produced and where it is consumed. This geographical separation weakens the link between local responsibility and global accountability (Wurtenberger *et al.*, 2006). Government-directed agro-environmental policies in Europe recognize that the use of public funds to subsidize food production, which underpinned the original Common Agricultural Policy, is no longer acceptable to a population with little understanding of how food is produced. Food subsidies are being replaced by the 'public good' of subsidizing farmers to remedy negative environmental and social impacts, which were the unforseen consequences of the original food security objective. A second lesson from the past has also been learned: the need for closer alignment between society's demand for environmental services and landholders' abilities to meet those demands (Hodge, 2001; Milne, 2005). This alignment is being addressed in many countries through increased consultation of stakeholders during the development of policy. Evidence from the UK suggests that farmers still think predominately in terms of production goals (Burton and Wilson, 2006) and policy development will have to consider how to stimulate the recognition of the new values which society is placing on agriculture (Burton, 2004).

So what are the key questions in the context of redesigning sustainable livestock systems that will deliver environmental, economic and social benefits in a rapidly changing climate of uncertainty? Who should take responsibility for the negative environmental and social impacts of a cheap food policy as demonstrated in Europe? How do we reconnect farmers and consumers within a global trading forum? These impacts and their societal implications affect today's farmers. We explore some of the issues that contribute to these questions and in so doing aim to provoke debate by comparing examples of livestock production systems in two very different economic contexts. We also explore the extent of impact on the environment and society and highlight some of the 'solutions' that have been identified in the academic literature.

Economic Drivers of Livestock Production Systems in the 21st Century

At the third Horizons in Livestock Science conference this paper was given in Australia, by a person from Scotland; two countries with strong historical links. Domesticated livestock farming in Australia owes much of its origins to practices which had evolved in the countries of the original settlers. Since the establishment of the Common Agricultural Policy (CAP) in Europe the economic drivers of these two continents have diverged. A comparison of the contrasting drivers between two different sectors within Australian and European livestock production systems provides an interesting contrast for study. Here we compare the UK dairy industry operating under the policies of the European Union with the Australian beef industry operating within the global free market system.

UK dairying

The recent history (post-World War II) of European agriculture emphasizes the changing economic priorities of livestock agriculture in response to broader social priorities that are interpreted through government intervention. The Common Agricultural Policy (CAP) was introduced to increase production in order to ensure a reliable food supply. The increase in food supply was achieved through the application of science and by decreasing production risks via a guaranteed market. Governments purchased surplus food, which was put into storage. The original justification for the intervention policy was to ensure food production was increased and food supplies were sufficient to meet community needs. People would never go hungry again. Government intervention thus created a market which provided the catalyst for enhanced productivity.

By the early 1980s the oversupply of food started to build up in intervention stores, creating the so-called 'milk lakes' and 'butter mountains'. The interventionist policy had successfully increased production through agricultural intensification. Increased mechanisation, improved genetics and more widespread use of agrochemicals underpinned the economic model. However, in the mid-1980s it became apparent that the oversupply of dairy products was becoming a problem. Subsidized food that entered the global market place created trade tensions. Within the EU there was increasing concern that farmers were receiving public money to produce food that was no longer needed. Within the UK to counteract these tensions, policies were adopted to restrict output, and in 1984 milk quotas were set and economic penalties imposed for overproduction (Colman, 2000). In the UK, farmers who had successfully streamlined their businesses for increased production now had to find new ways of running profitable businesses.

The introduction of quotas created a new market scenario with a shift from production maximization to either maximize efficiency or add value. The relatively rapid shift in emphasis created significant economic challenges particularly for those farmers that operated an output-driven economic model with large capital investment which enabled them to maximize production. High capital overheads cannot quickly be reduced. Although the new economic framework focused attention on inputs there was still a drive to increase output per unit area or per unit livestock. Quotas encouraged dairy farmers to explore their business options. At the farm level some producers tried to expand production to increase business efficiency. However, there were a number of dairy farmers who decided to leave the industry and others that added value through niche markets, including processing and organic production.

The UK dairy industry has seen a continued cost price squeeze with falling milk prices and increasing costs of production (Blackburn, 2006). Added to the economic challenges there has been an increasing burden of environmental and food safety compliance. Agricultural intensification policies within the EU created highly efficient (in economic terms) dairy production systems and succeeded in delivering cheap milk through supermarkets. However, mounting evidence indicates that this was at the expense of the environment (Aarts *et al.*, 2000). Johnes (Chapter 11 this volume) describes in detail the new challenge posed by implementation of the EU Water Framework Directive. Efficiency in the past was judged in terms of maximizing marketable outputs rather than minimizing wasteful ones in dung and urine and it took some time for the problems that these wastes created to be recognized.

More recently there is a forecast that falling milk supplies are leading to market forces acting on the milk price; in certain locations, processors and retailers are starting to have to pay a premium to secure milk supplies (Blackburn, 2006). History has shown that dairy farmers in the EU have the ability to respond to government policies that change the economic drivers. The effectiveness of economic liberalization coupled with greater environmental intervention in resolving environmental problems is yet to be seen; however, the challenge to meet the exigencies imposed by the EU Water Framework Directive will provide an interesting test.

Australian beef

In contrast to the policy-driven economic framework of the UK dairy industry the Australian beef industry has been largely driven by global market forces. The neoliberal framework explains industry structural changes largely independently of political intervention. Australia is one of the world's largest beef-exporting nations, selling its products to a diversified market including the USA, Asia and the EU (Pritchard, 2006). Australian beef producers have successfully managed to respond to the global market-driven needs. The growth in the export market has come about through a combination of increased beef consumption in Japan and Korea through the 1990s, an opportunity to sell low-grade beef to the USA and Australia's unique ability to address food safety concerns through its ability to combine a high volume of low input production with disease-free status for internationally controlled diseases such as bovine spongiform encephalopathy (BSE). Recent adoption of an electronically based ID system to enable rapid and accurate traceback has ensured the industry is able to guarantee the highest level of food safety and bio-security (Bailey and Britt, 2001). Bio-security is an increasingly important issue when trading in global food markets (Smith *et al.*, 2005). In response to global market demands, Australia has developed a diversified beef production system serving both the high and the low end of the market with both live and post-slaughter products. The opportunity to utilize large areas of northern Australia with tropically adapted Brahman (*Bos indicus*) cattle has maintained growth in output.

The ongoing expansion and growth in beef exports has relied on global markets remaining active. In looking forward, Pritchard (2006) suggests that ongoing expansion in Australian beef outputs will rely on continuing trade liberalization; he points to the problem of border restrictions particularly in trading with the USA and falling red meat consumption again in the USA. He also highlights some uncertainties in the continued expansion of the north-east Asian beef markets. Finally the emergence of new competitors from Latin America will put further pressure on Australian beef producers. Australian beef production systems have responded, and continue to respond, to global market demands. Future economic growth in the Australian beef sector will require the industry to align with global market drivers. The global free market is a political construct based on national interests. At the highest level, political motives represent an aggregated social driver. Efficiency of production and the ability to meet strict food safety requirements have maintained the Australian beef sector's competitiveness. Future global political drivers may be both more demanding and wide ranging and will include meeting environmental standards. Australian beef producers will need to develop strategies that are able to maintain economic success in response to these new demands.

Analysis of economic drivers

There are contrasting economic constraints between the UK dairy industry and the Australian beef industry. Nevertheless, farmers in both cases have adapted to changing economic drivers. Australia is focused on maintaining beef export markets and must remain globally competitive in an uncertain market; it is unclear how the industry will respond to global pressures on the cost price squeeze. In the UK, the dairy industry has had to respond to new economic drivers. The shift in policy away from production support to a supply–demand driven market required to meet increasingly stringent environmental and food safety concerns has been rapid and challenging. If livestock farmers are going to meet social and environmental goals they will have to be integrated within a sound economic performance model. In both the UK dairy and the Australian beef sectors, linking industry structural changes to economic performance creates a positive incentive for farmers. Multifunctional livestock farming systems are based on diversification within and between farms with a shift from quantity of production to quality of production (Evans *et al.*, 2002). We speculate that the contrasting Australian and UK economic frameworks shape social and environmental goals by encouraging different management behaviours. The diversity of economic options in the Australian example encourages a diversity of management systems through neoliberalized space (Peck and Tickell, 2002). The EU CAP initially aimed to create stable market conditions that guaranteed economic returns for the UK dairy industry. However, changes in the EU intervention policy have led to free market conditions. A restriction of production with falling milk prices and increasing costs of production is creating economic pressures for the UK dairy industry. Aligning the broad economic direction with community needs through a combination of government intervention and market alignment can create economic stability.

The Environment and the Impact of Livestock Production

It is important to recognize that natural ecosystems are in a continual state of flux due to a range of pressures. If modern agriculture had never existed, there would still have been natural changes that would have shaped the landscape. Throughout the last 3.5 million years there have been a series of ice ages which have covered vast areas of the planet under ice and influenced global climate. The palaeoecological records show that as the global temperature rose and the ice retreated, so natural succession occurred and previously barren areas became highly productive, species-rich environments (Lowe and Walker, 1997). Therefore change is an ongoing phenomenon and the natural environment has evolved diversity to continually adapt and exploit new opportunities that are created through change (Shugart, 1998).

Agriculture has accelerated the rate of change in the landscape and the environment in general. The change has been driven by a need to increase food production either to feed expanding national populations or to respond to international markets. The question we are addressing is; who is responsible for the cost of that impact? It could be those who initiated the change (which may be generations back), those who currently own or farm the land, or those who reaped the benefits of the outputs (i.e. policy makers and consumers). This section considers some of the factors which affect community perceptions and values in relation to shaping the landscape, and changing the environment, before exploring the implications.

Environmental values

There are wide and varied cultural differences in society's concerns about environmental degradation (Ignatow, 2006). The value society puts on the environmental landscape is in part linked to an aesthetic appeal. The appeal with the natural world has been termed biophilia (Wilson, 2003), in other words our environmental preferences are determined by our ancestral origins. Although there is still debate over the exact environmental conditions that shaped our evolution, there is general consensus that there is a direct link between our ancestral interactions with past environments and our current innate environmental preferences (Potts, 1998).

Good or bad environmental changes are commonly considered in relation to human values and perceptions (Ignatow, 2006). Previous agricultural practices which utilized natural resources for society's benefits have created what are now considered as resources of high environmental value, for example coppice woodland or species-rich grasslands, which encourage biodiversity. Therefore, understanding and quantifying community environmental landscape value is not easy. Our historical and cultural affiliations combined with our knowledge and understanding of environmental impacts will shape our individual core values. Whether these values are articulated through government policy or by consumer buying preferences, they are an important consideration for redesigning animal agriculture.

Who is responsible for landscape and environmental change?

Farmers are traditionally viewed as custodians of the land and their activities shape the environmental landscape (Firbank, 2005; Zurlini *et al.*, 2006). The broader community may therefore consider the farmer as having a responsibility to preserve the landscape in its current form or for converting it to a perceived 'better' form. In a previous section we considered how economic incentives were powerful drivers for change. Farmers have responded to the drive to produce cheap food through increased agricultural intensification, higher stocking rates and the use of larger amounts of agrochemicals. This intensification has led to increased pressure on the natural environment (Milne, 2005). Therefore as farmers farm, so the environment will be disturbed both visibly and invisibly, both locally and at a distance, with some impacts being transient and others longer term.

Farmers are responding to direct and indirect drivers of change and the responses can result in unexpected disturbances to the environment. The impacts of these disturbance events in relation to the broader community perception of environmental benefits can be both positive and negative (Milne, 2005; Ignatow, 2006; Young, 2006). For example an open landscape (a visible impact resulting from high stocking rates and low woodland regeneration) may be favoured by walkers as it provides easy walking with good views. These changes to the landscape can be assessed using visual projection techniques and there is evidence that current EU farmed landscapes with some wooded areas are generally viewed as most attractive (Hunziker and Kienast, 1999; Lothian, 1999; Miller, 2001).

On the other hand, many impacts are invisible and the consequential negative environmental impact can accumulate significantly before being recognized. Greenhouse gas emissions (e.g. nitrous oxide, methane and carbon dioxide) are an example of by-products of modern farming systems. Methane from livestock accounts for approximately 12% of the Australian national net greenhouse gas emissions (Howden and Reyenga, 1999). One of the proposed solutions is to feed animals better to reach their slaughter weight more quickly, which reduces the amount of methane output lost per unit of meat produced (Hunter *et al.*, 1993). This higher growth rate is perceived as having a net benefit to the environment, but unless the carbon emissions associated with delivery of the higher quality feed to the animals are taken into account, the net benefit at the global scale, which is the appropriate scale for considering impacts on the air, is over-represented.

Other invisible impacts are the transfer of nutrients and microorganisms to water; here the appropriate scale is the catchment. Johnes (Chapter 11 this volume) provided evidence of the negative impact of livestock farming on nutrient inputs to the UK's freshwater catchments. Modern farming methods can degrade the localized environment to a point that it is no longer useful and productive (Tilman *et al.*, 2002), while pollutants in water courses including nitrates and phosphates from organic and inorganic sources can lead to eutrophication in distant water courses (Pimentel *et al.*, 2004). Disturbed and polluted water courses reduce biodiversity and can cause human health problems, and net costs in environmental, societal and economic terms.

The environmental impacts presented above provide some specific examples of the impact of livestock farming on the environment. There are many other examples including loss of biodiversity, overgrazing, introduction of invasive species, air pollution, deforestation and soil erosion. Much of the damage accumulated before either the effect was recognized, or the relationship between cause and effect identified, with major examples highlighted by Diamond (2005). There is a wide body of literature on who should pay for the environmental costs of market-driven industries. The question being addressed in Europe is whether livestock farmers, who were responding to economic drivers imposed by policy makers, should receive payments from the public purse for agro-environment schemes. By offering these payments, it can be argued that society is recognizing and accepting its responsibility for the impact of past policies, and that such schemes are essential to protect the economic viability of agriculture during the transition from 'productivist' to 'multifunctional' farming systems, as referred to above. For example, payment for the adoption of organic farming methods can meet the broader needs of society and also ensure that economic production is either maintained or grows (Pimentel *et al.*, 2005).

Changing Societal Values Towards Agriculture of Livestock and the Environment

Underlying community values can influence the degree of acceptance of environmental change and impact. Although there is a perception that farmers are responsible for negative environmental impacts, farmers are in fact responding to a number of broader community drivers. Society's values are likely to change depending on whether it is a consumer of farm products or a consumer of environmental services or both. Community values create a cultural backdrop which farmers respond to. Farming is a business and, as with any business, customer preferences exert a major pressure on what enters the market place. The demand for cheap, abundant food has led to more intensive farming methods, which can degrade the environment and also fall below society's expectations of how animals should be treated. This section explores how some of these conflicts have arisen through changing relationships between people and livestock, which provides lessons for the context within which new livestock production systems should be developed.

Where we have come from

Containment and domestication of wild animals marked the shift from hunter-gatherer communities to the earliest forms of agricultural production (Dexter, 1969). Livestock domestication has been a global phenomenon and the co-evolution with humans emphasizes the partnership between livestock and human development (Bruford *et al.*, 2003). Domestication started with livestock playing an important role as a source of food and fibre, but in many countries, livestock still provide draught power and in many societies, livestock play an important cultural role (Anderson, 1997).

Thompson (Chapter 3 this volume) identified a changing value system where the use of animals has shifted from a partnership model, as described in the previous paragraph, to a utilitarian model. This shift reflects the changing dynamics between people and animals (Buller and Morris, 2003), resulting from trends which occur as economies develop, with evidence of a strong positive association between urbanization and increasing per capita income on the one hand and increased consumption of livestock products on the other (e.g. Cranfield *et al.*, 1998; Rae, 1998). At this point in time, this trend is most apparent in China, where meat consumption per capita grew from 16 to 54 kg between 1983 and 2003 as part of a rapidly expanding market for animal products (Delgado, 1995). This increase is not confined to China, however, but is a phenomenon in many developing countries, leading to a trend away from extensive (grass-based) to intensive (concentrate feed-based) livestock systems.

Such a rapid increase in demand can only be met by an increase in intensive production systems, and all of these factors (urbanization, increasing income and intensification of agriculture) inevitably lead (as has already happened in the developed world) to consumers having little understanding of the process of rearing and caring for animals (Holloway, 2001). Livestock production systems are not only fragmented they are also isolated. Community perception is as voyeur, observing from a distance, making value judgements through observation not participation (Holloway, 2001). Oscar Wilde captured the essence of the change when he wrote 'it seems to me that we all look at nature too much and live with her too little'.

Alston (Chapter 2 this volume) provides an in depth assessment of the important factors that maintain vibrant rural communities. The industrialization of farming has led to the amalgamation of smaller farms into larger, more economically efficient farming units (Burton and Walford, 2005). Rural communities are no longer bound together by the 'farm'. The rapid depopulation of rural areas coupled with the difficulties in maintaining a basic infrastructure of hospitals, schools, police, and communications is creating a rural/urban divide (Smithers *et al.*, 2005). The changing emphasis of social interactions means rural farming communities need to establish partnerships with urban or non-farming communities (Sharp and Smith, 2003). Regional rural identity is important in developing a local economic competitiveness (Amin, 1999). In trying to close the loop we consider how community interactions with farmers, in particular when purchasing farm produce, can begin to address some of the issues raised in the previous sections. We consider how greater connectedness between producers and consumers can contribute to ensuring that sustainable economic and environmental goals are met.

A growing disconnection

There is a growing body of literature that explores how the homogenization and disconnection of urban landscapes are creating an urban population with little understanding of the natural world (Miller, 2005). Studies have shown a diminishing appreciation of nature as the majority of people live in or near urban centres (Turner *et al.*, 2004). The number of people that are directly employed to work on farms continues to decline. Within the general population, environmental awareness is on the rise; however, the general population is unable to appreciate direct rural environmental issues (Pyle, 2003) that impact livestock farmers, for example the effect of the EU Water Framework Directive. Community connections are an important part of building societal engagement in rural communities (Miller, 2005) but in recent decades, public perceptions have been informed more by media coverage than either experience or by scientific fact.

The public image of agriculture was damaged in particular by food scandals (Hobbs and Kerr, 2006). Perhaps one of the most damaging food safety issues was the emergence of bovine spongiform encephalopathy (BSE), initially and to the greatest extent, in the UK. The increased incidences of *Escherichia coli* infections and global concerns about avian influenza have further unsettled an already mistrustful society. Consumer confidence is built around quality assurance, and the upsurge in the implementation of food safety standards around the world reflects the growing recognition that the public are seeking a guarantee that the food they are buying is good for their health, as well as animal welfare- and environmentally friendly (Weatherell *et al.*, 2003; Manning *et al.*, 2006). Although quality assurance standards aim to give consumers specific guarantees, they will only be sustainable if farmers can recoup the increased costs involved in meeting them. In some cases it is doubtful whether consumers fully understand what the standards actually guarantee, since they have not been developed in a way that promotes engagement between consumers and producers (Manning *et al.*, 2006). The lack of knowledge about specific quality assurance criteria is further evidence of the disconnection between consumers and producers. Retailers and producer groups set standards based on perceived consumer needs without making them fully aware of what is being done on their behalf. However, there is evidence that supermarkets are beginning to realize the importance of providing direct information to consumers through labelling, in-store information and via their websites.

Opportunities to reconnect

Developing informed consumers is about reconnecting them with food production; this is about building knowledge, which leads to mutual sustainability as links are formed between urban and rural communities (Smithers *et al.*, 2005). There is evidence that farmers are beginning to recognize the potential benefits of creating a much more direct connection with the consumer. Renting *et al.* (2003) differentiate between alternative food networks where farmers find alternative routes to the consumer as opposed to the more recent development of short food-supply chains. These short food-supply chains have a central goal of enabling farmers to have a direct interaction with consumers; Renting *et al.* (2003) identify three domains where this is occurring: organic farming, quality production and direct selling. In each of the three domains, food processing may form an integral part of adding value before the farm gate. Direct-selling methods include farm shops, farmers' markets, home deliveries and pick-your-own, in each case the farmer becomes more connected with the consumer, allowing a deeper understanding of the economic, environmental and social issues associated with the production of food.

Conclusions

In redesigning both animals and parts of the systems in which they are produced, scientists are providing new tools and knowledge for farmers; animal agriculture is moving from the industrial and chemical revolutions into the molecular revolution. The new era will undoubtedly lead to change with the emergence of new high-tech farming systems which have the potential to further increase the 'mining' of our natural resources. History has shown that exploitation of natural resources can eventually lead to societal collapse, where societies are too slow in recognizing and/or responding to potentially catastrophic problems. We posed the question as to whether the application of the sustainability principles of social, environmental and economic goals was feasible and highlighted areas of potential tension. Given that change is an ongoing and continuous process then perhaps building a sustainable future is about building resilience, i.e. fostering the capacity to adapt to change and absorb the impacts that drive change. Economic drivers are powerful incentives for change and farmers have proven willing and able to adapt to emerging economic opportunities. Factors that have a direct impact on farmers' incomes are likely to have the strongest influence on farmers' choices. Addressing environmental objectives has been shown to be more complex. As owners or tenants of the land, farmers have a key role to play in the delivery of environmental benefits. However, community concerns over environmental degradation cannot be easily translated into a farmer mandate for change. Consumers are not always able or willing to pay the increased prices for food which would result from widespread adoption of farming in an environmentally friendly way. One way forward which is gaining momentum is the trend to reconnect urban dwellers (i.e. the majority of consumers) with the food production process itself. This is an important lesson which needs to be fully considered in the introduction of any new technology-based livestock production systems.

Experience suggests that if the 'triple bottom line' (accounting for environment, economy and society) is ignored, then ultimately it results in societal collapse (Diamond, 2005). Scientists collectively have a responsibility to ensure that the implementation of scientific advances does not lead to catastrophes for farmers, the environment or the general public. Fulfilling social, environmental and economic roles is not only feasible, it is also essential to the successful implementation of re-designed livestock production systems.

References

Aarts, H.F.M., Habekotte, B. and Van Keulen, H. (2000) Nitrogen (N) management in the 'De Marke' dairy farming system. *Nutrient Cycling in Agroecosystems* 56, 231–240.

Amin, A. (1999) An institutionalist perspective on regional economic development. *International Journal of Urban and Regional Research* 23, 365–378.

Anderson, K. (1997) A walk on the wild side: a critical geography of domestication. *Progress in Human Geography* 21, 463–485.

Bailey, P.J. and Britt, A.G. (2001) Electronic identification and tracking of cattle in Victoria, Australia. In: Klopcic, M., Malinger, K., Siard, N. and Zgur, S. (eds) *Performance Recording of Animals – State of the Art, 2000.* Wageningen Academic Publishers, Wageningen, Netherlands, pp. 327–330.

Blackburn, N. (2006) *Industry Report on Milk Production Forecast 2007–2012*. Kite Consulting, Stafford, UK.

Blaxter, K.L. and Robertson, N. (1995) *From Dearth to Plenty: the Modern Revolution in Food Production*. Cambridge University Press, Cambridge.

Bruford, M.W., Bradley, D.G. and Luikart, G. (2003) DNA markers reveal the complexity of livestock domestication. *Nature Reviews Genetics,* 4, 900–910.

Buller, H. and Morris, C. (2003) Farm animal welfare: a new repertoire of nature-society relations or modernism re-embedded? *Sociologia Ruralis* 43, 216–237.

Burton, R.J.F. (2004) Seeing through the 'good farmer's' eyes: towards developing an understanding of the social symbolic value of 'productivist' behaviour. *Sociologia Ruralis,* 44, 195–215.

Burton, R.J.F. and Walford, N. (2005) Multiple succession and land division on family farms in the South East of England: a counterbalance to agricultural concentration? *Journal of Rural Studies* 21, 335–347.

Burton, R.J.F. and Wilson, G.A. (2006) Injecting social psychology theory into conceptualisations of agricultural agency: towards a post-productivist farmer self-identity? *Journal of Rural Studies* 22, 95–115.

Colman, D. (2000) Inefficiencies in the UK milk quota system. *Food Policy* 25, 1–16.

Cranfield, J.A.L., Hertel, T.W., Eales, J.S. and Preckel, P.V. (1998) Changes in the stucture of global food demand. *American Journal of Agricultural Economics* 80, 1042–1050.

Delgado, C.L. (1995) Rising demand for meat and milk in developing countries: implications for grasslands-based livestock production. In: McGilloway, D.A. (ed.) *Grasslands: a Global Resource*. Wageningen Academic Publishers, Wageningen, Netherlands, pp. 29–40.

Dexter, P. (1969) Fauna of Çatal Hüyük: evidence of early cattle domestication in Anatolia. *Science* 164, 177–179.

Diamond, J. (2005) *Collapse: How Societies Choose to Fail or Survive*. Penguin, Melbourne.

Evans, N., Morris, C. and Winter, M. (2002) Conceptualizing agriculture: a critique of post-productivism as the new orthodoxy. *Progress in Human Geography* 26, 313–332.

Firbank, L.G. (2005) Striking a new balance between agricultural production and biodiversity. *Annals of Applied Biology* 146, 163–175.

Hobbs, J.E. and Kerr, W.A. (2006) Consumer information, labelling and international trade in agri-food products. *Food Policy* 31, 78–89.

Hodge, I. (2001) Beyond agri-environmental policy: towards an alternative model of rural environmental governance. *Land Use Policy* 18, 99–111.

Holloway, L. (2001) Pets and protein: placing domestic livestock on hobby-farms in England and Wales. *Journal of Rural Studies* 17, 293–307.

Howden, S.M. and Reyenga, P.J. (1999) Methane emissions from Australian livestock: implications of the Kyoto Protocol. *Australian Journal of Agricultural Research* 50, 1285–1291.

Hunter, R.A., Sillence, M.N., Gazzola, C. and Spiers, W.G. (1993) Increasing annual growth-rates of cattle by reducing maintenance energy-requirements. *Australian Journal of Agricultural Research* 44, 579–595.

Hunziker, M. and Kienast, F. (1999) Potential impacts of changing agricultural activities on scenic beauty – a prototypical technique for automated rapid assessment. *Landscape Ecology* 14, 161–176.

Ignatow, G. (2006) Cultural models of nature and society – reconsidering environmental attitudes and concern. *Environment and Behavior* 38, 441–461.

Levy, S.S. (2003) The biophilia hypothesis and anthropocentric environmentalism. *Environmental Ethics* 25, 227–246.

Lothian, A. (1999) Landscape and the philosophy of aesthetics: is landscape quality inherent in the landscape or in the eye of the beholder? *Landscape and Urban Planning* 44, 177–198.

Lowe, J.J. and Walker, M.J.C. (1997) *Reconstructing Quaternary Environments*. Addison Wesley Longman, Edinburgh.

Manning, L., Baines, R.N. and Chadd, S.A. (2006) Quality assurance models in the food supply chain. *British Food Journal* 108, 91–104.

Miller, D. (2001) A method for estimating changes in the visibility of land cover. *Landscape and Urban Planning* 54, 91–104.

Miller, J.R. (2005) Biodiversity conservation and the extinction of experience. *Trends in Ecology and Evolution* 20, 430–434.

Milne, J.A. (2005) Societal expectations of livestock farming in relation to environmental effects in Europe. *Livestock Production Science* 96, 3–9.

Peck, J. and Tickell, A. (2002) Neoliberalizing space. *Antipode* 34, 380–404.

Pierce, J.T. (1994) Towards the reconstruction of agriculture – paths of change and adjustment. *Professional Geographer* 46, 178–190.

Pimentel, D., Berger, B., Filiberto, D., Newton, M., Wolfe, B., Karabinakis, E., Clark, S., Poon, E., Abbett, E. and Nandagopal, S. (2004) Water resources: agricultural and environmental issues. *Bioscience* 54, 909–918.

Pimentel, D., Hepperly, P., Hanson, J., Douds, D. and Seidel, R. (2005) Environmental, energetic, and economic comparisons of organic and conventional farming systems. *Bioscience* 55, 573–582.

Potter, C. and Tilzey, M. (2005) Agricultural policy discourses in the European post-Fordist transition: neoliberalism, neomercantilism and multifunctionality. *Progress in Human Geography* 29, 581–600.

Potts, R. (1998) Variability selection in hominid evolution. *Evolutionary Anthropology* 7, 81–96.

Pritchard, B. (2006) The political construction of free trade visions: the geo-politics and geo-economics of Australian beef exporting. *Agriculture and Human Values* 23, 37–50.

Pyle, R.M. (2003) Nature matrix: reconnecting people and nature. *Oryx* 37, 206–214.

Rae, A.N. (1998) The effects of expenditure growth and urbanization on food consumption in east Asia: a note on animal products. *Agricultural Economics* 18, 291–299.

Renting, H., Marsden, T.K. and Banks, J. (2003) Understanding alternative food networks: exploring the role of short food supply chains in rural development. *Environment and Planning A* 35, 393–411.

Sharp, J.S. and Smith, M.B. (2003) Social capital and farming at the rural–urban interface: the importance of nonfarmer and farmer relations. *Agricultural Systems* 76, 913–927.

Shugart, H.H. (1998) *Terrestrial Ecosystems in Changing Environments.* Cambridge University Press, Cambridge.

Smith, G.C., Tatum, J.D., Belk, K.E., Scanga, J.A., Grandin, T. and Sofos, J.N. (2005) Traceability from a US perspective. *Meat Science* 71, 174–193.

Smithers, J., Joseph, A.E. and Armstrong, M. (2005) Across the divide: reconciling farm and town views of agriculture–community linkages. *Journal of Rural Studies* 21, 281–295.

Tilman, D., Cassman, K.G., Matson, P.A., Naylor, R. and Polasky, S. (2002) Agricultural sustainability and intensive production practices. *Nature* 418, 671–677.

Todt, O. (2004) Regulating agricultural biotechnology under uncertainty. *Safety Science* 42, 143–158.

Turner, W.R., Nakamura, T. and Dinetti, M. (2004) Global urbanization and the separation of humans from nature. *Bioscience* 54, 585–590.

Weatherell, C., Tregear, A. and Allinson, J. (2003) In search of the concerned consumer: UK public perceptions of food, farming and buying local. *Journal of Rural Studies* 19, 233–244.

Wilson, E.O. (2003) *The Future of Life.* Vintage Books, New York.

Wurtenberger, L., Koellner, T. and Binder, C.R. (2006) Virtual land use and agricultural trade: estimating environmental and socio-economic impacts. *Ecological Economics* 57, 679–697.

Young, O. (2006) Vertical interplay among scale-dependent environmental and resource regimes. *Ecology and Society* 11, 27.

Zurlini, G., Riitters, K., Zaccarelli, N., Petrosillo, I., Jones, K.B. and Rossi, L. (2006) Disturbance patterns in a socio-ecological system at multiple scales. *Ecological Complexity* 3, 119–128.

Index